SWEETENERS AND DENTAL CARIES

(A Special Supplement to Feeding, Weight & Obesity Abstracts)

Evaluation of available and potential
new sweeteners as sugar substitutes in
development of non-cariogenic foods
and beverages.

The correct manner in which to refer to a paper from this
publication is as follows:

Author of paper, Title of paper
Proceeding 'Sweeteners and Dental Caries'
Eds. Shaw, J.H. and G.G. Roussos. Sp. Supp.
Feeding, Weight & Obesity Abstracts, 1978

pp —⁻–⁻, 1978

SWEETENERS AND DENTAL CARIES

(A Special Supplement to Feeding, Weight & Obesity Abstracts)

Proceedings of a workshop on evaluation of available
and potential new sweeteners as sugar substitutes in
development of non-cariogenic foods and beverages.

October 24-26 1977
The New England Center for Continuing Education
Durham, New England.

Edited by
James H. Shaw
and
Gerassimos G. Roussos

Presented and Sponsored by
National Caries Program
National Institute of Dental Research
National Institutes of Health
in collaboration with the Harvard School of Dental Medicine.

Information Retrieval Inc.
Washington D.C. and London.

International Standard Book Number 0-917000-05-6
Library of Congress Catalog Card Number 78-52278

Published as a special supplement to:
Feeding, Weight & Obesity Abstracts
Published quarterly by Information Retrieval Limited, London.
Published by Information Retrieval Inc.
1911 Jefferson Davis Highway, Arlington, Virginia, 22202.

Printed in the United States of America

Contents

Contents(continued)

Contents(continued)

Contents(continued)

List of Participants

Harold Amos
Professor of Microbiology
Department of Microbiology
    and Molecular Genetics
Harvard Medical School
25 Shattuck Street
Boston, Massachusetts 02115

Ray H. Anderson
Applied Research and Nutrition
James Ford Bell Technical Center
General Mills, Inc.
9000 Plymouth Avenue North
Minneapolis, Minnesota 55427

Kristina Arvidson
Assistant Professor
Department of Histology
Karolinska Institutet
S-104 01 Stockholm 60, Sweden

Karl M. Beck
Manager, Market Research & Development
Chemical and Agricultural Products
    Division
Abbott Laboratories
North Chicago, Illinois 60064

Lloyd M. Beidler
Professor of Biophysics
Department of Biological Sciences
Florida State University
Tallahassee, Florida 32306

Basil G. Bibby
Research Associate
Eastman Dental Center
800 East Main Street
Rochester, New York 14603

G. N. Bollenback
Scientific Director
The Sugar Association, Inc.
1511 K Street, N. W.
Washington, D. C. 20005

Robert G. Bost
Director of Food Products
Regulatory Affairs
G. D. Searle and Co.
P. O. Box 1045
Skokie, Illinois 60076

William H. Bowen
Chief, Caries Prevention and Re-
    search Center
National Caries Program
National Institute of Dental
    Research
Bethesda, Maryland 20014

Joseph G. Brand
Associate Member
Monell Chemical Senses Center
University of Pennsylvania
3500 Market Street
Philadelphia, Pennsylvania 19104

Myron Brin
Associate Director, Biochemical
    Nutrition
Hoffmann-La Roche, Inc., and
Adjunct Professor of Nutrition
Columbia University
Nutley, New Jersey 07110

Robert H. Cagan
Member, Monell Chemical Senses
    Center and
Associate Professor of Biochem-
    istry
University of Pennsylvania
3500 Market Street
Philadelphia, Pennsylvania 19104

Sidney M. Cantor
President
Sidney M. Cantor Associates
Haverford, Pennsylvania 19041

James P. Carlos
Associate Director
National Caries Program
National Institute of Dental
    Research
Bethesda, Maryland 20014

Robert B. Choate
Chairman
Council on Children, Media and
    Merchandising
1346 Connecticut Avenue, N. W.
Washington, D. C. 20036

William Cooley
The Proctor & Gamble Company
Winton Hill Technical Center
6110 Center Hill Road
Cincinnati, Ohio 45224

David Coursin
Director of Research
Research Institute
St. Joseph's Hospital
Lancaster, Pennsylania 17604

Alan L. Coykendall
Veterans Administration Hospital
555 Willard Avenue
Newington, Connecticut 06111

Guy A. Crosby
Director, Chemical Synthesis
Dynapol
1454 Page Mill Road
Palo Alto, California 94304

George Ev. Demetrakopoulos
Research Fellow in Microbiology and
     Molecular Genetics
Harvard Medical School
25 Shattuck Street
Boston, Massachusetts 02115

Jane Frances Emele
Director, Biological and Clinical
     Affairs
Warner-Lambert Company
201 Tabor Road
Morris Plains, New Jersey 07950

Joseph R. Fordham
Nutritionist
Market Development and Customer
     Service
Clinton Corn Processing Company
Clinton, Iowa 52732

Harold M. Fullmer
Director, Institute of Dental
     Research
School of Dentistry
The University of Alabama in
     Birmingham
The Medical Center
University Station
Birmingham, Alabama 35294

K. F. Gey
Head, Sweeteners Section
F. Hoffmann-La Roche and Co. AG.
and Vice-President for Research,
     Xyrofin
Grenzacherstrasse 124
CH-4002 Basel, Switzerland

Clarence C. Gilkes
Supervisory Dental Officer
Division of Surgical Dental
     Drug Products
Bureau of Drugs
Food and Drug Administration
5600 Fishers Lane
Room 18B08 Parklawn Building
Rockville, Maryland 20857

William C. Griffin
Associate Director
Product Development Department
Specialty Chemicals Division
ICI United States, Inc.
Wilmington, Delaware 98197

Michael Gumbmann
Research Leader
Toxicology and Biological Eval-
     uation Research Unit
Western Regional Research Center
U.S. Department of Agriculture
Berkeley, California 94704

Victor Herbert
Chief, Hematology and Nutrition
     Laboratory
Veterans Administration Hospital
130 West Kingsbridge Road
Bronx, New York 10468

L. Kenneth Hiller
The Proctor & Gamble Company
Winton Hill Technical Center
6110 Center Hill Road
Cincinnati, Ohio 45224

Robert M. Horowitz
Research Chemist
Fruit and Vegetable Chemistry
     Laboratory
U.S. Department of Agriculture
Pasadena, California 91101

William A. Hoskins
Manager, Contract Research
Research and Development Center
Foremost Foods Company
6363 Clark Avenue
Dublin, California 94566

George E. Inglett
Chief, Cereal Science and Food
    Laboratory
No. Regional Research Labora-
    tories
U.S. Department of Agriculture
1815 N. University Street
Peoria, Illinois 61604

Richard J. Jones
Director, Division of Scientific
    Activities
American Medical Association
535 North Dearborn Street
Chicago, Illinois 60610

Morley R. Kare
Director, Monell Chemical Senses
    Center and Professor of
    Physiology
University of Pennsylvania
3500 Market Street
Philadelphia, Pennsylvania 19104

Keatha K. Krueger
Diabetes Program Director
Extramural Programs
National Institute of Arthritis,
    Metabolism and Digestive
    Diseases
Bethesda, Maryland 20014

Donald A. M. Mackay
Vice President, Research and De-
    velopment
Life Savers, Inc.
40 West 57th Street
New York, New York 10019

Kauko K. Mäkinen
Professor in Biochemistry
Institute of Dentistry
University of Turku
Lemminkäisenkatu 2
SF-20520 Turku 52, Finland

Howard Moskowitz
MPI Sensory Testing, Inc.
770 Lexington Avenue
New York, New York 10021

Juan Navia
Senior Scientist
Institute of Dental Research,
    and Professor of Biochem-
    istry and Nutrition
School of Dentistry
The Medical Center
The University of Alabama in
    Birmingham
University Station
Birmingham, Alabama 35294

Charles S. Nevin
Assistant to the Vice-President
Wm. Wrigley Jr. Company
Research and Quality Control
410 North Michigan Avenue
Chicago, Illinois 60611

Ernest Newbrun
Professor of Oral Biology
School of Dentistry
University of California
3rd and Parnassus
San Francisco, California 94143

Abraham E. Nizel
Professor of Nutrition
School of Dental Medicine
Tufts University
136 Harrison Avenue
Boston, Massachusetts 02111

Peter P. Noznick
Director, Corporate Science and
    Technology
Beatrice Foods Company
Research Center
1526 South State Street
Chicago, Illinois 60605

Ellis Ratner
Staff Attorney
Bureau of Consumer Protection
Federal Trade Commission
Washington, D. C. 20580

William E. Rogers, Jr.
Scientific Coordinator for Grants
   and Contracts
National Caries Program
National Institute of Dental Re-
   search
Westwood Building, Room 549
Bethesda, Maryland 20014

Gerassimos G. Roussos
Project Officer
National Caries Program
National Institute of Dental Re-
   search
Westwood Building
Bethesda, Maryland 20014

Richard J. Ronk
Director, Division of Food and
   Color Additives
Food and Drug Administration
200 C Street, S. W.
Washington, D. C. 20204

Philip A. Rossy
BASF AG
Research and Development
6700 Ludwigshafen/Rhein
West Germany

J. Rozanis
Associate Professor of Bacter-
   iology and Immunology
Health Sciences Center
The University of Western Ontario
London, Ontario, Canada N6A 5C1

Gordon H. Schrotenboer
Secretary, Council on Dental Thera-
   peutics
American Dental Association
211 E. Chicago Avenue
Chicago, Illinois 60611

James H. Shaw
Professor of Nutrition
Harvard School of Dental Medicine
188 Longwood Avenue
Boston, Massachusetts 02115

J. Stuart Soeldner
Associate Director
Elliott P. Joslin Research Laboratory
1 Joslin Place
Boston, Massachusetts 02115

Charles D. Stone
Senior Research Scientist
Scientific Affairs Department
M&M/Mars
High Street
Hackettstown, New Jersey 07840

Philip A. Swango
Community Programs Section
Caries Prevention and Research
   Branch
National Caries Program
National Institute of Dental
   Research
Bethesda, Maryland 20014

John M. Talbot
Senior Medical Consultant
Life Sciences Research Office
Federation of American Socie-
   ties for Experimental Biology
9650 Rockville Pike
Bethesda, Maryland 20014

Stanley Tarka
Group Leader, Nutritional Sciences
Hershey Foods Corporation
Research Laboratories, P. O. Box 54
Hershey, Pennsylvania 17033

John Townsley
Chief, Caries Grant Programs
National Caries Program
National Institute of Dental
   Research
Westwood Building, Room 522
Bethesda, Maryland 20014

John W. Turner
Research Fellow in Pediatric
   Dentistry
Children's Hospital Medical Center
300 Longwood Avenue
Boston, Massachusetts 02115

H. van der Wel
Unilever Research
Vlaardingen/Duivan
Olivier dan Noortlan 120
Vlaardingen, The Netherlands

Bruce J. Walter
Corporate Staff Economist
CPC International, Inc.
Englewood Cliffs, New Jersey 07632

James M. Weiffenbach
Oral and Pharyngeal Development
    Section
Clinical Investigations Branch
National Institute of Dental
    Research
Bethesda, Maryland 20014

Sidney Weiss
Associate Director of Research
Colgate-Palmolive Research Center
909 River Road
Piscataway, New Jersey 08854

Ronald G. Wiegand
Director, Scientific Affairs
Chemical and Agricultural Products
    Division
Abbott Laboratories
North Chicago, Illinois 60064

Ferdinand B. Zienty
Manager, Research and Development
Food, Feed and Fine Chemicals
Monsanto Company
800 N. Lindbergh Boulevard
St. Louis, Missouri 63166

The scientific reporter for the conference was:

    Mr. Ira P. Maisel
    Leavitt Reporting Service
    28 Broad Street
    Weymouth, Massachusetts 02188

Introductory remarks: purpose and objectives of the workshop. Definition of the problem and charge to the workshop participants

James P. Carlos

National Caries Program, National Institute of Dental Research, Bethesda, Maryland 20014

It is my pleasure to welcome all of you on behalf of the National Caries Program and the National Institute of Dental Research. We are very pleased that many of our colleagues from industry and other countries are able to be here.

It is hardly necessary to say that this discussion of dietary sweeteners occurs at an exceptionally appropriate time since the subject currently is of intense scientific, commercial and public interest.

During these $2\frac{1}{2}$ days we will, hopefully, determine the present status and future potential of a large array of candidate sweeteners. To provide some background I will begin with a few words about where we have been in the recent past.

As many of you know, the National Caries Program conducts and supports research on caries prevention in four major Strategy Areas. Strategy I involves attempts to interfere with the microbiological agents involved in caries etiology. The second Strategy is concerned with methods to improve tooth resistance; the third, with altering harmful dietary factors; the fourth Strategy involves research on improved delivery and public acceptance of preventive methods.

An examination of our activities over the past five or six years leads to the somewhat disappointing conclusion that, of the four Strategy Areas, we have made least progress in devising ways to alter the dietary factors in caries--or, to be more specific--to limit the amount, form and frequency of sucrose ingestion by the American population.

When the Program began in 1971, several approaches to this problem were considered. The most obvious approach, health education, was also the one most quickly rejected. Our staff and our advisory bodies agreed that ample evidence existed to suggest that efforts to persuade the population to make major changes in their dietary habits to prevent caries, was unlikely to be a cost-effective undertaking.

Instead, we began research on the possibility that other sugars, particularly invert sugar, glucose and fructose could be substituted for sucrose in snack foods to good effect. This, as it turned out, proved to be a poor choice. Results from a series of experiments in rodents and primates, as well as microbiological studies, strongly indicated that simple sugars would also be sufficiently cariogenic to eliminate them as desirable candidates to replace sucrose.

Our attention, therefore, turned to non-sugar sweetening agents including, of course, the polyols and, at present, our entire research effort in Strategy Area III is devoted to the study of such compounds. Thus far this effort has been very modest--approximately $800,000 worth of research during the past year. Even so, and I trust that you will find this as surprising as I do, the National Caries Program is the only program at the NIH which is involved in research on dietary sweeteners. As a result of the current controversy over saccharin the effect has been, rather inadvertently, to thrust a dental research activity into the limelight. Hence my remark on the timeliness of this Conference. We like to think that research on dental disease frequently produces information which is useful and applicable to other areas of biomedical investigation. Certainly research on sweeteners is a prime example of this, especially if sweeteners can be developed which are not only noncariogenic but noncaloric as well.

In addition to the scientific challenges in developing safe and acceptable dietary sweeteners, the effort is strongly influenced by several other factors. Foremost among these is the intense commercial interest and involvement in this field, a factor which is both a blessing and a problem. A blessing, because without commercial investment no new sweeteners, no matter how attractive from a scientific standpoint, will ever reach the marketplace and the consumer. But also, a possible problem because marketing means advertising, and advertising--sometimes--means overenthusiastic and premature claims, to put it kindly. Excesses can, and have occurred, to the considerable embarrassment of many scientists working in this field. It is certainly not our role to judge the ethics of the marketplace, but as we move forward with development of new sweeteners, prudence in our public announcements of results would seem the wisest course. I should add that we are especially pleased with the heavy representation from industry at this Conference, since no useful dialogue on sweeteners would be possible otherwise.

A second major factor in influencing this area of research and development is the need to satisfy the extensive requirements of the Food and Drug Law before any new sweetener can be made available to the public. Many persons in this room know far better than I how expensive, how time consuming, and how frustrating that process invariably becomes. My own limited experience convinces me that there is no easy solution and no shortcuts. So, in addition to my plea for prudence in sweetener research I must also counsel "patience"--and plenty of it!

There is no doubt that our discussions at the Conference will be stimulating and productive. The program, the speakers, and the guests are as comprehensive and well-balanced as one could possibly hope for. The publication which will result cannot fail to be the definitive statement on sweeteners and caries.

I should like to congratulate and thank, in advance, Jim Shaw, Gerry Roussos, and the others who have helped to organize the Conference, and who will assemble and edit the proceedings.

Once again, welcome to all of you and our thanks for your willingness to join us in what I expect will be an exceptionally important series of discussions.

SESSION I.

PERSPECTIVES ON SWEETENERS

Moderator:

M.R. Kare

Director, Monell Chemical Senses Center and
Professor of Physiology
University of Pennsylvania
3500 Market Street
Philadelphia, Pennsylvania 19104

Biophysics of taste receptors

Joseph G. Brand

Monell Chemical Senses Center, University of Pennsylvania, Philadelphia, PA 19104

SUMMARY

The ability of the animal to monitor his surroundings is critical to survival. Taste is one of the senses directly involved with gauging the acceptability of food for purposes of nutritional maintenance and poison avoidance. Our knowledge of the peripheral mechanisms in taste perception is not complete enough to provide a detailed biophysical explanation of this phenomenon. Nevertheless several stages in this process have been hypothesized and some are demonstrable.

The taste receptor is a differentiated epithelial cell synaptically contacting sensory nerve fiber(s). A collection of these cells and other presumably supporting cells forms the multicellular structure known as the taste bud. Taste information is carried by three cranial nerves, VII, IX, and X. A qualitative and quantitative message is perceived after complex processing in the brain such that the stimulus is recognized as either sweet, sour, salty, or bitter, or some combination of these four. The neural basis for this specificity is still a matter of conjecture.

Both neurophysiological and biochemical experiments point to the existence of several steps in taste receptor functioning for transduction of the chemical message. First, the stimulus binds to the taste receptor cell membrane at the apex of the taste bud. This binding then presumably leads to conformational changes in the receptor that initiate plasma membrane conformational changes in the taste cell below the level of the tight junctions. These conformational changes of the plasma membrane result in changes in membrane conductances that permit the influx and efflux of ions which initiate the receptor potential. The receptor potential generates the release of synaptic vesicles to the innervating sensory nerve fiber to initiate the neural spike. Evidence to confirm the existence of several of these steps has accumulated and is presented.

INTRODUCTION

Survival of all forms of life depends upon their ability to sense the immediate environment and react in an appropriate advantageous manner to the stimulus gradients therein. The quality of our lives is enhanced through our appreciation of the environment, and information concerning our surroundings is monitored through our senses. We are primarily visual and auditory animals, and our knowledge of the mechanisms of these senses is, though incomplete, fairly extensive. Less is known of our chemical senses, taste and smell. Yet both are intimately involved in our selection of a hospitable environment, both external and internal.

The search for food for nutritional maintenance is a major environmental encounter for the animal. The sense of taste aids in monitoring the acceptability of food. A decision on whether or not to ingest a food is largely dependent on the nutritional needs of the moment and the results of the chemical analysis of the food that is carried out in the oral cavity. Taste in this context thus serves as a monitoring device, gauging the acceptability of a food. Other functions of taste and general oral stimulation are also well appreciated[1-6]. The unique association of the taste of a food with gastrointestinal consequences following ingestion of that food is also well documented. The many studies to date show a central role for taste as the conditioned stimulus in this form of learning[7,8].

In the oral cavity chemicals are recognized following their contact with the epithelial cells. Recognition of chemicals may take the form of non-specific irritations such as the so-called common chemical sensation. Alternatively, recognition may be more specific, permitting quality discriminations beyond simple preference or aversions. In taste one usually delineates at least four primary quality sensations in man: sweet, sour, salty and bitter[9]. Quality recognition for taste may also be logically paired with chemical structure so that if unique receptors exist for each recognized quality, one might anticipate some molecular similarity among all chemicals that elicit a single quality sensation. This hypothesis has received most attention for the sweet modality[10].

Taste quality recognition is a function of the cells of the taste buds. Perception of the sensation is a result of complex processes in the brain. The mechanism of the initial contact and recognition steps is largely unknown, and the nature of the transductive step is still a matter of conjecture. While it is often assumed that taste quality specificity resides in the peripheral taste cell membrane receptors, this concept has never been rigorously shown biochemically. Indeed the very existence of unique receptor entities capable of recognizing chemical stimuli in a quality-specific manner has only been inferred. Many experiments point to the existence of such templates, particularly in the recognition of the sweet sensation, but none have been unequivocally demonstrated. Knowledge

of these receptors and their function can aid in the design of stimuli of
uniform taste quality and desired impression.   Studies on the biochemical
and biophysical mechanisms of taste at the peripheral level are designed to
achieve this level of understanding.

TASTE BUD STRUCTURE

An appreciation of the morphology of the taste bud unit aids in
placing in perspective critical assessments of biochemical experiments.

Taste buds are localized in the buccal cavity and throat in regions
of the tongue, soft palate, and pharynx.   In the human tongue, taste buds
can be localized to three specialized structures: the fungiform papillae,
clustered in the anterior two-thirds of the tongue; the foliate grooves,
located in the lateral lingual epithelium; and the circumvallate papillae,
located in the posterior portion of the tongue. The 8 to 10 circumvallate
papillae in the human are arranged in a V-shape along the posterior dorsal
surface of the tongue.   Each one contains from 100 to 300 taste buds.
The exact number of taste buds in all of these structures is variable from
individual to individual, but does seem to be age-dependent[11-13].   A
recent review of the topography of the tongues of many species is
available[14].

The sense of taste is mediated by three separate sensory nerves[15].
The chorda tympani branch of the VIIth cranial nerve (facial) innervates
taste papillae on the anterior two-thirds of the tongue, although a
posterior field was reported in the calf[16].   The glossopharyngeal nerve
(the IXth cranial nerve) innervates taste buds in the posterior field of
the tongue, that is, those of the circumvallate papillae and the foliate.
The vagus nerve (Xth cranial nerve) innervates taste buds in the soft
palate and throat.   The fact of vagal innervation suggests a mechanism
for taste that directly influences digestive functions of the gut[1,2].
Sensory neural innervation is essential for taste bud integrity, both
morphological[17,18] and apparently functional[19].

The taste bud is a multicellular, pear-shaped epithelial structure,
the apex of which is exposed to the environment of the oral cavity
(Figure 1).   Taste bud cells are differentiated epithelial cells of
markedly different morphology than the undifferentiated cells that
surround the bud[20,21].   Taste bud cells have the appearance of
specialized epithelial receptors innervated by sensory nerve fibers; yet
in many morphological studies, only a few cells within the bud have ever
been shown to have classical synaptic relationships with innervating

nerves[21].   Cells of the taste bud are constantly being renewed with a
turnover rate in the rat fungiform papilla estimated at 8 to 13 days[22].
Cholinesterase activity has been demonstrated histochemically in the taste
bud[23], and denervation decreased the intensity of this staining according
to the normal course for nerve degeneration[24].   Recent histofluorescence
studies also suggested the possibility that a biogenic monoamine is
present in the rabbit foliate taste bud cell[25] and, possibly, in the
frog[26,27].

Fine structural observations of the taste bud have been reported for
several species over the past decade and a half.   Those of the rat
fungiform[20], the rabbit foliate[21], the frog[28], and aquatic species[29,30]
have been reported regularly.   These studies are generally concerned with
discerning the cell types within taste buds, the concentration and type
of subcellular organelles within these cell types, the nature of the cell-

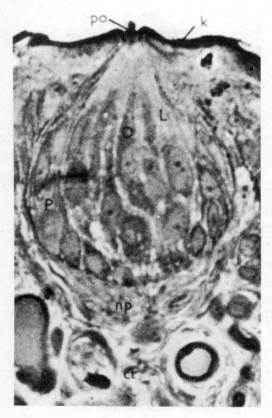

Figure 1. A longitudinal
section through a taste bud of
a rat fungiform papilla. Above
the level of the connective
tissue (ct) is a nerve plexus
(np). Cells within the taste
bud are here designated as
three in type: the peripheral
cell (P) and two spindle-shaped
cells, one with dark cytoplasm
(D) and the other with light
cytoplasm (L). The pore area
(po) of the bud is exposed to
the oral environment and often
contains a plug of unknown
composition. Keratinized
epithelium (k) surrounds the
bud pore. This figure and
general description are taken
from Farbman[20].

nerve contact, and the overall relationships among cells. Each worker
has recognized certain specific attributes of unique cell types and used
these to form classification schemes of the taste bud cells. Attempting
to synthesize this information, it is possible to report some fine
structural details that have bearing on the present problem.

The bud is composed of several morphologically distinct cell types,
some of which reach the bud apex and some of which exhibit microvilli at
these apical ends. The two main types of cells are the so-called dark
cells and light cells. Based on its location and extension to the apex,
the dark cell is considered by Farbman[20] to be the receptor cell, and the
light cell is suggested to have a role in maintenance of the taste bud
pore. Murray[21] considers a specialized light cell, his so-called "Type
III cell" to be the receptor cell, since he claims a classical synaptic
relationship between this cell and an innervating fiber. Other cell
types include the basal and peripheral cells that could be either
structural entities of the bud or cells in transitional forms.

Regardless of the exact classification of these cells and their
relationship to nerve fibers, their organization and gross morphology
suggest certain mechanistic explanations for the taste process. Taste
stimuli interact with the apical portions, possibly the plasma membranes,
of the taste cells after being solvated by saliva and the material
sometimes present in the taste pit. This interaction sets off a
transductive event that may encompass the entire taste bud. Farbman
observed tight junctions between Type I cells (dark cells) and suggested
that these could be used for cell-cell contact as channels of low
electrical resistance necessary for the spread of electrotonic potential
among all cells of the bud[20]. These junctions could also act as barriers
to free diffusion of stimulus molecules from the exterior surface of the
tongue into the lower regions of the taste bud. Whether some of these
tight junctions are also gap junctions remains to be definitively
demonstrated. Electrotonic coupling among cells implies the presence of
gap junctions. West and Bernard have recently demonstrated electrical
coupling in Necturus taste cells[31]. The transductive event then spreads
until it culminates in the initiation of a spike potential at the
innervating sensory fiber. In taste, therefore, the receptor potential
is initiated in specialized epithelial cells at a level probably below
the tight junctions. The generator potential initiated in the sensory
nerve is a result of synaptic vesicular release by the taste cell. The
neural spike train then proceeds to the brain where further neural

processing occurs, ultimately resulting in a sensory experience. Thus at least four major stages are involved in the perception of the taste message -- reception at the taste cell periphery, transduction through the taste bud, generation and propagation of the spike, and complex processing of the message by the central nervous system (CNS) resulting in a quality specific impression. The information arriving at the CNS is invariant, yet the affective behavioral response can be modified by context and experience.

ACTIVITY IN THE SENSORY NERVES

Neurophysiological experiments using both extracellular and intracellular techniques have provided information on the neurological events in taste perception, both transduction and transmission, and indirectly about the nature of the peripheral receptor events that initiate these neural responses. Recordings from the sensory nerves either as a bundle of fibers or from individual fibers have shown that chemical stimuli placed on the tongue can trigger neural firing very rapidly after onset of the stimulus [∿ 20 and 50 milliseconds (msec) for sodium chloride in the rat][32]. Firing patterns in a single fiber are dependent on the nature of the stimulus applied, its concentration, and its temperature[33-35].

Recording neural activity from single fibers of gustatory nerves is a more specific technique than recording from the whole nerve bundle. It has been found that each fiber generally responds best to a single quality stimulus. Frank has suggested naming fibers in terms of their best response characteristic[36]. Thus she would label some as sucrose (sweet)-best fibers or sodium chloride (salty)-best fibers. It is interesting that few fibers that are sucrose-best also respond to the bitter stimulus quinine. This is true for chorda tympani fibers as well as glossopharyngeal fibers[37]. Single fibers, though showing this specificity, innervate more than a single receptor cell or even a single taste papilla in mammals. Thus their activity is a measure of the summated activity of many receptor elements. It would be interesting to determine if these categories of best-stimulus labeling are upheld when fibers are tested with varying concentrations of stimuli. To date, most of the best-stimulus naming of fibers have used only a single concentration of stimuli that represent the four basic tastes.

Two theories have been proposed to account for generation of the quality-specific message in the CNS. The across-fiber pattern theory assumes that a unique pattern will emerge for each quality if one could summate all the fiber firings in all neurons of a particular taste nerve. If all fibers could be tested with most available stimuli, presumably certain stimuli-response patterns would group together based on these electrophysiological data. These groups would then represent the basic taste qualities for that particular nerve and species. Erickson points to behavioral and electrophysiological data that support this theory[38]. However, clear taste quality sensations can be elicited by stimulating only a single papilla. Presumably a single papilla is not innervated by all sensory fibers, though probably it is innervated by more than one fiber. With this limitation in mind, therefore, it is apparent that at least the entire pattern across all fibers is not necessary for simple quality recognition.

A theory based on the hypothesis of Müller in 1848 suggests that sensory receptors transduce only one sensation, independent of their manner of stimulation. Thus in taste one may assume that the separate taste modalities must have specific fiber pathways to the CNS. The picture in taste is complicated by the fact that the receptors are non-neural. Nevertheless, Bekesy's behavioral experiments tend to support this theory[39]. Electrophysiological results are, however, puzzling. Were this theory correct, taste fibers should elicit only one sensation, regardless of how they are stimulated. However, it has often been shown that fibers respond to more than one type of chemical stimulus. Even so, they generally respond to a single stimulus quality best, and to others only to a lesser degree. Wang[40] has suggested an explanation for this apparent confusion. When, for example, a sucrose-best fiber responds also to sodium chloride, it might not be responding to the saltiness of sodium chloride but to the sweetness of sodium chloride. Bekesy[41] suggested that there may be no monogustatory stimulus for taste; that, in fact, all chemical stimuli elicit more than one pure quality.

The two theories may not be mutually exclusive. While specific fibers may exist for each quality stimulus (in the neo-Müllerian sense), complex tastes may be discriminable only by the pattern of response across all fibers[42]. Thus we can differentiate between sucrose and saccharin even though the dominant impression from each is sweet.

Electrophysiological studies have provided the impetus for two theories of taste receptor mechanisms. In 1954, Beidler[43] described taste neural response in terms of an adsorption theory, wherein the adsorption of the stimulus on the receptor matrix was directly relatable to the summated, steady-state electrophysiological response up to a maximum. This maximum response was assumed to be directly related to the saturation of the receptor sites on the taste bud membrane. In spite of the assumptions inherent in a formulation such as this and the non-linearity of some taste-response data[44-45], this theory of taste stimulation continues to have impact in the field and remains a reasonable working hypothesis for peripheral taste mechanisms.

Beidler's formulation considers only the tonic response characteristics in the electrophysiological record and does not deal with the phasic responses. However, at least the rat can make taste discriminations within the time period of the phasic response. Halpern and Tapper[46] claimed that the rat is able to make behavioral quality discriminations within 250 msec, and Halpern[47] reported rejection of aversive stimuli by rats within 50 msec of sampling. Theories formulated to explain the use of the phasic portion have assumed that response is based on the rate of reaction of the stimulus with the receptor. As Beidler[44] pointed out, rate formulations such as those of Renqvist[48] and Lasareff[49] predict zero steady state response by being proportional to the net rate of the interaction which is obviously not the case in taste. Other kinetic theories such as those for drug action of Paton[50] and for taste of Heck and Erickson[51] assume response is proportional only to the rate of formation of the stimulus-receptor complex, that is, the rate of adsorption of the stimulus. Faull and Halpern[32], however, found that the rate theories do not predict well the relationship between concentration and phasic response magnitude. Smith and Bealer[52] reported that the initial transient portion seen in a chorda tympani discharge is a result of the rate of stimulus onset. Were Heck and Erickson's theory[51] correct, this transient response in the nerve should be preceded by a transient in the receptor potential. The magnitude of this transient should be proportional to the rate of stimulation.

A direct test of this prediction by Balnave and Wolbarsht[53] using iontophoretically applied current pulse stimuli with intracellular recordings from taste cells revealed no transient potentials in these cells. Peak depolarizations corresponded with the peak of the iontophoretic current pulse. They concluded that the receptor potential is only a function of stimulus intensity and not a function of the rate of rise of the stimulus. A similar conclusion was reached by Sato[54] using chemical stimulation. These results are contrary to the kinetic prediction of Heck and Erickson[51]. Smith[55] has recently suggested that the chorda tympani response is sensitive to the amplitude of the receptor potential as well as to its rate of change. Thus, the nerve "adds a differential to the taste receptor input" and, "...the rate sensitivity of the chorda tympani nerve can be accounted for under an occupation model..." as proposed by Beidler[43]. The transient, thus, probably originates at the synaptic level in taste. Regardless of the generation of these neural events, both the phasic and the tonic components appear to be behaviorally important. It has been suggested that the phasic response can be used by the rat for quality discriminations while the tonic component is necessary for concentration discriminations[39].

No theory yet stated adequately predicts the wealth of information that must be present in the neural train. An even more complex puzzle results when one attempts to decipher the peripheral steps responsible for the initiation of the neural spike. Many of the assumptions on the nature of these steps are derived from intracellular neurophysiological recordings.

PERIPHERAL MECHANISMS IN TASTE

Intracellular (taste cell) receptor potentials generally show a time course to peak depolarization that is relatively slow, and in the majority of cells examined, responses were found to more than one taste quality. Because of the junctions among cells of the taste bud, it is possible that chemically-evoked responses from a single taste bud cell cannot be recorded using an in situ mammalian preparation[40]. The question of whether or not each taste cell responds to only one quality stimulus might not be answerable without an isolated cell preparation.

From the few studies carried out to date using the intracellular recording technique, some facts can be presented concerning the generation of the receptor potential. Depolarizing receptor potentials recorded from rat, hamster and frog taste cells have shown that these cells respond to multiple stimuli and that the sensitivity of the cells is different from one cell to the next. The receptor potential is also proportional to stimulus intensity[56-58]. Ozeki[59] examined conductance changes of the rat taste bud cells during chemical stimulation and found positive conductance changes for sodium chloride, sucrose and hydrochloric acid stimulation. These papers and a recent publication by Sato and Beidler[60] point to a hypothesis of taste cell transduction from the interaction of the stimulus with the receptor site (resulting in the depolarization of the taste cell) to, via a synaptic mechanism, the generation of the neural spike. Sato and Beidler[60] assume three steps in this process: (a) the adsorption of the taste stimulus to the receptor matrix of the taste cell membrane, (b) conformational changes in the receptor matrix that initiate changes in the taste cell membrane structure at sites distant from the receptor matrix and (c) changes in ionic permeability of the taste cell plasma membranes resulting in generation of the receptor

potentials.   The receptor potential presumably then initiates the release
of synaptic vesicles that generate the neural spike potentials.

A diagram of these steps is presented in Figure 2.   In the case of
salty, sweet and acid stimuli, but perhaps not bitter[59], the influx of
extracellular sodium represented by step C is brought about by a lowering of
plasma membrane resistance that accompanies the membrane conformational change
in step B.   This increase in ionic permeability initiates the receptor
potential that is recorded as a depolarization.   Sato and Beidler's study[60]
presents good evidence for the central role of sodium in this generation
process.   The decrease in conductance for quinine stimulation of the rat
taste cell reported by Ozeki[59] probably involves other membrane mechanisms as
well as the movement of ion(s) other than sodium.   Ozeki[61] suggests that

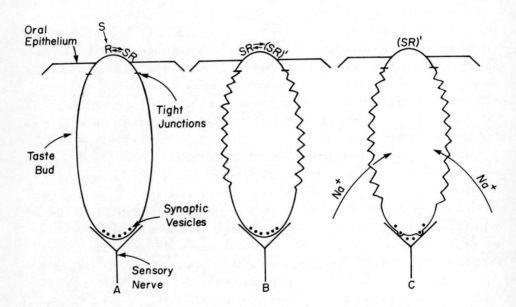

Figure 2.   Three hypothesized stages in the development of the receptor
potential: (A) interaction of the stimulus, S, with membrane-bound receptor,
R, to form a stimulus-receptor complex SR; (B) conformational change in SR
to (SR)' brought about by the interaction of S with R.   This change to (SR)'
initiates a change in plasma membrane conformation of taste cells, probably
below the level of the tight junctions; (C) membrane conformational changes
result in lowered membrane resistance and the consequential influx of
extracellular ionic species, probably sodium.   This influx of sodium
generates the receptor potential which induces synaptic vesicular release to
the innervating sensory nerve.   The generator potential ensues.
In this diagram, the entire taste bud is shown to be responding as a
unit.   Whether this actually occurs has not been definitively demonstrated.
The figure is drawn after a description given by Sato and Beidler[60].

potassium movement (presumably out of the taste cell) may accompany this particular change. Step C of Figure 2 is observed when recording from microelectrodes placed in the taste cell. Evidence for the existence of steps A and B of Figure 2 is also beginning to accumulate.

Attempts to detect step A of Figure 2, the interaction of the stimulus with the receptor, have generally taken the form of attempting to measure the putative receptor-stimulus complex. The literature in this field is not vast, yet already several hypotheses concerning the nature of this complex have been tested. Sweet, salty and bitter stimuli have most often been used in these studies. For the purposes of the present discussion, only studies reporting the interaction of sweet stimuli with taste tissue will be reviewed. Studies of the peripheral reception of salty[62-63] and bitter[64-65] stimuli will be alluded to only from a general mechanistic position.

The early work of Dastoli and Price[66] has often been cited as evidence for sweet receptor isolation. They assumed the receptor was a solubilizable protein that could be removed from its membranous surroundings by homogenization techniques. Their assay of activity depended on this protein being able to exhibit considerable conformational deformability upon interaction with sweet stimuli. Many of their conclusions and the assumptions inherent in them have since been shown to be in error. The protein fraction was isolated using primarily lingual epithelium with very little taste bud-related material in the initial tissue fraction. This procedure assumes that taste receptors for sweet are ubiquitous. While widespread distribution, in fact, may eventually be shown to be the case, it cannot be proven through a negative argument. From the viewpoint of determining the authenticity of taste receptors, it is necessary to perform studies using a tissue fraction highly concentrated in material from the taste bud, and to rigorously compare these results with those from lingual tissues known to be devoid of taste buds. Experiments of Koyama and Kurihara[67], Price and Hogan[68], Price[69] and Ostretsova et al.[70] have shown that the protein fraction claimed to be the "sweet-sensitive protein" is very heterogeneous. The original observations of Dastoli and Price[66], in addition, were not replicable in at least two published attempts[70-71]. Criticism has been made of the relative strengths of the reported binding constants and the comparison of human psychophysical data to binding data from bovine tissues[72].

The specific binding of taste-active chemicals and correlations between the observed biochemical results and the behavioral counterparts were first

investigated by Cagan[72].   Bovine tongue papillae, one type containing taste
buds, the other devoid, were homogenized in a defined manner and concentrated.   The ability of these two different papillae homogenates to bind
radioactively-labeled sugars was compared.   Sugars bound more extensively
to homogenates that originated from taste papillae.   In addition, this study
showed that while sucrose, a highly preferred sugar behaviorally[73], bound
more to taste bud papillae homogenates, lactose, a sugar to which calves are
indifferent[73], bound to a lesser extent.   The assay used in this study
involved centrifugation of the homogenate ^{14}C-labeled sugar mixture, followed
by resuspension of the homogenate in non-labeled stimuli, centrifugation, and
counting of the pellet.   The assay conditions may not be very selective in
distinguishing between sweet stimuli actually bound and that entrapped.   The
background binding of the stimuli in epithelial control tissues was
considerable and poses a unique problem in eventual determination of binding
constants[74].

Ostretsova et al.[70] used equilibrium dialysis in experiments aimed at
determining whether a membrane fraction of bovine papillae had the capacity
to bind sweet stimuli.   Their results suggested, though did not rigorously
show, binding specificity of ^{14}C-glucose by the taste papillae membranes when
compared to control membranes isolated from epithelial tissue.   A recent
report by Lum and Henkin[75] supports the results of these latter experiments
and contains relative binding constants for sweet carbohydrates and two
synthetic sweeteners to bovine tissue.   The two synthetic sweeteners had
binding constants lower than those reported for sucrose, fructose, and
glucose.   Since the calf is known to show very little behavioral preference
for the synthetic sweeteners, yet avidly selects sucrose, fructose and
glucose[76], strength of binding of the stimulus with the taste receptor
tissue probably represents a behaviorally relevant step in the transduction
of taste information.

These results immediately imply certain tentative conclusions about the
initial receptor event in taste.   First, binding specificity has repeatedly
been found to tissue homogenates, not to the soluble fraction of the
homogenate.   This fact suggests that the receptor for sweet is in the non-
soluble phase, presumably membrane-bound.   Also it appears that this
membrane-bound moiety is present either uniquely or at least to a greater
degree in tissue containing taste buds.   Secondly, the correlation between
the binding results and the behavioral experiments in the same animal lends
credence to the idea of peripheral control in specificity which had not been
rigorously shown before.   The notion of peripheral control is in agreement

with the earlier studies of Oakley[77] on cross regeneration of the sensory nerves in taste. He found sensory neural activity to be a function of the region of innervation and not of the nerve innervating that area. Cagan also examined the temperature stability of this binding and found, again in agreement with behavioral and electrophysiological experiments, that the binding was reasonably temperature-insensitive though it could be inactivated by prolonged heating[72,74].

These results strongly imply that peripheral receptors are responsible for the specificity of the reception of the sweet message. Significant problems are inherent in this approach, however. The tissue used in these experiments is not highly concentrated in taste bud material. The stimuli employed, primarily carbohydrate sweeteners, are not active at sufficiently low levels and do not show sufficiently high binding constants to make receptor isolation using these stimuli a reality in the near future. Attempts to concentrate taste tissue from mammalian tissue have been described[78-81] and these approaches are worth further study. Using taste stimuli active at concentrations lower than the sweet carbohydrates is also possible, and one may initially look to the bitter stimuli to fulfill this criterion. The calf apparently is indifferent to saccharin and some other synthetic agents reported to be sweet by humans[76], but is sensitive to low concentrations of stimuli reported as bitter by humans[82-83]. The stimulus-receptor interaction might also be observed using an animal heretofore unique to taste biochemistry, the channel catfish. This animal has a high density of taste buds on its barbels and is very sensitive to dissolved amino acids. Initial biochemical results have demonstrated taste-related specific stimulus binding in a partial membrane fraction from this animal[19].

The nature of the transductive event and the mechanisms underlying it are more obscure than those of the peripheral receptive event. The possibility that cyclic AMP might be involved in the transduction led to preliminary attempts to demonstrate taste stimulus-related changes in the level of this second messenger. Early reports designed to observe this interaction using sweet stimuli[74] and bitter stimuli[74,84] were encouraging. However, a more detailed test for the involvement of the cyclic AMP system in taste transduction failed to demonstrate the classical synergism between adenylate cyclase activation and the phosphodiesterase inhibition necessary for evoking this mechanism[85]. The possibility that phosphodiesterase may be involved in the transduction of bitter stimuli sensations should be further explored, since many stimuli that taste bitter to humans are phosphodiesterase inhibitors.

Recent experiments by Mooser[86] and Mooser and Lambuth[87] have demonstrated that protein modification reagents can inhibit the integrated chorda tympani electrophysiological responses to taste stimuli. These reagents complexed primarily sulfhydryl groups in one study[86] and carboxyl groups in another study[87]. Data from the experiments including time to inhibition and ether:water partition coefficients suggested that the sulfhydryl blockers acted at sites within the membrane or even in the cell, while the carboxyl reagents acted at sites at the cell surface. The reagents were applied to the tongue surface under aqueous conditions while the taste nerve was being stimulated in a steady state. After reaction of the reagents, subsequent stimulation by sodium chloride, sucrose and hydrochloric acid was inhibited. They suggest that a transduction step was being inactivated in each case because (1) the kinetics of inactivation were independent of the concentration of sodium chloride used to preadapt the electrophysiological response and independent of the degree of receptor cell stimulation, and (2) the protein modification reagents inhibit responses due to all modalities tested, sweet, sour and salty. No bitter stimuli were used. These results do not rule out the possibility that other non-competitive events may be responsible for the inactivation.

The results are, for example, compatible with the inhibition of an allosteric effector site. Such complexing could decrease the binding of the stimulus for the active receptor site. No independent measures of receptor site-stimulus association were described in Mooser's reports[86-87], although it is possible to perform such analyses[19,65,70,72,75], at least in principle. Also no data are presented to indicate whether the modification reagents are themselves stimuli for the chorda tympani response. The kinetics of inactivation of the preadapted sodium chloride response presented by Mooser[86-87] may be confounded by the possibility that the reagents are inhibiting their own response, in addition to the response of the stimuli. The fact that complete receptor-transductive inhibition was never attained, suggests to the authors that a multistate theory may be useful in explaining taste reception. This hypothesis is similar to the allosteric mechanism alluded to above. A recent report by Mooser[88] presents kinetic evidence for this multistate receptor theory in salt stimulation in the rat.

The multistate receptor theory discussed by Mooser and Lambuth[87] and Mooser[83] may find support in the earlier interpretation of L-alanine binding to taste tissue from the channel catfish[19]. Here the authors speculate that they are measuring not the initial (presumably very rapid) complexation between stimulus and receptor but a subsequent slower step wherein the initial complex is converted to a kinetically more stable form. Based on their kinetic studies, they suggest a type of negative cooperativity within the receptor-stimulus complex that is compatible with a multistate receptor complex. That is, Mooser and Lambuth[87] may not be inhibiting the initial step, but rather this conversion step, the first step in transduction. The residual activity remaining despite apparently complete modification may be due to the ability of the complex to convert via another mechanism, less effective than the unmodified one, or the ability of the complex to convert despite the modification, but to a lesser degree than when unmodified.

The actual observation of step B in Figure 2, the membrane conformational change that initiates the change in permeability has not, as yet, been observed definitively. Our research has been concerned with this step as well as with the initial receptor event. We reported[89] and later extended and characterized[80] the isolation of a cell suspension derived from bovine taste tissues. Based on morphological observations, we concluded that our cell suspension contained primarily taste bud cells. Oxygen uptake experiments with this fraction showed that the cellular material was still viable although somewhat leaky to agents that enhance mitochondrial oxidation. The uptake of the fluorescent probe, 8-anilino-1-naphthalene sulfonate (ANS), onto these cells was described.

Fluorescent probes, in the context used here, are fluorogenic organic compounds that exhibit low fluorescence yield in polar solvents but that increase their yield and sometimes shift their emission maximum, in less polar media. Since this initial definition, many other types of fluorogenic reagents have been synthesized, many for use in specialized applications in membrane biophysics[90-91]. The fluorescent probe ANS is a polarity-sensitive probe[92] that has been used to report biophysical properties of membranes and to follow changes in membrane-associated events, including conformational changes[91].

We have found that ANS reports a fairly apolar environment when titrated onto a taste bud cell suspension (Figure 3). This apolarity is in agreement with earlier estimates of the high lipoidal content of taste papillae[93]. However, this latter study was based on extracting entire papillae, not just taste bud cells. Our fluorescent results, therefore, support the contention that taste bud cells as well contain a large amount of lipid. Whether this lipid, or functional groups specific to it, play a role in taste reception is still a matter of conjecture[93-94].

Subsequent chemical stimulation of the taste bud cells labeled with ANS produced changes in the fluorescence emission of the probe only for certain

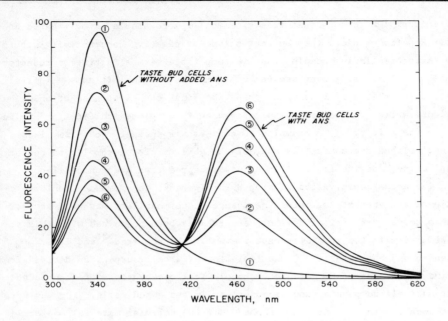

Figure 3.   Titration of a suspension of taste bud cells by the fluorescent probe, 8-anilino-1-naphthalene sulfonic acid, magnesium salt (ANS).   The total concentration of ANS in the cuvette at each step is: curve 1, zero; curve 2, 10 µmol; curve 3, 20 µmol; curve 4, 30 µmol; curve 5, 40 µmol; curve 6, 50 µmol.   The fluorescence near 340 nm is due to aromatic amino acid fluorescence of the cells, that near 465 nm is due to the bound ANS. Excitation was with 285 nm light.   An emission filter designed to transmit light efficiently only above 310 nm was interfaced between the cuvette and the photoanalyzer to eliminate reflected and scattered excitation light. Temperature was 25.0° C.   Taken from Brand and Cagan[80].

bitter tasting stimuli.   The intensely bitter stimulus, Bitrex[R] (denatonium benzoate), was a prime example of this effect (Figure 4).   Using this preparation of cells in the fluorescence arrangement previously described[80], the introduction of Bitrex leads to enhancement of ANS fluorescence plus a slight change in peak emission maximum.   This enhancement is dependent upon the concentration of Bitrex.   The change in peak intensity could be the result of a redistribution of the probe within the cell or of a change in the equilibrium between bound and free ANS leading to more ANS being bound at newly-created sites.   Whether this change in ANS fluorescence is indicative of a conformational change in the taste bud cell membrane cannot, at this point, be determined.

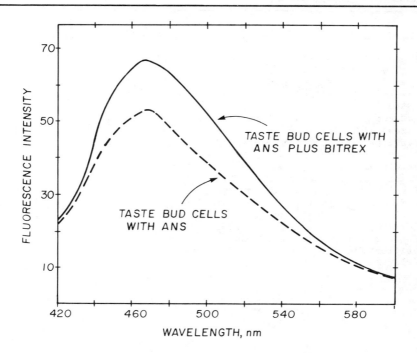

Figure 4. Change of ANS fluorescence in a suspension of taste bud cells after the addition of the bitter stimulus, Bitrex[R]. ANS concentration was 50 μmol; total cell protein was 120 μg. Bitrex was added to a final concentration of 0.67 mmol. Excitation wavelength was 375 nm. A 390 nm filter was interfaced between the cuvette and the photoanalyzer. Temperature was 25.0°C. The cells were prepared and suspended in buffered saline[80].

ACKNOWLEDGMENTS

The research from our laboratories described herein was supported in part by Research Grant No. NS-08775 (to Dr. Robert H. Cagan) and Postdoctoral Fellowship No. NS-54195 (to the author) from the National Institute of Neurological and Communicative Disorders and Stroke, USPHS. For critically reading the manuscript, the author thanks Dr. Peter Balnave, Dr. Rudy Bernard, Dr. Robert Cagan, and Dr. John Teeter. The author thanks Mr. Douglas Bayley for technical assistance.

REFERENCES

1. Kare, M.R. in Olfaction and Taste III, edited by C. Pfaffmann, Rockefeller University Press, New York, pp. 586-592, 1969.
2. Naim, M. and Kare, M.R. in The Chemical Senses and Nutrition, edited by M.R. Kare and O. Maller, Academic Press, New York, pp. 145-162, 1977.
3. Nicolaidis, S. Ann. N.Y. Acad. Science 157, 1176-1200, 1969.
4. Kakolewski, J.W. and Valenstein, E.S. in Olfaction and Taste III, edited by C. Pfaffmann, Rockefeller University Press, New York, pp. 593-600, 1969.

5. Hommel, H., Fischer, U., Retzlaff, K., and Knofler, H. Diabetologia 8, 111-116, 1972.
6. Fischer, U., Hommel, H., Ziegler, M., and Jutzi, E. Diabetologia 8, 385-390, 1972.
7. Rozin, P. and Kalat, J.W. Psychol. Rev. 78, 459-485, 1971.
8. Garcia, J., Hankins, W.G., and Rusiniak, K.W. Science 185, 823-831, 1974.
9. McBurney, D.H. Chem. Senses Flav. 1, 17-28, 1974.
10. Birch, G.G. Critical Reviews in Food Science and Nutrition, 57-95, 1976.
11. Arey, L.B., Tremaine, M.J., and Monzingo, F.L. Anat. Record 46 (Suppl. 1), 9-25, 1935.
12. Mochizuki, Y. Okajimas Folia Anat. Japan 15, 595-608, 1937.
13. Mochizuki, Y. Okajimas Folia Anat. Japan 18, 355-369, 1939.
14. Bradley, R.M. in Handbook of Sensory Physiology IV, Part 2, edited by L.M. Beidler, Springer-Verlag, New York, pp. 1-30, 1971.
15. Oakley, B. and Benjamin, R.M. Physiol. Rev. 46, 173-211, 1966.
16. Bernard, R. Am. J. Physiol. 206, 827-835, 1964.
17. Guth, L. in Handbook of Sensory Physiology IV Part 2, edited by L.M. Beidler, Springer-Verlag, New York, pp. 63-74, 1971.
18. Zalewski, A.A. Exp. Neurol. 22, 40-51, 1968.
19. Krueger, J.M. and Cagan, R.H. J. Biol. Chem. 251, 88-97, 1976.
20. Farbman, A.I. J. Ultrastruct. Res. 12, 328-350, 1965.
21. Murray, R.G. in The Ultrastructure of Sensory Organs, edited by I. Friedmann, North-Holland Pub. Co., Amsterdam, pp. 1-81, 1973.
22. Beidler, L.M. and Smallman, R.L. J. Cell Biol. 27, 263-272, 1965.
23. Ellis, R.A. J. Histochem. Cytochem. 7, 156-163, 1959.
24. Zalewski, A.A. Exp. Neurol. 25, 429-437, 1969.
25. Nada, O. and Hirata, K. Histochemistry 43, 237-240, 1975.
26. Hirata, K. and Nada, O. Cell Tiss. Res. 159, 101-108, 1975.
27. DeHan, R.S. and Graziadei, P. Life Sci. 13, 1435-1449, 1973.
28. Graziadei, P.P.C. and DeHan, R.S. Acta Anat. 80, 563-603, 1971.
29. Grover-Johnson, N. and Farbman, A.I. Cell Tiss. Res. 169, 395-403, 1976.
30. Ovalle, W.K. and Shinn, S.L. Cell Tiss. Res. 178, 375-384, 1977.
31. Bernard, R. personal communication, 1977.
32. Faull, J.R. and Halpern, B.P. Science 17, 73-75, 1972.
33. Zotterman, Y. in Progress in Brain Research, 23, Sensory Mechanisms, edited by Y. Zotterman, Elsevier Publ. Co., New York, pp. 139-154, 1967.
34. Sato, M. in Handbook of Sensory Physiology IV Part 2, edited by L.M. Beidler, Springer-Verlag, New York, pp. 116-147, 1971.
35. Sato, M. in Contributions to Sensory Physiology 2, edited by W.D. Neff, pp. 223-251, 1967.
36. Frank, M. J. Gen. Physiol. 61, 588-618, 1973.
37. Frank, M. in Olfaction and Taste V, edited by D.A. Denton and J.P. Coghlan, Academic Press, Inc., New York, pp. 59-64, 1975.
38. Erickson, R.P. in Olfaction and Taste I, edited by Y. Zotterman, Pergamon Press, New York, pp. 205-213, 1963.
39. Bernard, R. in Olfaction and Taste V, edited by D.A. Denton and J.P. Coghlan, Academic Press, New York, p. 68, 1975.
40. Wang, M.B. in Research in Physiology, edited by F.F. Kao, K. Koizumi and M. Vassalle, Aulo Gaggi Publisher, Bologna, pp. 483-488, 1971.
41. Bekesy, G. von J. Appl. Physiol. 19, 1105-1113, 1964.
42. Pfaffmann, C. Chem. Senses Flav. 1, 60-67, 1974.
43. Beidler, L.M. J. Gen. Physiol. 38, 133-139, 1954.
44. Beidler, L.M. Prog. Biophys. Biophys. Chem. 12, 107-151, 1962.
45. Tateda, H. in Olfaction and Taste II, edited by T. Hayashi, Pergamon Press, New York, pp. 383-397, 1967.

46. Halpern, B.P. and Tapper, D.N. Science 171, 1256-1258, 1971.
47. Halpern, B.P. in Olfaction and Taste V, edited by D.A. Denton and
    J.P. Coghlan, Academic Press, New York, pp. 47-52, 1975.
48. Renqvist, Y. Skand. Arch. Physiol. 38, 97-201, 1919.
49. Lasareff, P. Arch. Gesamte Physiol. 194, 293, 1922.
50. Paton, W.D.M. Proc. Roy. Soc. Ser. B. 154, 21, 1961.
51. Heck, G.L. and Erickson, R.P. Behav. Biol. 8, 687-712, 1973.
52. Smith,D.V. and Bealer, S.L. Physiol. Behav. 15, 303-314, 1975.
53. Balnave, P.A. and Wolbarsht, M.L. J. Gen. Physiol., in press, 1977.
54. Sato, T. Experientia 32, 1426-1427, 1976.
55. Smith, D.V. VI International Symposium on Olfaction and Taste, Paris,
    Abstract, 1977.
56. Kimura, K. and Beidler, L.M. J. Cell. Comp. Physiol. 58, 131-140,
    1961.
57. Akaike, N., Noma, A. and Sato, M. Proc. Japan Acad. 49, 464-469, 1973.
58. Tateda, H. and Beidler, L.M. J. Gen. Physiol. 47, 479-486, 1964.
59. Ozeki, M. Science 228, 868-869, 1970.
60. Sato, T. and Beidler, L.M. J. Gen. Physiol. 66, 735-763, 1975.
61. Ozeki, M. J. Gen. Physiol. 58, 688-699, 1971.
62. Kamo, N., Miyake, M., Kurihara, K. and Kobatake, Y. Biochim. Biophys.
    Acta 367, 1-10, 1974.
63. Miyake, M., Kamo, N., Kurihara, K. and Kobatake, Y. J. Membrane Biol.
    22, 197-209, 1975.
64. Koyama, N. and Kurihara, K. Biochim. Biophys. Acta 288, 22-26, 1972.
65. Brand, J.G., Zeeberg, B.R. and Cagan, R.H. Intern. J. Neuroscience
    7, 37-43, 1976.
66. Dastoli, F.R. and Price, S. Science 154, 905-907, 1966.
67. Koyama, N. and Kurihara, K. J. Gen. Physiol. 57, 297-302, 1971.
68. Price, S. and Hogan, R.M. in Olfaction and Taste III, edited by C.
    Pfaffmann, Rockefeller Univ. Press, New York, pp. 397-403, 1969.
69. Price, S. Nature 241, 54-55, 1973.
70. Ostretsova, I.B., Safarian, E. Kh., and Etingof, R.N. Proc. Acad. Sci.
    U.S.S.R. 223, 1484-1487, 1975.
71. Nofre, C. and Sabadie, J. Comptes Rendus 274D, 2913-2915, 1972.
72. Cagan, R.H. Biochim. Biophys. Acta 252, 199-206, 1971.
73. Kare, M.R. and Ficken, M.S. in Olfaction and Taste I, edited by Y.
    Zotterman, Pergamon Press, New York, pp. 285-297, 1963.
74. Cagan, R.H. in Sugars in Nutrition, edited by H.L. Sipple and
    K. McNutt, Academic Press, New York, pp. 19-36, 1974.
75. Lum, C.K.L. and Henkin, R.I. Biochim. Biophys. Acta 421, 380-394, 1976.
76. Kare, M.R. in Handbook of Sensory Physiology IV, Part 2, edited by
    L.M. Beidler, Springer-Verlag, New York, pp. 278-292, 1971.
77. Oakley, B. J. Physiol. 188, 353-371, 1967.
78. Lo, C.H. Biochim. Biophys. Acta 291, 650-661, 1973.
79. Uehara, S. J. Gen. Physiol. 61, 290-304, 1973.
80. Brand, J.G. and Cagan, R.H. J. Neurobiol. 7, 205-220, 1976.
81. Lum, C.K.L., Whittaker, N.F. and Henkin, R.I. Biochim. Biophys. Acta
    421, 353-361, 1976.
82. Goatcher, W.D. and Church, D.C. J. Animal Sci. 31, 373-382, 1970.
83. Chalupa, W. and Brand, J.G. unpublished observations, 1977.
84. Kurihara, K. FEBS Letters 27, 279-281, 1972.
85. Cagan, R.H. J. Neurosci. Res. 2, 363-371, 1976.
86. Mooser, G. J. Neurobiol. 7, 457-468, 1976.
87. Mooser, G. and Lambuth, N. J. Neurobiol. 8, 193-206, 1977.
88. Mooser, G. VI International Symposium on Olfaction and Taste, Paris,
    Abstract, 1977.
89. Brand, J.G. and Cagan, R.H. Soc. Neurosci. 3rd Annual Meetings,
    San Diego, Abstr. 18.1, p. 173, 1973.

90. Conti, F. in Annual Review of Biophysics and Bioengineering Vol. 4, edited by L.J. Mullins, W.A. Hagins, L. Stryer and C. Newton, Annual Reviews, Inc., Palo Alto, pp. 287-310, 1975.
91. Radda, G.K. and Vanderkooi, J. Biochim. Biophys. Acta 256, 509-549, 1972.
92. Penzer, G.R. Eur. J. Biochem. 25, 218-228, 1972.
93. Kurihara, K., Koyama, N. and Kurihara, Y. in Olfaction and Taste IV, edited by D. Schneider, Wissenschaftliche Verlagsgesellschaft, Stuttgart, pp. 234-240, 1972.
94. DeSimone, J.A. and Price, S. Biophys. J. 16, 869-881, 1976.

The biological and cultural role of sweeteners

Lloyd M. Beidler

Department of Biological Sciences, Florida State University, Tallahassee, Florida 32306

TASTE AND SWEETENERS

The pleasure associated with intake of sugars is observed with many animals. Since carbohydrates are rich in calories and are reasonably abundant in nature, it is often proposed that the pleasure of sweets is related to the search for an energy source. If the positive response to sugars is so prevalent in the animal kingdom, then one might expect that a mechanism of taste stimulation exists that is common to many different animal species. Much evidence suggests that such is the case.

TASTE MECHANISMS

Man has over 10,000 taste buds that are contained mainly on the dorsal surface of the tongue, although a number are distributed over the palate and elsewhere in the oral cavity. The highest taste bud density is found with the circumvallate papillae at the tongue posterior. A lesser number, but of greater importance in response to sweets, is associated with the fungiform papillae. Each of these contains 0 to 7 taste buds and each taste bud contains about 50 cells. Each cell lives about 10 days and then is replaced by a new one[1].

Taste buds appear early in human development and are already present in the four-month human fetus. The total number of taste buds remains rather constant from early youth to old age. Many individual taste cells within these taste buds respond to sweet stimuli but with varying sensitivity. Those taste cells very responsive to sugars may also be responsive to other stimuli but with lesser magnitude. The neural coding of tastes has been extensively studied by others but will not be discussed here[2].

How do sugars actually interact with taste cells? This subject can be studied using three major but quite distinct methods. First, the molecular structure of the taste stimulus can be related to the magnitude of the response it elicits. Second, the electrical neural activity of the

taste cell or nerve may be recorded and its magnitude and temporal characteristics related to the stimulus. Third, psychophysical measurements
of the human taste sensations can be studied and some offer limitations on
the possible transduction and neural mechanisms. Such observations are
very important, for example, in determining the relative magnitudes and
qualities of tastes associated with stimuli to be correlated in structure-
function analysis. All three methods have advantages and disadvantages,
and each is limited in the nature of the data revealed.

Electrophysiological research suggests that the sweet molecule adsorbs
to specific sites on or in the cell membrane of the taste microvilli. The
binding strength is usually only a few kilocalories per mole[3]. The membrane
molecules undergo a change in conformation that ultimately leads to a
change in electrical potential difference across the membrane. This potential change excites the taste nerve and results in neural impulses that are
transmitted toward the brain. The resultant taste sensation depends upon
which taste cells have been stimulated, to what extent, and the previous
history of stimulation. Taste mixtures can result in taste inhibition,
additivity, or synergism.

In order to perform electrophysiological experiments, a portion of the
taste nerves must be isolated. Usually this procedure is done with experimental animals by surgically freeing the chorda tympani taste nerve that
innervates the taste buds on the anterior two-thirds of the tongue. This
can be accomplished in man rather simply by merely lifting the eardrum and
placing electrodes on the chorda tympani nerve in the middle ear. A number
of experiments utilizing neural recordings in man have been undertaken by
Zotterman and Diamont in Sweden[4]. The magnitudes of neural responses to
sugars were relatively similar to the intensity of sweetness as determined
by the same subjects before surgery. Their experiments with sweet inhibitors and taste modifiers provided insights to the mechanisms of both.

If the taste stimulus adsorbs weakly to the taste cell membrane, then
there should be some common elements of the various stimuli adsorbed at any
given receptor site. At first glance it would appear that various sweet
stimuli--sucrose, chloroform, beryllium chloride, saccharin, etc.--have
little structural relationship. However, Shallenberger studied their
properties more carefully and concluded that each can form two hydrogen
bonds with a possible receptor site and that these bonds would be about
3.0 Å from one another[5]. Later, additional steric considerations were added to the requirements for an adequate sweet substance.

One might conclude from the above analysis that all sweet stimuli interact with the same receptor site on the taste cell membrane. A number
of experiments indicate that this assumption is not true. Dual sites have
been most clearly demonstrated with the fly where a "glucose" site is separate from a "fructose" site and a disaccharide may interact with both
simultaneously[6]. Similar multiple sites are known for man which greatly
complicates any structure-function study.

THE NEED FOR SWEETENERS

As noted previously, many animals respond positively to sweet substances. Such a response has been demonstrated in electrophysiological
studies of taste buds of foetal sheep[7]. Saccharin injected into the amni-
otic fluid of a pregnant woman increases the rate of swallowing in a 4-
month human foetus. Thus, operable taste buds are present before birth and
the day-old newborn already expresses a positive response to sweet stimuli.
This positive response is maintained for many years and children as well as
some adults continue to crave sweetness and often results in an increase
in caloric intake with an accompanying obesity. Sweet craving may be
thought of as a strong biological drive and therefore difficult to control.
Many years ago the source and variety of food was limited and man had little
problem with this drive, resulting in an excess sugar intake. If sugar is
offered in unlimited amount, such as to a sugar cane cutter, a sustained
sucrose intake of as high as 500 gm per day is observed and a transient ingestion of 2 to 3 lbs. in less than 10 minutes is not uncommon[8]. The recent unlimited access to sugar has produced problems in our eating habits.

The Arabs introduced sugar cane to the Western world about 1,000 years
ago but only the wealthy had real access to it until the 15th century.
Columbus brought sugar cane to the Caribbean countries where it flourished
with the help of imported African slaves. As countries increased their
per capita income, and as world sugar production increased, the per capita
sugar consumption soared. Today, the average American consumes about 125
pounds per year. Japanese consumption is presently rising as their economy improves, even though their traditional meals contain little sugar.

The great importance of sweetness to man is emphasized by the fact
that a baby at birth will prefer sweet substances to those with other
tastes as well as the fact that many animals also prefer sweetness. The
close relationship between sugar consumption and sugar availability suggests that cultural factors may not be of great importance in the craving
for sweetness as are those that determine the overall menu of a particular

culture. What other factors influence man's behavior toward sweet substances?

The genetic basis for taste discrimination and taste preference has
not been studied for most taste stimuli. However, the ability to detect a
certain group of bitter stimuli (those with a N-C=S bond) is genetically
determined[9]. Also, genotype is a major determinant of individual differences in saccharin preference in mice[10]. Thus, our genetic background may
influence our taste preferences.

The early environment in which a young animal develops can also alter
the way a sensory system is organized and thus the way it functions. To
what extent do the early patterns of a child's food consumption contribute
to the later taste preference of the adolescent? Will a child given access
to many sweet candies and drinks have a greater craving for sweet things
as an adolescent? The answers are not known. However, ample evidence indicates that experience plays a role in the establishment and modification
of food preference. In particular, young rats with much experience with
sucrose show a difference in sucrose preference in adult life as compared
to those with no early sucrose experience[11]. Also, it has been reported
that the flavor of the mother's milk may be a determinant in the preference
of foods with similar flavors in later life[12].

NEW SWEETENERS AND MODIFIERS

If a strong biological drive for sweetness exists, then it will be
difficult to circumvent it. An alternative approach is to offer sweeteners
that have none of the disadvantageous properties of sugars and are less expensive. The accidental discoveries of the sweetness of calcium cyclamate
and sodium saccharin allowed man to satisfy his craving for sweetness without increasing his caloric intake. Since some predict that over 80% of
cancers are induced by environmental factors including food additives, and
since some believe that certain carcinogens in mixtures are additive, there
is a trend to limit food additives including cyclamates and saccharins.
What are the alternatives?

One might examine other more undeveloped cultures where sugar is still
not too available. In recent years, this search led to the recognition of
a number of naturally occurring sweet materials. In addition, inhibitors
and modifiers of the sweet sensation have been studied. Finally, a few new
noncarbohydrate molecules have been found to be sweet and can be synthesized. None of these by themselves have properties that allow them to be
substituted for saccharin or sucrose in all foods. However, each has potential application for the use with a restricted number of foods.

It is interesting to consider the impact of our culture and mores on the potential use of a new sweetener. Society would like a continued use of foods with a pleasant sweetness. Industry, however, imposes more stringent restrictions and it is reluctant to change old habits. Industry would like a new "saccharin" which has sweetness without bitterness, sweetness without aftertastes, and a product thermally stable and with a long shelf life. Ideally the sweetener should also have a high osmotic pressure to act as a preservative and glass-like physical properties so that it could be used to make candies without sugars. Since none of the new sweeteners (dihydrochalcones, aspartame, monellin, miraculin, etc.) have all these properties, industry is resistant to their use.

The product most developed in recent years, miraculin, illustrates the difficulty. About $7,000,000 was spent over about a five-year period to develop a product to be sold to a consumer with culturally oriented tastes and to an industry with a culturally oriented methodology. The product includes a large glycoprotein of about 44,000 molecular weight that tastes sweet if the pH of the tongue surface is low[13]. Thus, since miraculin binds strongly to the taste bud and is active over a period of many minutes to an hour, the sweetness can be turned on and off at will be merely changing the pH. Since low pH elicits sourness, one might also expect a sour taste. Actually the sweetness experienced is a tarty or fruity sweet as expected for a mixture of sourness and sweetness. Electrophysiological recordings with both humans and monkeys showed that miraculin stimulates the sweet receptors at low pH. That is, miraculin acts like any other sweetener and does not change the responsiveness of other taste cells to any other taste nor modify the taste cells like a drug. Unfortunately, the early literature referred to the modification of the taste sensation (psychological, not physiological, modification), leading some to conclude wrongly that miraculin is a drug[14].

Miraculin was sold on the market for a short time with favorable reactions by thousands of diabetics. This compound has the distinct advantage that it is not necessary to eat it every time you would like sweetness. It can be taken once before a meal and sweetness can be obtained by using a sour substance. Miraculin has no effect on bitter, salty, or other sweet substances[14]. Thus, a meal can be made without sugars or continued use of artificial sweeteners, but miraculin does require a remaking of the menu. Diabetics are highly motivated and are willing to change their eating habits previously culturally determined. Other people, however, do not desire to change.

A certain number of special products can be formulated. Chewing gum with miraculin and an acid releasing substance will double the normal flavor retention time. Patents have been granted for use with such foods as yogurt and soft drinks. In all cases, miraculin cannot simply be added to a product but the product must be completely redesigned, necessitating much development and much expense.

A change in our eating habits as well as a change in food formulation must be expected if present food additives, whether saccharin or nitrites, are to be either eliminated or restricted. Neither society nor industry is happy about such changes.

Application of our knowledge concerning taste mechanisms may help to solve the sweetness problem in the future. Three possibilities are apparent: other cultures can be search for natural substances used as sweeteners; structure-function relationships can be studied to synthesize new sweet tasting molecules; or inhibitors or other modifiers of sweet taste can be introduced. Brief examples of each will be given and others in this conference will expand on these subjects.

Inglett pioneered in the search for natural sweeteners and initiated studies on miracle fruit and other sweet proteins. Stevioside was used by certain populations in Paraguay. Glycorrhizin is used today as a licorice sweetener. The Indians of Venezuela add to their jellies the leaves of a fern that grows high in the mountains. Lists of economic plants around the world include a large number of other sweeteners.

Once the chemical structure of a sweetener is established, many variations and analogs can be synthesized. In recent years the dihydrochalcones and dipeptides have been particularly studied. Increased sweetness, decreased bitterness, faster onset, less aftertaste, etc. are examined. The most interesting and of great potential use are the proteins discovered to have a sweet taste.

The group of substances that have been little studied for possible commercial use are the sweet inhibitors. Gymnemic acid has been known for almost 100 years and was sold in Germany by Merck for control of diabetes shortly after World War I. It can almost completely abolish sweet sensations for over an hour. It would be of particular use for those who crave sweets between meals. The structure and pharmacological properties probably prohibits its introduction to the market. A number of other inhibitors exist, particularly disulfide inhibitors.

None of the above rational approaches have really helped much in the development of sweeteners or inhibitors. The main sweeteners such as saccharin, cyclamate and aspartame are the results of accidental discoveries. However, the science of taste and its mechanisms is progressing rapidly and will contribute to our future success in increasing the probability of guessing the correct structures that elicit sweetness.

REFERENCES

1. Beidler, L. M. and Smallman, R. L., J. Cell Biol. 27; 263, 1965.
2. Frank, M., J. Gen. Physiol. 61; 588, 1973.
3. Beidler, L. M., J. Gen. Physiol. 38; 133, 1954.
4. Diamant, H., Oakley, B., Ström, L., Wells, C. and Zotterman, Y. Acta Physiol. Scand. 64; 67, 1965.
5. Shallenberger, R. S. and Acree, T. E. Chapter 12 in Handbook of Sensory Physiology, Vol. IV, L. M. Beidler, Ed., Springer-Verlag, New York, N. Y., 1971.
6. Morita, H. and Shiraishi, A. J. Gen. Physiol. 52; 559, 1968.
7. Bradley, R. M. and Mistretta, C. M. J. Physiol. 231; 271, 1973.
8. Campbell, G. D. Personal communication.
9. Kalmus, H. Chapter 9 in Handbook of Sensory Physiology, Vol. IV, L. M. Beidler, Ed., Springer-Verlag, New York, N. Y., 1971.
10. Pelz, W. E., Whitney, E. and Smith, J. C. Physiol. Behav. 10; 263, 1973.
11. Greenfield, H. Ph.D. Thesis, University of London, 1970.
12. Galef, B. G. and Henderson, P. W. J. Comp. Physiol. Psychol. 78; 213, 1972.
13. Kurihara, K. Chapter 16 in Handbook of Sensory Physiology, Vol. IV, L. M. Beidler, Ed., Springer-Verlag, New York, N. Y., 1971.
14. Bartoshuk, L. M., Gentile, R. L., Moskowitz, H. R. and Meiselman, H.L. Physiol. Behav. 12; 449, 1974.

The psychology of sweetness: historical trends and current research

Howard R. Moskowitz

MPI Sensory Testing, Inc., 770 Lexington Avenue, New York, NY 10021

AN OVERVIEW

The psychology of sweetness is a multi-disciplinary body of findings. In contrast to other branches of psychology, the study of sweetness perception is an amalgam of many different, often disjointed, research inputs. The array of references in the bibliography of this report gives quiet testimony to the diversity of research efforts and the array of research talent.

One might well inquire as to whether underneath all of the research cited here there might lie one or a parsimonious few general "organizing principles". If so, could we, perhaps, develop a predictive model that would account for the diversity of findings and also unify the inputs of all the contributing disciplines. Physiologists, chemists, psychologists, and other basic researchers look for such unifying principles as keystones to the growth of their respective sciences. Unfortunately, no such general principles yet have been found to explain molecular structure vs. quality and intensity of sweetness, or hedonics of sweetness, to name just two areas. We are still far from having integrative models to account for (if not to explain) sweetness perceptions. Nor, perhaps, ought we expect to have such principles soon. As science advances in all areas touching on the sense of sweet, new methods and findings emerge more quickly than do theories to account for them.

In prospect, therefore, we have an interesting future awaiting us. We can look forward to an increase in knowledge about sweetness perception, and, unfortunately, perhaps to a consistent gradual reduction in the number of theories that valiantly try to account for the data.

INTRODUCTION

The taste sensation of sweetness plays an unusually important role for man. Studies of neonates suggest that there is an innate preference for sweetness, which is independent of learning[1,2]. Children in our American culture are accustomed to consume sweet foods in quantity, be they cereals, candies or cakes and cookies. Culturally, sweetness is associated with good things. There is an old Jewish custom to coat a page of a young child's first Bible text in school with honey, so that the words of learning will taste sweet to him.

Underlying these aspects of sweetness lies a tremendous and still-growing volume of information on how we respond to sweet-tasting materials. That information is the basis for a psychology of sweetness.   The data derive from several disciplines, including psychology itself, food science, anthropology, sociology and economics, in addition to the biological sciences.

This paper considers three aspects of the psychology of sweetness: Quality, Intensity, and Hedonics (liking/disliking).   The data presented derive both from scientific studies with model systems and from applied studies using consumable foods.   Much of the data that researchers have obtained with model systems exhibit well-behaved functions relating concentration to perception of sweetness, and to judgments of hedonic tone, but those taste stimuli are too simple to be considered foods.   In contrast, research on real food items uncovers the complexities in sweetness perception which are encountered when the food stimulus provokes taste responses, flavor responses, and cognitive expectations.   A comparison of the two types of data, model systems and real foods, in the three aspects of quality, intensity and hedonics, can provide a good balanced perspective.

QUALITY OF SWEETNESS

A century ago, Fick[3] proposed four primary tastes, sweet, salty, sour and bitter.   However, other researchers before and after Fick suggested that there are many other tastes in addition to these four.   Table 1 shows a partial listing of the suggested primary tastes which have appeared during the course of the last several centuries[4].

The list of tastes in Table 1 comprises other sensory characteristics as

Table 1.   Qualities of taste (taste primaries)[4]

Bravo (1592)	Linnaeus (1751)	Haller (1751)	Fick (1864)	Bekesy (1964)
sweet	sweet	sweet	sweet	sweet-bitter
bitter	bitter	bitter	bitter	-
acid	acid	acid	sour	sour-salty
saline	saline	saline	salty	-
sharp	sharp	sharp	-	-
insipid	insipid	insipid	-	-
pungent	-	pungent	-	-
harsh	-	harsh	-	-
fatty	fatty	-	-	-
-	nauseous	putrid	-	-
-	-	spiritous	-	-
-	-	aromatic	-	-
-	astringent	-	-	-
-	viscous	-	-	-
-	aqueous	-	-	-
-	-	urinous	-	-

As well as: metallic (various authors), alkaline (various authors) and MSG "taste" ("Vollmundigkeit") (various authors)

well, which are more properly classified as texture and smell. Researchers living in past centuries were not fortunate to have well-purified chemicals with which to work, so that in their sensory description of stimuli they often measured effects of contaminants which stimulated texture and smell impressions as well as taste sensations. In addition, the fine distinctions between smells and tastes were not made.

In addition to these categorizations, sweetness itself comprises many nuances, as Greeves[5] suggested in his description of the tastes and smells of the materia medica in use 150 years ago. Greeves' list of sweetness nuances comprised tactile and smell sensations, as well as taste sensations. Table 2 shows Greeves' listing for sweetness.

With the advent of organic chemistry, the discovery of saccharin in 1879 and of many other artificial sweeteners, the problem of describing the quality of sweetness reared itself again. Numerous adjectives were used to describe the sweet taste, especially by Cohn[6], who in his compendium "Die Organischen Geschmackstoffe" literally tried to describe the taste of sweeteners in the published chemical literature. Cohn's description of sweetness emphasized the recognition of many side-tastes that could differentiate one sweet-tasting material from another.

The commercial development of sweeteners, starting with invert sugar in the 1920's, continuing with the analysis and production of cyclamates, and continuing into the current development and evaluation of such sweeteners as aspartame and dihydrochalcones, forced researchers to continue their study of sweetness quality in order to produce a product that consumers find to be an acceptable substitute for sugar. To this end various studies, both formal and informal, were undertaken to describe the sweet quality of various materials. For instance, Rader et al.[7] through high refinement and crystallization of saccharin, attempted to ascertain the "true" quality of saccharin, in order to determine whether the bitter taste is inherent in the saccharin molecule, or whether the bitter taste derives from the impurities inadvertently included during crystallization. The true taste of saccharin comprises a bitter note, even under the purest syntheses.

Subsequent work on the quality of sweeteners has focused upon the

Table 2. Greeves' (1828) categories of sweetness in the materia medica

Category	Characteristics
1. amylaceous	intermediate between insipid and sweet (e.g. starch)
2. mucous-unctuous	similar to amylaceous, but with an unctuous or oily character (e.g. gums)
3. faint	slight bitter + sweet
4. frugous	flavor of fruit + sweet
5. sweet-spicy	sweet + warmth, heat

description of sweetness sensation, often with a meager vocabulary. Few panelists possess an adequate and communicable list of descriptor terms with which to describe taste perceptions. As a result, two primary approaches in quality evaluations have had to be followed:

Approach 1

Consumers (or panelists) are forced to use a series of reference chemical standards to reflect various taste qualities (e.g., types of sweeteners). Then, in the evaluation of a new, previously uninvestigated compound, experimenters can train the panel to point to the chemical which produces a similar quality. This approach requires extensive training and by virtue of its cost and investment finds only a moderate use in large-scale screenings. Its major use is with small, expert screening panels, for quick evaluation of taste impressions (e.g., initial go - no-go decisions).

Approach 2

Consumers evaluate tastes using their own vocabulary to describe taste quality. The consumers may use a structured questionnaire, wherein they are first instructed to rate new sweeteners in terms of degree of sweetness and to add descriptions of other tastes (i.e., how sweet, sour, salty or bitter the sweeteners taste), as well as describe the sensations in their own terms. This method can be used with many panelists, is cost-efficient in terms of the quantity of data obtained, and from the large matrix of data (many panelists, many sweeteners, with varying levels of concentration), the experimenter can begin to build a descriptor profile of quality.

Both of the foregoing methods have shown the following general results for artificial sweeteners:

(1) The sweeteners provoke different qualities of taste compared to sucrose or other carbohydrate sweeteners.

(2) The nature of this taste is hard to quantify. Some of the descriptors used include bitter, and harsh, as well as the more nebulous word "off-taste" (for lack of a better term).

(3) Quite often impure artificial sweeteners provoke smell sensations as well, but this is probably due to the volatiles that are included (e.g. especially the licorice smell of glycyrrhizin).

(4) The sweetness of artificial sweeteners is not described as a "clean" sweetness, whereas the sweetness of sucrose is "clean".

(5) The time course of the sweetness/taste impression of artificial sweeteners also differs from that of sucrose, thus leading to a perception of

quality difference. Artificial sweeteners possess a more "ragged", slower-decaying sweetness sensation over time. In contrast, carbohydrate decline once the sweetening agent is removed from the mouth.

Lest it be thought that the sucrose sweetness sensation is the optimum sweetness, one should note that an "ethereal" and almost perfect sweetness was evoked and reported by Bekesy[8], in his stimulation of the single papillas of the tongue by electrical current.

In actual food products the differing tastes of sweeteners show up in reports both of modified sweetness sensation, as well as modified flavor and texture. For instance, the substitution of equally sweet, high-fructose corn syrup for sucrose or invert sucrose in a beverage may emerge as a report of modified flavor. Although the sweetener, and only the sweetener, has been changed, the panelist perceives a change in the overall complex of sensations, and attributes this to the flavoring agents, rather than to the sweetening agents.

MEASURES OF SWEETNESS DEGREE OR INTENSITY

Psychological measurement of sensations, sweetness notwithstanding, has evolved slowly and with many tortuous side-paths over the past century. To the non-psychologist it appears difficult to assign numbers to sensations with the same ease that one assigns numbers to more measurable aspects of physical reality, such as force and mass. Yet the past century has seen the rise of sensory measurement systems which exhibit the same type of mathematical rigor and usefulness as physical measurements do.

The methods commonly used to measure degree of sweetness are: (1) threshold, (2) equal-sweetness matches, (3) category scaling, and (4) ratio scaling.

Thresholds

Threshold measurement aims to determine what is the lowest concentration of a sweetener that the individual can either detect (called the detection threshold) or recognize (called the recognition threshold). The only response which the individual makes in detection is to state "yes" or "no" (or, it is there vs. it is not there). In recognition, the individual must describe the quality that he or she detects. These two thresholds are quite different entities. Sodium chloride, for instance, exhibits a concentration (around 0.005 or less M) where it begins to taste sweet. That abnormally low level is its sweetness recognition threshold, even though its typical taste is salty[9].

Table 3 shows a compilation of threshold values for a variety of
sweetener substances.   Note that this is only a partial listing.   Many
thresholds have been published by the American Society for Testing and
Materials, in their Series DS48[10].

Some cautions are warranted in interpreting taste thresholds:

(1) The threshold is not a fixed point but instead must be considered
a range.   There is no single threshold, as there is a single temperature
level at which water boils.   The publication of various threshold values
represents averages, often from a quite widely-dispersed group of individual
threshold values.

(2) The thresholds represent a physical point (or range) at which the
individual says that he or she cannot detect or recognize a stimulus.   Thus,
the individual subject is actually measuring with a limited response scale.
The responses are limited to yes or no.   The threshold values thus

Table 3. Thresholds for sucrose and other sweeteners

A. Sucrose thresholds

Compound	Threshold* % Wt./Vol.		Source**
sucrose			
males	1/199 D		Bailey and Nichols, 1887
females	1/204 D		" " "
sucrose	0.17 D		Richter and Campbell, 1940
	0.55 D	1.30 R	Fabian and Blum, 1943
	0.51 D		Janowitz and Grossman, 1949
	0.35 D		Schutz and Pilgrim, 1959
	0.28 D		Yensen, 1959; Furchgott and Fried-man, 1960, cited by Linker, Moore and Galanter, 1964
		0.685 R	Crocker and Henderson, 1932; Berg, Filipello, Hinreiner and Webb, 1955
	0.275 D	0.54 R	Cooper, Balish and Zubek, 1959
		0.274 R	Pangborn, 1959
	0.35 D	0.35 R	Kelty and Mayer, 1971

B. Other sweetener thresholds

Compound	Threshold		Source**
saccharin	1.6×10^{-5} M		Williams, 1970
sodium saccharin	2.0×10^{-5} M		
glucose	0.721 – 1.621%		Pfaffmann, 1959
DL-erythritol	0.81 of sucrose		Dermer, 1946
fructose	0.80	"	"
ethylene glycol	0.77	"	"
pentaerythritol	0.91	"	"
glycerol	0.92	"	"
L-arabitol	2.17	"	"
dulcitol	1.35	"	"
D-mannitol	1.75	"	"
DL-sorbitol	1.85	"	"
D-xylose	2.00	"	"
maltose	2.22	"	"
phamnose	3.125	"	"
galactose	3.22	"	"
quebrachitol	3.22	"	"
raffinose	4.54	"	"

* D = detection threshold, R-recognition threshold.
** See references for Table 3 at end of general reference list.

represent <u>physical</u> <u>concentrations</u> which evoke precisely the same sensory
response.   One cannot measure relative sensory intensity using thresholds.
One can say that sweetener A is 500 times more <u>potent</u> than sweetener B in
evoking a sensory response, if the threshold for A is 1/500 that of B.   One
cannot state, however, that sweetener A is 500 times sweeter than B.   The
panelist has never directly assessed sweetness.   Although seemingly
academic, this difference between potency and sensory intensity is critical,
for many claims are made about <u>relative</u> <u>sweetness</u>.   What should be claimed is
relative potency of compounds.

<u>Some Major Results Concerning Sweetness Thresholds</u>

(1)   The thresholds for carbohydrate sweeteners are generally many times
higher than the thresholds for the so-called artificial sweeteners, even when
molecular weight is accounted for.   Thus, artificial sweeteners are usually
more potent.

(2)   Many chemicals (especially sodium chloride) exhibit sweetness
thresholds, at an atypically low concentration, where they do not exhibit
their standard, prototypical tastes.   This indicates that there is a range
of thresholds provoked by tastants, and that a taste can go through many
different thresholds.   Even such complicated stimuli as monosodium glutamate
and inosine monophosphate exhibit a concentration where they taste sweet.

(3)   There is a wide range of inter-individual variation in thresholds,
both for sweetness in particular, and for taste perception in general.   It is
not unusual to find a 10-fold range in threshold.   As far as can be determined, the threshold distribution is approximately normal.   There do not
seem to be multi-mode thresholds for sweetness of chemicals as there are for
the bitterness of some alkaloids, such as phenyl thiocarbamides.

(4)   Sweetness thresholds increase with age[11].   This means there is a
diminished ability to detect sweetness as we age.   Unfortunately, sufficient
data have never been published on the exact nature of the sensitivity shifts,
nor the extent of the shifts for a representative variety of sweeteners.

(5)   The threshold can be increased, often quite substantially, if the
solvent medium for taste is not pure aqueous solution.   If the solvent
background provokes another taste, then sweetness threshold increases (e.g.,
in the presence of sodium chloride, simultaneously dissolved in solution[12]).
The degree of threshold increase depends upon the nature and concentration of
the other material in the solution (how strong it is, whether it exceeds its
own threshold), and the nature of the sweetener being tested.   Sucrose may
be less affected than glucose by some taste agents, but more affected by

other taste agents.

(6) The threshold can be modified by temperature variations. Usually, stimuli nearest body temperature (34°-37° C) exhibit the lowest threshold (i.e., we are most sensitive to them[13]).

(7) Thresholds will increase if the solvent medium is not liquid, but rather a foam or a gel[14,15]. Furthermore, thresholds can be increased if the solution is viscosified (i.e., by the addition of sodium carboxymethyl cellulose or other non-tastable viscosifying agents).

(8) The threshold can be modified by psychological means that have little to do with sensory processes. For instance, in threshold determination, when there is a reward for correctly guessing if a taste is present, and no punishment for stating that a taste is present when in fact the sample is a blank (called a false alarm), then most individuals will often state that they detect a taste since there is no penalty for incorrect guessing. This guessing artificially lowers the threshold value. The apparent threshold level can be increased, in contrast, if severe penalties are levied for an incorrect guess that a taste is present when in reality it is absent, but no penalty is levied if the respondent fails to detect when a taste is present. In that situation the rational individual will reduce the number of guesses and become more conservative.

(9) Studies have shown that taste concentrations at equal number of threshold units above threshold are not necessarily equally sweet. That means that threshold values cannot be used as reference bases, multiples of which base can be used to scale perceived sweetness. A study by Lemberger[16] showed this very nicely. Saccharin and sucrose thresholds were measured for a group of individuals. Concentrations at ten times threshold were not equally sweet, although they lay in the same multiple of threshold. Thus the threshold value is not necessarily a fixed, equal reference point for stimuli, on which a sensory scale of sweetness (or other taste) intensity can be based.

Equal Sweetness Matches

If thresholds are abandoned as units of measurement for sweetness, then the next higher form of measurement, at least in a historical sense, is equal sweetness determination. The experimenter first presents to the panelist a reference sweetener at a fixed level (e.g., 1% or 10% sucrose). Then the panelist is presented with a series of comparison sweeteners, which range in concentration. The test requirement is to find a comparison sweetener which tastes as sweet as the reference. A number of major results have been published using this method:

(1) Equal sweetness matches between sugars usually are in equivalent or overlapping concentration ranges. This means that for two sweeteners, e.g., sucrose and lactose, rarely is the ratio of concentration more than 10:1 for equal sweetness[17,18].

(2) For a comparison of artificial and carbohydrate sweeteners, the ratio of equally sweet concentrations is more in the order of 100's or 10's to 1. Table 4 shows some of the relative concentrations needed to produce equal sweetness matches, for a variety of sweeteners.

(3) Experimenters who have used equal sweetness matching have often assumed that the ratio of concentrations needed to produce the equal sweetness match defines relative sweetness, rather than defining relative potency. As one might expect, this confusion leads to such claims as "saccharin is 180-220 times sweeter than sucrose". Such conclusions are probably incorrect as measures of relative sweetness just as ratios of thresholds are incorrect.

(4) If experimenters determine the equal sweetness matches at several concentrations, then they can plot "equal sweetness contours". Cameron[17] did this for many sugars, as well as amino acids. Pangborn[18] developed the equal sweetness contours for sucrose vs. glucose, fructose and lactose, respectively. The contours turn out to be straight lines if the experimenter plots the logarithm of sucrose concentration vs. the logarithm of the equally sweet test concentration, for the following conditions:

(a) If the equal sweetness match is done within the middle range of concentrations of each sweetener (especially for artificial sweeteners such as saccharin, whose high concentrations provoke bitterness as well as sweetness[19,20].

(b) If the sweetness match is done with glucose as a reference instead of sucrose, Cameron reported[21,22] that equal sweetness matches with sucrose as the reference sweetener often were non-linear, whereas the same

Table 4. Concentrations as sweet as 10% sucrose (% by weight)

raffinose	45.4	D-glucose	14.4
rhamnose	30.3	glycerol	12.7
lactose	25.6	invert sugar	15.4
dulcitol	24.3	D-fructose	8.8
maltose	21.7	glycyrrhizin	.00066
D-sorbitol	19.6	stevioside	.00033
D-mannose	16.9	Ca cyclamate	.0029
galactose	15.8	dulcin	.001
D-xylose	14.9	saccharin	.00033
D-mannitol	14.5		

equal sweetness matches tended to remain linear when glucose was the reference sweetener. Quite often the equal sweetness matching function exhibits a slope of 1.0 in log-log coordinates.

(c) If instead of equal sweetness matches between sugars, one of the sweeteners is an artificial sweetener or an amino acid, then the straight line relation in log-log coordinates may either become a curve (e.g., with amino acids[17]), or the slope may dramatically shift from 1.0 to around 1.7 (as it does with saccharin: cf. Magidson and Gorabachow[19] and Taufel and Klemm[20], whose equal sweetness matches were replotted by Moskowitz[23]). Although equal sweetness matches do not yield a true sensory measure of perceived sweetness, the observation that the matching function fails to exhibit a 1.0 slope in log-log coordinates means that the sweetness of the two substances must grow sensorically at two different rates. The relative rate of sweetness growth can be assessed. For instance, in the case of sweetness matches between sucrose and saccharin[23] the function was: saccharin $= K(sucrose)^{1.7}$. A ten-fold change in the concentration of sucrose required a $(10)^{1.7}$ or approximately a 50-unit change in the concentration of saccharin to maintain equal sweetness. Hence, saccharin sweetness grows at a far slower rate than sucrose sweetness (albeit both rates are in absolute values as yet undetermined).

Category Scaling

With category scaling the experimenter requires the subject to choose a number from a set of discrete, equally-spaced numbers to represent sensory intensity. The scale may comprise any number of points, depending upon the whim and experiences of the particular experimenter. The scale points are usually assumed to reflect equal intervals of sensory intensity.

Category scaling in the chemical sense can be traced back a century ago to the French physiologist Corin[24] who instructed a group of panelists to assign an appropriate descriptor from an array of graded verbal descriptors to each of various concentrations of acids. The descriptor terms themselves were assumed to reflect equal increments in sensory intensity (just as, for instance, the sensory levels of "no sourness", "barely sour", "mildly sour", "moderately sour", etc., can be assigned sensory intensities on a graded scale).

The results of a comprehensive category scaling study were reported by Schutz and Pilgrim[25]. The subjects in that study were presented with various pure sweeteners at different aqueous concentrations, and told to select the appropriate category from a graded numerical scale to reflect the sweetness. Among the salient findings in that study were:

(1) Judged sweetness vs. concentration described on different curves, depending upon the sweetener. For instance, alanine, glycerol, maltose and glycine all described concave upwards curves for category value vs. log concentration, whereas fructose, sucrose, dulcin, saccharin and cyclamate all described concave downwards curves. This curvature reflects different rates of sweetness growth with sweetener concentration. As a general rule, carbohydrate sweeteners, which describe the concave upwards curve, grow more rapidly in sweetness than do those describing a concave downwards curve.

(2) At the same concentrations, different sweeteners are assigned different category values. Thus, sweeteners are not equally sweet at matched concentrations, which confirm previous equal sweetness ratings for sweeteners at disparate concentrations.

(3) The logarithmic nature of the functions means that equal ratios of concentrations produce almost equal additive increments in rated sweetness. The functions found by Schutz and Pilgrim were almost, but not quite, 100% logarithmic in nature. A perfect logarithmic function would be linear in log-linear coordinates. Thus, the prototypical sweetness equation for these data would have to be written as:

$$\text{sweetness} = k \left[\log (\text{concentration} + c) \right] + k'$$

The additive constant c is necessary to correct for the curvature. It is not clear whether c has any psychological meaning for the data, other than as a correction and linearizing factor.

Despite the attractiveness of the category scale as a measuring method for suprathreshold sweetness, category scales are subject to several biases. Stevens and Galanter[26] reviewed some of them extensively, in a detailed analysis of category scales for a dozen perceptual continua. Among the problems and biases which they uncovered are the following:

(1) When using a category scale, the subject often may hesitate to use the extreme scale points for fear of running out of numbers. This is known as the "category end" effect. It can effectively truncate a 9-point scale to a 7-point scale, and reduces the usefulness of the highest and lowest categories of any fixed scale. The severity of the end effect varies with (a) the conservatism of the subjects (more conservative subjects will avoid the end points more often), and (b) the number of available categories (the effect is lessened if the subject has available a larger number of inter-mediate categories).

(2) The number of available categories determines the slope of the intensity functions. Subjects tend to match their distribution of category assignments to the available range of categories. Thus for a fixed range of

sugar concentrations, the function may appear steeper if the subject assigns numbers using a 9-point scale, and flatter if the subject assigns numbers on a 7- or 5-point scale, respectively. Unfortunately, from laboratory to laboratory there seems to be relatively little consistency in sweetness functions, due to the use of different category scales, different concentrations of sweeteners, etc.

(3) The subjective differences in sweetnesses on the category scale are unequal. For example, differences between the first and second category may be smaller than differences between the second highest and highest categories (e.g., on a 1-9 point scale, the differences between 1 and 2 are smaller than the differences between 8 and 9). Table 5 shows the psychological distances between adjacent categories in a typical category scale as quantified by the method of magnitude estimation (see below).

Despite the problems associated with the category scale, the scale continues to be a useful measuring instrument for at least two reasons:

(a) Consumers (unpracticed subjects) can use it to assign numbers to reflect perceived sweetness, thus allowing for a truer measurement of the subjective sweetness sensation.

(b) The dose-response curve between category value and physical concentration of sweetness can be used to compare the rate of growth of different sweeteners (assuming, of course, that the same size scale in terms of numbers of categories was used).

Ratio Scaling (Magnitude Estimation)

The ratio-scaling method for developing sweetness scales originated

Table 5. Typical sweetness categories scaled by magnitude estimation[*] and comparison to actual sweetened beverages rated on same scale

Category	Mean Numerical Rating
extremely sweet	183.5
very sweet	135.1
moderately sweet	95.1
slightly sweet	58.1
barely sweet	24.6
not at all sweet	0
% Sucrose in Beverages (W/V)	
22.5	129.8
15.0	100.5
10.0	70.9
7.5	47.2
5.0	33.8

* Panelist (1) evaluated various sweetened products by magnitude estimation (beverages containing 22.5%, 15%, 10%, 7.5% and 5% w/v sucrose, respectively), (2) assigned numbers in the same magnitude estimation framework to match other hypothetical beverages that are "extremely sweet", etc.

three decades ago, from the pioneering research of Beebe-Center[27] at Harvard University who was interested in erecting ratio scales of taste intensity, so that (a) various concentrations of sucrose would be assigned sensory scale values called "gusts", and (b) these gusts would exhibit ratio properties (viz., a taste of 80 gusts was to be subjective _twice_ as strong as a taste of 40 gusts). Beebe-Center's gust scales were erected for the four basic tastes, with sucrose serving as the typical sweetener[28].

Stevens[29] reported the outcome of a series of experiments on sensory scaling which were later to allow the more rapid development of ratio scales for sweetness. In the preliminary study, he showed subjects lights of different luminances and asked them to assign numbers to the lights to reflect perceived brightness. No restrictions on the numbers were made, except that ratios of numbers assigned (first called absolute judgments, but then called magnitude estimates) were to reflect, as closely as possible, ratios of brightness. The function thus obtained appeared concave -- downwards, meaning that large changes in measured physical luminance of the light produced small changes in the numbers that the subject assigned. In linear coordinates, this relation turned out to be curvilinear and concave downwards but in logarithmic coordinates the curve straightened out. The equation governing the data was a power function, of the form:

$$ME = k(PL)^n \text{ or } \log(ME) = n(\log PL) + \log k \tag{1}$$

(ME = magnitude estimates = numerical intensity ratings; PL = measured physical luminance; k = multiplier constant; n = exponent)

In research on taste sweeteners, the application of the magnitude estimation procedure has illustrated some important differences among various sweetening agents in aqueous solution. The major findings are:

(1) Sweeteners exhibit different dose-response functions, as indexed by the exponent $\underline{n}$. For sweeteners, $\underline{n}$ is usually larger if the sweetener is carbohydrate-derived (e.g., sugars, polyhydric alcohols) than if the sweetener is "artificial" (viz. saccharin, cyclamate). Table 6 lists some representative power function exponents for different sweeteners, evaluated in water solutions.

(2) At higher concentrations, artificial sweeteners lose some of their perceived sweetness. As a result, dose-response relation as presented in log-log coordinates flattens out, and the curve may even proceed downward, indicating that higher concentrations are less sweet than lower concentrations. This effect is most noticeable for the taste of saccharin[30].

(3) Carbohydrate sweeteners may exhibit parallel dose-response

Table 6. Parameters of the power function $S = kC^n$ relating sweetness and concentration [31]

A. SUGARS			
Sugar	n Least- Squares Exponent	k Molarity Intercept	k Percentage Intercept
experiment I			
glucose	1.31	29.21	.65
dulcitol	.48	4.43	1.51
glycerol	1.14	11.53	.93
ribose	1.35	20.61	.52
D-xylose	1.31	16.71	.48
experiment II			
glucose	1.09	22.18	.94
arabinose	1.22	17.71	.84
fructose	.99	39.92	2.26
sorbose	1.04	23.46	1.15
experiment III			
glucose	1.25	24.83	.66
galactose	1.16	17.65	.61
inositol	1.35	22.34	.45
mannitol	1.26	20.19	.52
experiment IV			
glucose	1.31	27.33	.61
maltose	1.30	24.79	.25
sorbitol	1.37	14.81	.28
sucrose	1.03	59.38	1.56
experiment V			
glucose	1.05	23.82	1.13
lactose	1.21	35.28	.48
raffinose	1.02	24.74	.38
rhamnose	1.54	38.20	.59
experiment VI			
glucose	1.73	31.69	.21
adonitol	1.87	15.61	.09
D-arabitol	2.30	21.06	.04
L-arabitol	1.37	13.75	.33
cellobiose	1.31	18.54	.18
erythritol	1.69	20.81	.30
melibiose	1.17	13.46	.19
melezitose	2.35	99.14	.01
trehalose	1.27	18.57	.18
experiment VII			
glucose	1.70	27.01	.26
α-1-fucose	1.37	14.87	.32
lyxose	1.63	17.78	.21
ribitol	1.50	21.01	.35
turanose	.84	15.02	.77
xylitol	1.48	33.54	.60
L-xylose	1.46	17.97	.36
experiment VIII			
glucose	1.39	26.20	.47
2-deoxy-galactose	1.82	6.15	.62
6-deoxy-galactose	1.55	26.27	.34
6-deoxy-glucose	1.28	16.50	.46
2-deoxy-mannose	.65	4.94	.80
2-deoxy-ribose	2.61	5.38	.006
β-D-fucose	1.45	16.57	.29
glucoheptose	1.70	67.97	.42
mannoheptulose	2.17	78.10	.12
perseitol	1.26	26.03	.55
sedoheptulose	1.67	125.02	.90
experiment IX			
glucose	.90	18.66	1.38
tagatose	1.63	21.73	.20

B. ARTIFICIAL SWEETENERS [40]		
Sweetener	N	K (Molarity Intercept)
sodium cyclamate	0.75	$10^{1.41}$
sodium saccharin	0.45	$10^{1.81}$
saccharin	0.47	$10^{1.90}$

functions in log-log coordinates, but these sweeteners are separated by a constant distance (corresponding to a constant ratio in linear coordinates). This separation means that although the sweeteners describe parallel functions, one carbohydrate sweetener at a fixed concentration is always a constant percentage sweeter or less sweet than another.

(4) Sucrose and fructose exhibit anomalous growth laws vis-a-vis concentration. Both of these sweeteners exhibit curvature (concave downwards) in log-log coordinates. The curvature is most evident at the moderate and high concentrations where sweetness is maximum[31]. The reason for such curvature is not known. A similar phenomenon occurred in Cameron's equal sweetness matches[17] where the curvature was quite evident for sucrose and fructose, but not evident for glucose. The curvature may be related to the ability of the sweetener molecule to penetrate into and escape from the sweet receptor site on the tongue.

(5) In some instances, where a molecule excites dual tastes (viz., saccharin, cyclamate), perceived sweetness conforms to one type of growth law vs. physical concentration, whereas total sensory intensity is governed by a different, and often steeper, growth law[23].

(6) The mode of sweetener application to the tongue can affect the response function. If the taste material is sipped, swirled about in the mouth, and expectorated, then the relation between magnitude estimation and physical concentration is a power function of higher exponent than if the sweetener is flowed over the extended tongue[32]. The ratio of power function exponents for perceived sweetness in sipping vs. flowing may be a factor of three or more. The reason for the dramatic difference also is unknown.

Translatability of Sensory Functions to Foods

Much of the foregoing data, ranging from threshold determinations to psychophysical power functions, have been accumulated for model systems, which are most often a tasteless aqueous solvent plus sweetener. To what extent are these laws applicable to complex foods? Several studies suggest that if the concentration of the sweetener is varied systematically in a food, with all other factors (or at least most other factors) maintained constant, then the experimenter can recover a reasonable dose-response function for sweetness in the food. The following examples typify the results with simple foods:

(1) For cherry-flavored beverage, varying in sucrose concentration, the function re-establishes itself over a sucrose range of 3% - 15% w/v[33]. In addition, if in the study the sucrose and the flavoring agent are both

varied simultaneously, to produce a grid of products, and if the subjects evaluate the full grid, then one can obtain typical-appearing psychophysical sweetness functions, one function for varying sucrose levels against a single fixed flavor level[34].

(2) For other foods (e.g., puddings, carbonated soft drinks), the psychophysical functions can also be obtained, by the magnitude estimation method, when the experimenter a priori varies the concentration of sweetener in the food (whether the sweetener be sucrose or artificial sweeteners). However, the parameters of the sensory function may not be precisely parallel or duplicate those obtained for sucrose or model aqueous systems. The mixture effects, with suppression of one taste by another in the food, and the added effect of texture and aroma all interact to diminish the degree of direct, immediate translatability.

Other Aspects of the Intensity of Sweetness

The foregoing portion of the paper concerned the measurement of simple sweetness. The sweetness perception can be modified in different ways to change its intensity.

(1) Mixtures of tastes can be made, in which the sweetener is mixed either with another sweetener agent (homogenous mixtures), or else the mixture can comprise a sweetener and a non-sweet-tasting chemical (heterogeneous mixtures).

(2) The sweetener can be presented in different media, and at different viscosities.

(3) The sweetener can be presented at different temperatures.

Taste Mixtures

More than 120 years ago a short note appeared in a pharmacy magazine, discussing the effect that sugar and coffee had on the bitter taste of quinine sulfate[35]. That paper suggested the well-known masking effect. In contrast, food technologists have, in the past century, known that mixtures of sweeteners can be used to enhance the sweet taste of foods. Here sweetness is increased, rather than reduced, in the mixtures.

During the past half century we have been privy to several series of studies on taste mixture, especially with regard to the effects that the mixture played on sweet taste. The major results have shown that when similar tasting sweet substances are mixed, the result is an additive result -- i.e., the mixture tastes stronger than either component alone. In contrast, when dissimilar tasting substances are mixed together, then the mixture shows suppression.   In both instances, however, it is not clear what is the underlying quantitative law of taste mixture.

Similar Tasting Mixtures of Sweeteners

(1) Cameron[17] studied mixtures of sugars with each other, and mixtures of sugars plus amino acids.   The principal finding was that if all of the sweetener components were expressed in terms of the concentration of glucose that is equally sweet (i.e., for sucrose and glucose, one would determine the concentration of glucose as sweet as the sucrose), then summation of such "glucose equivalents" produced a total glucose level which, when experimentally tested, is as sweet as the mixture (i.e., additivity).   Such summation and perfect arithmetic additivity failed when (a) the reference sugar was fructose or sucrose rather than glucose, or (b) one of the sweeteners was an amino acid, rather than a carbohydrate sweetener.

(2) Kamen[36] investigated mixtures of cyclamate and sucrose, using category scales as the measuring instrument.   The empirical findings of that study were that the mixtures were significantly sweeter at moderate level than would be expected on the basis of simple arithmetic additivity.One must, however, realize that the non-linearities of the category scale make it impossible to deduce any truly general quantitative law of sweetness summations.

(3) Stone and Oliver[37] investigated various mixtures of sugars and artificial sweeteners, and found less than 100% additivity when sweetness was quantified by magnitude estimation.   In a later experiment, using the same method of magnitude estimation in a study of glucose and fructose and their mixtures, Stone, Oliver and Kloehn[38] reported partial synergism.   However, in neither paper did the authors offer a suggestion about a possible explanation or even descriptive model for sweetness additivity.

(4) Yamaguchi et al.[39] investigated equal sweetness matches between sugars and artificial sweeteners.   In some instances the mixtures were synergistic, in other cases the mixtures were additive.   She and her colleagues developed an ad hoc model for predicting the mixture sweetness from the component sweetness.   The model comprised several terms, which could reflect and thus account, in an empirical sense, total summation, partial summation, and even synergism.   Like Cameron's model, Yamaguchi's model for sweetness summation was couched strictly in terms of equal sweetness matches.   As a consequence, we can only guess about the actual psychological scale that the panelists used, and the actual degree of sensory synergism.   We do, however, know the percentage under- or overprediction in terms of concentrations of sweeteners.

(5) Moskowitz[40, 41] and Moskowitz and Klarman[42] integrated Cameron's model of sweetness additivity with the magnitude estimation model (which allows subjects to assign numbers with ratio scale properties to sweeteners and sweetener mixtures). They considered two types of summation that could occur (note that these are hypotheses or models of what could happen in sweetness mixtures).

Model 1. The gustatory system adds together the <u>concentrations</u> of the two sweeteners in the mixture, and then transforms the total or overall concentration to a sweetness impression. The summation of concentrations is done in terms of summing equivalents of concentration (e.g., glucose equivalents). This transformation parallels the summation proposed by Cameron, who first transformed the component sweetener concentrations to equivalents of glucose and afterwards summed those equivalents. Here, once the summation of glucose equivalents is made, the total sum (viz. a higher level of glucose) is then transformed, via the appropriate psychophysical equation, to a magnitude estimation of perceived sweetness. Model 1 can be expressed by the predictor equation:

$$\text{mixture sweetness} = k_X(C_X \tilde{} A + C'_X \tilde{} B)^{N_x} \qquad (2)$$

(A = concentration of sweetener A; B = concentration of sweetener B; C_X = concentration of the reference sweeteners X (e.g., glucose) which is as sweet as A; C'_X = concentration of the reference sweetener X which is as sweet as B; N_x = power function exponent for reference sweetener X (viz. the glucose exponent) relating concentration of X to perceived sweetness; k_x = coefficient of the power function for sweetener X).

Model 1 allows a straightforward prediction of perceived sweetness for mixtures. Any sweetener (including sweeteners not used in the study) can be used as a reference, so long as the equal sweetness levels are known, and as long as the psychophysical function for reference sweetener X is known (transforming concentration of X to perceiving sweetness).

Model 2. The gustatory system adds together subjective sweetness impressions treating the sweeteners in the mixture as separate sources of sweetness information. The tongue (or brain) sums together the perceived sweetnesses arithmetically (rather than summing concentrations) to determine mixture sweetness. Model 2 is expressed by the considerably simpler equation:

$$S_{ab} = S_A + S_B = k_a(C_a)^{n_a} + k_b(C_b)^{n_b} \qquad (3)$$

(S = sweetness; A, B = separate sweeteners A and B; ab = mixture; k = coefficient for the power function relating concentration (C) to sweetness;

n = exponent of the power function).

In both models there is considerable deviation between predicted and obtained sweetnesses. Both models consistently underpredict the sweetness of mixtures of pairs of sugars (when one of the sugars is either glucose or fructose, and the other sugar is sucrose, sorbitol, etc.)[41]. The models fail to predict the taste ingensity of mixtures of artificial sweeteners with each other[42]. The models predict sugar mixtures satisfactorily, but do worse in predicting the sweetness of artificial sweeteners (e.g. saccharin, cyclamate) mixed with sugar[40].

Despite the proposed models which predict additivity, and the empirical observations on sweetness mixtures obtained during the past fifty years, we still do not have an adequate theoretical treatment of what occurs when two sweeteners are tasted together. It may turn out that a simple biological model of mixtures, such as a pharmacological model for drug action, will prove more adequate in the long run than the empirical approaches suggested above. What is needed is a more coherent approach to sweetener mixtures, which will begin with a series of alternative models based on theory, proceed with a set of test sweeteners, and use sensitive sensory scaling methods. Through standardized experimental conditions, researchers should be able to better discern the nature of sweetness additivity, and use the rules of summation to find mixtures that exhibit synergistic, enhanced sweetness.

Mixtures of Different Tastes

When sweet and bitter substances are tasted together, each taste quality is diminished. This observation was made 120 years ago, as noted above. Since then, numerous studies have shown that in heterogenous mixtures comprising two or more tastes, the taste intensity of each component is almost invariably reduced (mutual suppression).

Précis of such investigations are given below:

(1) A century ago, the mutual suppressing effect of sugar and coffee on quinine was noticed[35], but there was no quantification of the extent of the effect.

(2) Various studies by pharmacists were concerned with the nature of sweetening agents plus flavors (as syrups) as potential disguisers of the bitter tastes of drugs[43]. In most instances, a sweetener was added to a flavoring agent in order to form a flavoring complex which would suppress the unpleasant bitter taste. Interest was focused upon the extent of bitterness reduction, but with little or no quantification. The observations were made by pharmacists, who were interested in the practical and

immediate usage of the information.

(3) Determinations of threshold revealed that the threshold for sweetness is increased (sensitivity to sweetness is diminished) if the background solution has a taste other than sweet or insipid (tasteless)[12].

(4) Category scaling methods of suprathreshold sweetness reveal a suppression of the sweet taste in the presence of other taste materials. However, the precise decrement in perceived sweetness depends upon the concentration of sweetener, the type of sweetener, and the concentration of the other taste material[44-47]. At very low concentrations of sodium chloride, however, sugar sweetness often is enhanced, rather than suppressed. The reason for this enhancement is not yet known, although one possibility is that at very low concentrations of salt, the sweetness in the mixture suppresses the saltiness of salt. It is also known that sodium chloride at extremely low concentrations (less than 0.01 M) tastes sweet[48]. Quite possibly, the mixture of sugar sweetness and low salt levels diminishes the saltiness (due to mutual suppression), down to the level where sodium chloride tastes sweet, and thus adds to the sugar sweetness.

(5) Equal intensity matches between sucrose-salt mixtures and pure unmixed sucrose solutions show that the shift in sweetness obeys a power law. Beebe-Center et al.[49] instructed subjects to match the sweetness of mixtures of sucrose and salt to unmixed sucrose sweetness. The equation that they obtained could be expressed as follows:

$$\log \left(\frac{C}{C_o}\right) = -0.159 \, M^7 \tag{4}$$

(where C_o, C represent, respectively, the sucrose solution in the mixture and the like sample sucrose solution that matches it, both in grams/100 cc solvent; M = the concentration of NaCl, also in grams/100 cc solvent)

Equation 4 shows that the mixture sweetness is proportional both to the concentration of the sweetener and to the concentration of added sodium chloride.

(6) Studies using ratio scaling methods (e.g., magnitude estimation) show again that in mixtures the sweetness is diminished[50]. However, it is impossible to predict the degree of suppression of sweetness, and of the other tastes separately, at least from the published studies. Moskowitz[50] found the following empirical equations to relate the total taste intensity of a mixture (e.g., for a sucrose plus NaCl mixture, the total sum of perceived mixture sweetness and mixture saltiness):

$$\text{mixture taste} \atop \text{(sweet + salty)} \approx 0.59 \; \text{(unmixed sweet + unmixed salty)} \quad R^2 = 0.90 \tag{5a}$$

mixture taste $\overset{\sim}{=}$ 0.49
(sweet + sour)   (unmixed sweet + unmixed sour) $R^2=0.97$ (5b)

mixture taste $\overset{\sim}{=}$ 0.46
(sweet + bitter)   (unmixed sweet + unmixed bitter) $R^2=0.94$ (5c)

These equations suggest that there is a systematic reduction in the total taste intensity, but that the relative decrements of the two tastes may not necessarily be the same.

Bartoshuk[51] suggested that the total taste intensity of mixtures could be modeled by a vector model, similar to that proposed by Berglund, Berglund and Lindvall[52] to account for odor mixture. In an intriguing and critical experiment, Bartoshuk mixed together 1-, 2-, 3- and 4-component mixtures. The perceived sweetness diminished with increasing number of other tastants in the mixture, but the total taste intensity (with sweetness as one portion) soon reached an asymptote. It appeared that with additional taste qualities in the mixture suppression was set up, so that the new component suppressed the other components to which it was added, and they in turn suppressed the added taste. That equilibrium maintained constant total taste, but diminished the mixture sweetness.

Quite a bit of work remains before we have an adequate picture of the details in hetero-qualitative masking. The evidence for suppression continues to reappear from study to study. The parameters of the suppression, however, appear to be unique to the types of sweeteners, their levels, the types of the other tastes added, and their concentration.

Effect of Taste Medium on Sweetness

Much of the basic research on taste perception has been done with simple aqueous media, in which the taste material has had no problem diffusing to the receptor. If the viscosity or the physical characteristics of the medium change, however, then taste intensity diminishes. Several studies have suggested that the medium can elevate threshold or diminish suprathreshold intensity. Among the most important published studies are:

(1) Type of medium. Mackey and Valassi[15] evaluated sucrose (as well as other taste) thresholds in gels, foams and liquids, and found that thresholds were highest in gels, intermediate in foams and lowest in liquids.

(2) Type of liquid. Mackey[14] determined thresholds for saccharin dissolved in aqueous solution, in water that had been viscosified with methyl cellulose to equal the viscosity of mineral oil, and mineral oil itself. Thresholds were lowest in water, intermediate in viscosified water, and highest in oil. The type of medium and its chemistry therefore are critical.

(3) <u>Viscosifying agents</u>. Arabie and Moskowitz[53], Moskowitz and Arabie [54], Stone and Oliver[55], and Vaisey et al.[56], all found that sweetness as well as other tastes diminish in strength if the taste material is water plus a viscosifying agent (e.g. a vegetable gum). In some instances there may be an aberrant and unexpected increase in reported sweetness[55]. To some extent, this increase may be ascribed to the confusion of viscosity with highly concentrated and super-sweet sugar solutions, such as those found in canned fruits.

(4) <u>Theoretical foundations</u>. Weinheimer[57] discussed the possible physical mechanism underlying a reduction in the intensity in viscosified solutions. He suggested that the viscosifying agent may retard diffusion of the sweet molecule into the receptor site, thus reducing perceived sweetness. According to Weinheimer, the data of Moskowitz and Arabie[54] for sweetness vs. concentration can be predicted from (1) diffusion of the molecules to the taste bud in a viscosified system and (2) a relatively quick reaction mechanism (e.g., first-order kinetic equation). The relation between sensory sweetness and concentration at the surface of the tongue can be expressed with high correlation by a power function of exponent 1.0, once the appropriate chemistry is modeled. The model takes into account the two steps (diffusion, reaction) with the viscosifying agents increasing the time required for diffusion.

Effects of Temperature on Sweetness

A jar of sweet fruit preserves taken out of the refrigerator tastes less sweet (depending upon the sugars involved) than the same sweet preserves tasted at room temperature. Experimenters have investigated the sensory effects of changing temperature upon sweetness. The results are often contradictory, with some investigators suggesting that sweetness increases with temperature, whereas others suggest that sweetness decreases. Table 7 presents a list of results on the sweet taste, showing the sweetener used, the scaling method (suprathreshold, threshold measures), and the findings.

One should note that there is the anomalous intermediate temperature of 37° C, corresponding to body temperature. Maximal taste sensitivity and sweetness intensities seem to occur around this region, suggesting that stimuli at body temperature convey maximal taste information. Superimposed upon this maximum peak is a trend for shifts in taste away from the body temperature peak, in both directions.

Table 7. Relation between temperature and sweetness: a historical perspective [94]

Reference	Conditions	Results
Luchtmann, 1758	immerse tongue in water at 40-41° C or in ice water	could not recognize sweetness of sugar
Kiesow, 1896	immerse tongue in water of either 0° or 50° C. In other studies, Kiesow investigated intermediate temperature of solutions	no taste sensation, even of concentrated stimuli. Solution temperature had little effect (e.g., at 5° solution, sensitivity was as great as at 30°C)
Schreiber, 1893	measured thresholds of sucrose at 3 temperatures	threshold for sucrose: at 30° and 40° = 0.1 M at 0° C = 0.4 M
Chinaglia, 1915	(a) used a U-shaped tube on tongue to pass heated or cold water. Measured thresholds at 5-50° C at 5° intervals	little temperature effect was found
	(b) measured reaction time to taste stimuli at 17° C, 37° C, and 50° C	for 10% sucrose, reaction time is slower as the temperature diminishes
Komura, 1920	sprayed 200-500 cc of solution on tongue for 5 minutes. Solutions were 10°, 20°, 30°, 40°, respectively	thresholds for sucrose (grams sucrose/like) 10° = 12.30 G/L, 20° = 6.67 G/L, 30° = 5.66 G/L, 40° = 5.33 G/L
Goudriann, 1930	flowed taste solution of varying temperatures onto tongue. Stimuli were at 10°, 15-21°, 40° C	for 10% sucrose, as the temperature rises, the sweetness rises
Hahn & Gunther, 1932	U tube was used to flow solution over tongue. Limited temperature to 17-47° to limit pain	for sucrose, temperature drops down with temperature increase (from 17° to 32°) and then rises again
Hahn, 1936	evaluated adaptation curve by measuring threshold vs. elapsed time since the continuing stimulation was first applied	there are different adaptation curves, depending upon temperature
Shallenberger, 1963	increase temperature and rate sweetness	sweetness increases with higher temperature
Stone, Oliver & Kloehn, 1969	magnitude estimates of sweetness of glucose and fructose plus their mixtures at 5°, 22° and 50°	no temperature effect
Moskowitz, 1973	5 concentrations of glucose tested at 25°, 30°, 35°, 40°, 45°, 50° by magnitude estimation	power functions fitted to sweetness were unaffected by temperature. Maximum overall sweetness around 37° C

Overview of Intensity Measures of Sweetness

Researchers have extensively studied sweetness intensity perception. As Desor[58] noted, sweetness is the "special sense", because it is a desired taste and is attractive to consumers. The commercial use of high amounts of sweeteners in foods has brought the measurement of sweetness into a clear perspective.

The research on sweetness above suggests the following general trends:

(1) There are different effective ranges of sweetener concentrations, depending upon molecular structure.

(2) Psychophysical (dose-response) relations describe different curves depending upon molecular structure.

(3) Sweetness can be modified by changing either (a) chemical structure, (b) concentration, (c) medium in which the sweetener is placed, (d) adding opposing or similar taste qualities to produce suppression or additivity, or (e) modifying the temperature of the taste material. In modifications c-e, sweetness can be diminished or enhanced in known and often quantifiable ways. Unfortunately, we do not as yet have a good way of predicting the intensity of sweetness given the chemical formula of the sweetener (modification a). In this regard, some of the work proposed by Dravnieks[59] to correlate odor intensity with chemical structure by a "building block model" may eventually be applied profitably to the prediction of suprathreshold sweetness intensity. The translation of chemical structure to sweetness via a pragmatic application of Dravnieks' approach awaits the adept chemist and psychophysicist.

SWEETNESS HEDONICS

As a result of its taste quality, sweetness provokes pleasure. In many languages sweetness is either directly associated with goodness and pleasure, or at least connotes goodness and pleasure. An extensive series of studies

Table 8. Frequency of appearance of the most characteristic features composing facial responses to gustatory stimuli in tested neonates[*]

	Termborn Normal		Anencephalic
	Below 20 Hrs n=75	3 - 7 Days n=100	1 - 19 Days n=4
sweet			
retraction of mouth angles	61	87	4
clear expression of satisfaction resembling "smile"	58	73	3
eager sucking with licking of the upper lip	74	97	4
sour			
pursing of lips	75	98	4
wrinkled nose	58	73	2
quick repeated blinking	67	70	2
increased salivation	61	65	4
flushing	57	64	1
bitter			
"arch-like" lips with depression of mouth angles	73	96	4
protrusion of flat tongue	59	81	3
salivation and spitting	57	87	4
general expression of "anger" and "dislike"	59	86	3
vomiting	34	52	3

* Data from Steiner[60].

has suggested that sweetness is innately pleasing, even to neonates who have had no cultural experience with sugar[60]. In a study of the facial reflexes of neonates who were presented with taste solutions of various qualities (as well as odorous stimuli of various qualities), Steiner discovered a reflex, called the gustofacial reflex. The reflex could be further divided into those movements which appeared to be associated with a rejection and withdrawal (provoked by such stimuli as bitter quinine and asafoetida), and movements that appeared to be associated with acceptance (given to the odor of milk and the taste of sugar). Table 8 provides the schematic outline of the gustofacial reflex. Nowlis[61] further suggested that tongue reflexes to sweet material parallel the degree of sweetness which adults exhibit.

Additional work by Desor et al.[62] suggests that there are hedonic differences between age groups and between races, with younger children preferring sweeter stimuli, and blacks preferring sweeter things than whites. Culture differences for pleasantness of the sweet taste did not, however, show up in a cross-cultural study reported by Moskowitz et al.[64], who investigated the sweetness perception and sweetness hedonics of glucose solutions of varying concentration with illiterate Indian laborers from the Karnataka region of India. The hedonics of sweetness were similar to that of the U.S. population.

Relatively little information exists on other classificatory variables for taste (e.g. sweetness) hedonics. Questions such as the potential existence of correlations between income and sweet preference (or taste preference in general), age, sex, nationality, etc., have never been answered in the public literature. Gemmill[65] has suggested even perhaps that a national index be constructed for the "sweet tooth" of nations. This index would reflect natural inclinations towards sweet products by people in various countries. Of course, such an index would have to take into account the various income and other characteristics of the population, as well as cost and availability of sweeteners.

Dose-Response Relations and Hedonics

Engel[66] suggested that there is a point on the sweetness continuum which is maximally pleasant. He based his initial assumptions upon the pioneering work of Wundt[67] who, a century ago, had postulated that the hedonic tone of all stimuli vs. sensory intensity conforms to an approximately inverted U- or L-shaped curve. Engel polled the population to find out the percentage finding varying concentrations of sucrose to be pleasing, displeasing or neutral. Engel did not inquire about the degree of pleasure or displeasure.

As expected, at very low and almost tasteless concentrations (around 1%), most individuals found the taste to be neutral. At higher concentrations an increasing proportion of individuals found the taste to be pleasing, and at still further higher concentrations stabilization appeared in the percent finding sucrose to be pleasing (at around 9%).

Engel's pioneering work lay dormant for several decades. In the early 1970's, Moskowitz and his associates at the U.S. Army Natick Laboratories began an extensive series of studies to more precisely map out the relation between the degree of pleasantness and the concentration of sweeteners. They based their initial approach[31] upon a pilot study reported by Kocher and Fisher[68] who had used the method of magnitude estimation to relate pleasantness and sweetness, respectively, to concentration.

The basic results of those early series of studies are:

(1) For glucose, pleasantness increases with concentration, usually at a slower rate than perceived sweetness. The increase is usually governed by a power function, whose exponent is approximately half that of the sweetness exponent. For instance, if glucose sweetness is proportional to (concentration of glucose)$^{1.3}$, then glucose pleasantness is proportional to (concentration of glucose)$^{0.65}$.

(2) For glucose, peak pleasantness occurs at a fixed concentration, around 1.0 M. Subsequent increases in concentration produce diminishing function.

(3) The hedonic functions for sucrose also exhibit an inverted U-shaped function, but with a much flatter sloping section. Changes in sucrose sweetness and concentration do not provoke as drastic a change in sucrose pleasantness.

(4) For most sugars, the peak point in pleasantness is at a concentration which is as sweet as 1.0 M glucose.

(5) If foods are sweetened with sucrose, then the break point tends to occur at the sweetness level equivalent to 1.0 M glucose. However, as the food becomes increasingly texturized (e.g. pudding, cake), departures from that clear breakpoint may occur[33].

(6) The peak point for sugar pleasantness (glucose solution) is the same in different regions of the world. Indian medical students, tested in Bangalore, with a category scale for pleasantness and with glucose solutions, exhibit the same breakpoint at 1.0 M glucose as do Westerners[69]. Illiterate laborers from the Karnataka region of India who assigned line lengths to match sweetness pleasantness also showed a peak pleasantness around 1.0 M glucose[64]. Their sourness function was normal as well, but in contrast to

Westerners, these laborers found the taste of citric acid to increase in pleasantness with concentration, until an extremely high sourness level was reached (corresponding to 0.03 M citric acid).

(7) Artificial sweeteners exhibit different hedonic functions, depending upon the sweetener[30,70]. The sweeteners, saccharin and cyclamate (both sodium and calcium salts) may peak in pleasantness at intermediate concentrations, but the sweetness of those intermediate concentrations does not necessarily equal the sweetness imparted by 1.0 M glucose. Nor, in fact, is the peak level of pleasantness of aqueous saccharin or cyclamate solutions equal to that imparted by sucrose.

(8) Aspartame exhibits an inverted U-shaped function for pleasantness, with a peak pleasantness at approximately 0.08% (equivalent to the sweetness produced by 9%-10% glucose)[71].

(9) Cola-flavored beverages, artificially sweetened with mixtures of cyclamate, saccharin and aspartame exhibit a peak pleasantness rating at a sweetness equal to that of approximately 14% sugar. This peak comes at a sweetness higher than the peak sweetness that would be most acceptable in a sweetened (sugar) cola[72].

In most of the foregoing studies, with the exception of Moskowitz et al.[33,72], the taste stimuli were constrained artificially to be simple, model systems. As Pangborn[73] points out, however, the context of the testing situation may not necessarily reflect what actually appears in the real world when we are confronted with food products. Just because a few ml of sweetened water are given highly acceptable ratings, more than lower or higher concentrations, does not necessarily mean that in free-eating situations we would ordinarily choose them. We might, in fact, quickly satiate on the sweeter products, and choose for long-term consumption the less sweet, more tart ones. The translatability of laboratory hedonic studies into real-world food eating situations is a moot question, open to validating research. Halpern (personal communication) has also found that sugar solutions tested by themselves are hedonically pleasing. If sweetened fruit-flavored beverages are tested in the same session, then sugar water becomes displeasing.

Are All Sweeteners Equally Pleasant?

Several studies have shown that sweeteners are not equally pleasant. Artificial sweeteners tend, even for matched sweetness, to be less pleasing than sugar or caloric sweeteners[70,71]. Chappell[74] had subjects rank order five sugars that were equal in concentration (25% sucrose, fructose, lactose, glucose and maltose) in terms of pleasantness and found that sucrose was the most pleasant, whereas maltose was the least pleasant tasting.

The reasons for the differentness of pleasantness vary. Among some reasons are:

(1) Some sugars stimulate unpleasant side tastes. For instance, mannose would be unpleasant relative to sucrose because mannose provokes a bitter taste[75]. Occasionally, however, the side taste may just be a presence or an absence of the sensation, without a clear registration of what the sensation is. Maltose possesses a clear side taste, undefinable in quality but noticeable. Moskowitz[76] illustrated the increase in the qualitative dissimilarity between sugars, due in part to the development of side tastes.

(2) Artificial sweeteners also provoke bitterness (as do saccharin, cyclamate) or off-tastes (aspartame). The clear bitterness leads to a rejection of the sweetness. In order to optimize acceptance, a way must be found to counter the bitter or off-tastes, by mixing the artificial sweetener with a carrier. Chemicals such as glucono-delta-lactone or specific mixtures between saccharin and cyclamate (in a 1/10 ratio by weight) effectively suppress the bitter or off-taste.

(3) Viscosity plans an important role, but seemingly a neglected role as well. The textural characteristics of sweeteners (viz. their "body") are functional properties that may be pleasing to the palate. Artificial sweeteners in beverages do not impart body. Saccharin or cyclamate sweetened syrups may have to be artificially viscosified by the addition of the appropriate vegetable gum. The functional properties of the gum used to impart body should be such that they are not felt to be slimy, but rather, ideally, impart the same sense of viscosity that sucrose-sweetened syrups impart[77].

(4) Finally in real food systems, sucrose and carbohydrate sweeteners play several roles. They sweeten the food but also, as Sjöström et al.[78] suggested, sugars broaden the flavor of food, and act as melding agents to integrate the various flavor notes of the food. Artificial sweeteners simply provide the sweetness needed, without the flavor blending and broadening. As a result, the sweetness of artificially sweetened food may be perceived simply as another sensation, perhaps a competing one.

Effects of Body State on Pleasantness

Three decades ago Mayer-Gross and Walker[79] found that the level of blood glucose in the body (which fluctuates daily) could influence the selection of sugar solutions on the basis of pleasantness. Subjects exhibiting high blood glucose levels preferred a relatively bland solution containing 5% sucrose, whereas individuals with low blood glucose preferred the much-sweeter 30% sucrose.

Since those pioneering studies, many studies have been run to assess how body state (as well as health and disease) influence the sensory pleasantness of stimuli in general, and sweetness in particular. Among the major results are the following:

(1) Studies evaluating hunger. Cabanac[80] proposed that the internal body state regulates the hedonic characteristic of sweetness. He named this internal regulation "alliesthesia". When individuals lost weight (so that they were in a depleted body state), they appeared to prefer the sweeter-tasting stimuli. In parallel studies, Moskowitz et al.[69] found that for Indian medical students who fasted overnight there was no change in the perceived sweetness nor pleasantness of glucose vs. concentration. Glucose pleasantness still exhibited maximum pleasantness at 1.0 M, and the ratings for degree of pleasantness at both the peak and non-peak pleasantness points did not differ significantly in hungry vs. non-hungry conditions.

(2) Studies evaluating satiety. Satiety is the inverse of hunger. Cabanac and Duclaux[81] found that by artificially sating an individual through a glucose load, one could diminish the taste pleasantness of a glucose solution. This finding indicated that taste hedonics is intimately related to satiety. Moskowitz et al.[69] reported, in contrast, that after ingesting an oral glucose load (1.5 g/kg body weight) the panelists found glucose pleasantness to increase, rather than to decrease (as Cabanac found). In addition, the satiating effect of glucose load eliminated, in a strong and statistically significant way, the breakpoint in glucose pleasantness which so often occurs at 1.0 M. Although at first glance these two results seem to differ sharply, they may not differ quite as dramatically. In both instances the subjects failed to show a discrimination between sweetness as information and sweetness as hedonics. In Cabanac's study one may speculate that the subjects simply mislabeled sweetness as unpleasantness. In Moskowitz's study, the subjects mislabeled sweetness as ratings of pleasantness. Both groups may have been scaling perceptions of sweetness, but whereas Cabanac's subjects classified their perceptions as unpleasant, the Moskowitz subjects classified the same perceptions as pleasant. A further discussion of this classification theory of hedonics is presented by Moskowitz[82]. Note that in either set of experiments, only the hedonic character of the sweet stimuli is changed, not the sensory quality.

Virtually no researcher has found _strong_ evidence for either threshold or suprathreshold changes with hunger or satiety[83,84].

(3) _Studies evaluating obesity_. Obese individuals (or, in fact, overweight individuals in general) may show systematically different hedonic responses than do normal weight individuals. Among the earliest studies suggesting the difference was a study reported by Pangborn and Simone[85] who found that overweight individuals gave higher hedonic ratings to desserts than did normal weight individuals. This bias to uprate sweeter things does not indicate a direct preference for sweeter things, since in a pairwise comparison of desserts the overweight individuals simply like the desserts more on their own scale. Subsequent work by various experimenters has shed more light on the issue. Grinker and Hirsch[86] found that the pleasantness of graded sucrose concentrations diminished with increasing concentration when obese individuals evaluated them, as if to indicate that the obese subjects find sweeter things less pleasant than more bland tastes. Moskowitz[87] suggested that in hedonic evaluations of sweet substances, obese individuals exhibit aberrant functions for hedonics vs. concentration because they are unable to translate sensory perceptions of sweetness to hedonic evaluations of those perceptions. Consequently, the psychophysical functions relating sweetness to concentration is identical to, or inverse of, the psychophysical functions relating pleasantness to concentrations. The sweetness and hedonic functions are parallel _if_ the obese individual categorizes the sweet taste as being pleasant. Then, the obese person would assign sweetness ratings but mistakenly categorize or classify those ratings as pleasantness ratings. In contrast, if the obese person categorizes sweetness as unpleasant or "bad" for him or her, then the ratings of sweetness will appear but disguised as ratings of unpleasantness. The parallelism for sweetness vs. pleasantness (or unpleasantness) has been shown in work on obese subjects by Abramson and Moskowitz[88], Rodin et al.[89], and by Thompson et al.[90].

(4) _Studies evaluating other metabolic disorders_. Abramson and Moskowitz[88] investigated the hedonic tone and the perceptual sweetness functions for sweetened solutions of Kool Aid as evaluated by normals, by individuals undergoing renal dialysis, and by obese individuals. The renal dialysis patients exhibited the hedonic functions for sweetened beverages that would be seen for normals, although the renal group exhibited heightened appreciation of the taste of salted soups.

(5) _Studies evaluating intolerance for the sweet taste or inability to_

recognize sweet. An early study by Davidenkov[91] reported several cases in the Soviet Union wherein individuals could not tolerate the sweet taste of sugar. However, that report did not provide any particulars about either the psychophysics of the tasting situation, or the methods for hedonic measurement. More recently, two studies have come forth on the problem of impaired sweetness perception. Henkin and Shallenberger[92] reported two males suffering from congenital idiopathic hypoparathyroidism. Both reported such sugars as fructose, glucose, galactose, as well as artificial sweeteners, to taste bitter, salty or sour, but not sweet. Billimora et al.[93] found that diabetics who could not adequately handle sweets biologically found the taste unpleasant, whereas those who could handle the sugars found the taste pleasant. In both sets of studies there may be unpleasant sensory effects that occur during the tasting of sugars, which may provoke displeasure and failure to tolerate the taste. Much more work remains to be done in this area.

REFERENCES

1. Maller, O. and Desor, J. in Fourth Symposium on Oral Sensation and Perception, ed. J. Bosma, p. 279, 1973.
2. Steiner, J. Ann. N. Y. Acad. Sci. 237, 229, 1974.
3. Fick, A. in Lehrbuch der Anatomie und Physiologie der Sinnesorgane, M. Lahr, Schaunburg & Co., p. 67, 1864.
4. Boring, E.G. Sensation and Perception in the History of Experimental Psychology, Appleton-Century-Crofts, N. Y., 1942.
5. Greeves, A. An Essay in the Varieties and Distinctions of Tastes and Smells and on the Arrangement of the Materia Medica, Edinburgh, 1828.
6. Cohn, G. Die Organischen Geschmackstoffe, Siemenroth, Berlin, 1914.
7. Rader, C. P., Tihanyi, S.G. and Zienty, F.B. J. Food Sci. 32, 357, 1967.
8. Von Békésy, G. J. Applied Physiol. 19, 1105, 1964.
9. Hüber, R. and Kiesow, F. Zeit Physikal Chemie 27, 601, 1898.
10. Stahl, W.S. (ed.) Compilation of Odor and Taste Thresholds, American Society of Testing and Materials DS 48, 1973.
11. Zubek, J.P. Med. Ser. J. Can. 15, 731, 1959, cited in Principles of Sensory Evaluation of Food, edited by M. A. Amekine, R. M. Pangborn, and E. B. Roessler, p. 58, 1965.
12. Anderson, R. J. Microfilm Abstracts 10 (4), Academic Press, N. Y., 287, 1950.
13. Hahn, H. Klinische Wochenschrift 15, 931, 1936.
14. Mackey, A. Food Res. 23, 258, 1958.
15. Mackey, A. and Valassi, K. Food Technol. 10, 238, 1956.
16. Lemberger, F. Plügers Arch. Ges. Phys. 123, 293, 1908.
17. Cameron, A.T., Rept. 9, Sugar Research Foundation, 1947.
18. Pangborn, R.M. J. Food Sci. 28, 726, 1963.
19. Magidson, O. J. and Gorabachow, S. W. Berichte der Chemische Gesellschaft 56B, 1810, 1923.
20. Taüfel, K. and Klemm, B. Zeitschrift fur Untersuchung der Lebensmittel, 50, 264, 1925.
21. Cameron, A.T. Canadian J. Res. E. Med. Sci. 24, 45, 1944.

22. Cameron, A.T. Canadian J. Res. E. Med. Sci. 23, 139, 1945.
23. Moskowitz, H.R. Percept. & Psychophys. 8, 40, 1970.
24. Corin, J. Arch. Biol. (Liege) 8, 121, 1888.
25. Schutz, H.G. and Pilgrim, F.J. Food Res. 22, 206, 1957.
26. Stevens, S.S. and Galanter, E. J. Exper. Psychol. 54, 377, 1957.
27. Beebe-Center, J.G. J. Psychol. 28, 411, 1949.
28. Beebe-Center, J.G. and Waddell, D. J. Psychol. 26, 517, 1948.
29. Stevens, S.S. Science 118, 576, 1953.
30. Moskowitz, H.R. in Chemical Senses and Nutrition II, Academic Press, N. Y., in press.
31. Moskowitz, H.R. Amer. J. Psychol. 84, 387, 1971.
32. Meiselman, H.L. Percept. & Psychophys. 10, 15, 1971.
33. Moskowitz, H.R., Kluter, R.A., Westerling, J. and Jacobs, H.L. Science 184, 583, 1974.
34. Moskowitz, H.R. J. Applied Psychol. 56, 60, 1972.
35. Quevenne, M. Pharm. J. & Pharmacist 7, 352, 1847-1848.
36. Kamen, J. Food Res. 24, 279, 1959.
37. Stone, H. and Oliver, S.M. J. Food Sci. 34, 215, 1969.
38. Stone H., Oliver, S.M. and Kloehn, J. Percept. & Psychophys. 5, 257, 1969.
39. Yamaguchi, S., Yoshikawa, T., Ikeda, S., and Ninomiya, T. Agric. & Biol. Chem. 34, 181, 1970.
40. Moskowitz, H.R., J. Exp. Psychol. 99, 89, 1973.
41. Moskowitz, H.R. in Sensation & Measurement: Papers in Honor of S.S. Stevens, D. Reidal, Dordrecht, p. 379, 1974.
42. Moskowitz, H.R. and Klarman, L. Chem. Senses and Flavor 1, 411, 1975.
43. Purdum, W.A. J. Am. Pharm. Assoc. 31, 289, 1942.
44. Pangborn, R.M. Food Res. 25, 245, 1960.
45. Pangborn, R.M. J. Food Sci. 26, 648, 1961.
46. Pangborn, R.M. J. Food Sci. 27, 495, 1962.
47. Pilgrim, F.J. in The Physiological and Behavioral Aspects of Taste, p. 66, 1961.
48. Dzendolet, E. Percept. and Psychophys. 3, 65, 1968.
49. Beebe-Center, J.G., Rogers, M.S., Atkinson, W.H. and O'Connell, D.N. J. Exp. Psychol. 57, 343, 1959.
50. Moskowitz, H.R. Percept. & Psychophys. 11, 257, 1972.
51. Bartoshuk, L.M. Phys. & Beh. 14, 643, 1975.
52. Berglund, B., Berglund, U. and Lindvall, L.T. Acta Psychol. 35, 255, 1971.
53. Arabie, P. and Moskowitz, H.R. Percept. & Psychophys. 9, 410, 1971.
54. Moskowitz, H.R. and Arabie, P. J. Text. Stud. 1, 502, 1970.
55. Stone, H. and Oliver, S.M. J. Food Sci. 31, 129, 1966.
56. Vaisey, M., Brunon, R. and Cooper, J. J. Food Sci. 34, 397, 1969.
57. Weinheimer, R. Master's Thesis, Dept. of Chem. Engineering, Carnegie Mellon University, Schenley Park, Pa., 1976.
58. Desor, J. Cereal Foods World, Feb., p. 69, 1976.
59. Dravnieks, A. Paper presented at the annual meeting of the American Chemical Society (Agricultural and Food Chemistry Division), San Francisco, Calif., Sept., 1976.
60. Steiner, J. in Fourth Symposium on Oral Sensation and Perception: Development in the Fetus and Infant, ed. J. F. Bosma, Govt. Printing Press, p. 254, 1973.
61. Nowlis, G. in Fourth Symposium on Oral Sensation and Perception, ed. J. Bosma, p. 292, 1973.
62. Desor, J., Maller, P. and Turner, R.E. J. Comp. Physiol. Psychol. 84, 496, 1973.
63. Desor, J., Greene, L. and Maller, O. Science 190, 686, 1975.

64. Moskowitz, H.R., Sharma, K.N., Kumaraiah, V., Jacobs, H.L. and Sharma, S.D. Science 187, 1217, 1975.
65. Gemmill, G. personal communication, 1975.
66. Engel, R. Pflügers Arch. ges. Phys. 64, 1, 1928.
67. Beebe-Center, J.G. The Psychology of Pleasantness and Unpleasantness, Van Nostrand, N. Y., 1942.
68. Kocher, E.C. and Fisher, G.L. Percept. & Motor Skills 28, 735, 1969.
69. Moskowitz, H.R., Kumaraiah, V., Sharma, K.N., Jacobs, H.L. and Sharma, S.D. Physiol. & Beh. 16, 471, 1976.
70. Moskowitz, H.R. and Klarman, L. Chem. Senses and Flavor 1, 423, 1975.
71. Moskowitz, H.R. and Dubose, C.N. J. Can. Inst. Food Sci. Tech. 10, 126, 1977.
72. Moskowitz, H.R., Wolfe, K. and Beck, C. Food Prod. Development, in press.
73. Pangborn, R.M. in Chemical Senses and Nutrition II, ed. M.R. Kare and O. Maller, Academic Press, N.Y., in press.
74. Chappell, G.M. J. Sci. Food Agric. 4, 346, 1953.
75. Pangborn, R.M. and Gee, S.C. Nature 191, 810, 1961.
76. Moskowitz, H.R. J. Food Sci. 37, 624, 1972.
77. Szczesniak, A.S. and Farkas, E. J. Food Sci. 27, 381, 1962.
78. Sjöström, L.B., Cairncross, S.E. and Caul, J.F. Food Technol. 11 (9), 20, 1957.
79. Mayer-Gross, W. and Walker, J.W. Brit. J. Exper. Pathol. 27, 297, 1946.
80. Cabanac, M. Science 173, 1103, 1971.
81. Cabanac, M. and Duclaux, R. Science 168, 496, 1970.
82. Moskowitz, H.R. Taste and Development: The Genesis of Sweet Preference (ed. J. Weiffenbach), Govt. Printing Office, Washington, D.C., 1977, p. 282.
83. Furchgott, E. and Friedman, M.P. J. Comp. Physiol. Psychol. 53, 576, 1960.
84. Meyer, D.F. J. Comp. Physiol. Psychol. 45, 373, 1952.
85. Pangborn, R.M. and Simone, M. J. Amer. Diet. Assoc. 34, 924, 1958.
86. Grinker, J. and Hirsch, J. in Physiology, Emotion and Psychosomatic Illness, Elsevier, Amsterdam, p. 349, 1972.
87. Moskowitz, H.R., unpublished manuscript.
88. Abramson, R. and Moskowitz, H.R. Paper presented to the Eastern Psychological Association, N.Y., N.Y., April, 1976.
89. Rodin, J., Moskowitz, H.R., Fleming, B. and Bray, G.A. Physio. & Beh., in press.
90. Thompson, D., Moskowitz, H.R. and Campbell, R. J. Appl. Physiol. 41, 77, 1976.
91. Davidenkov, S. J. Hered. 31, 5, 1940.
92. Henkin, R.I. and Shallenberger, R.S. Nature 227, 965, 1970.
93. Billimore, F.R., Shetty, H.B. and Rindani, T.H. Ind. J. Med. Res. 17, 329, 1963.

REFERENCES FOR TABLE 3

1. Bailey, E.H. and Nichols, E. Publ. American Assoc. Advanced Sci. 10 138, 1887.
2. Berg, H.W., Filipello, F., Hinreiner, E. and Webb, A.D. Food Technol. 9, 23, 1955.
3. Cooper, R.N., Balish, I. and Zubek, J.P. J. Gerontol. 14, 294, 1959.
4. Crocker, E.C. and Henderson, F. Am. Perf. & Essential Oil Rev. 27, 156, 1932.

5. Dermer, O. Proc. Oklahoma Acad. Sci. 27, 9, 1946.
6. Fabian, F.W. and Blum, H.B. Food Res. 8, 179, 1943.
7. Furchgott, E. and Friedman, M.P. J. Comp. Physiol. Psychol. 53, 576, 1960.
8. Kelty, M. and Mayer, J. Am. J. Clin. Nutr. 24, 197, 1971.
9. Linker, E., Moore, M.E. and Galanter, E. J. Exper. Psychol. 67, 59, 1964,
10. Pangborn, R.M. Amer. J. Clin. Nutr. 7, 280, 1959.
11. Pfaffmann, C.M. Handbook of Physiol., Vol. I, 507, 1959.
12. Richter, C.P. and Campbell, K.H. Amer. J. Physiol. 128, 291, 1940.
13. Williams, R.A. J. Comp. Physiol. Psychol. 70, 113, 1970.
14. Yensen, R. Quart. J. Exp. Psychol. 11, 221, 1959.
15. Janowitz, H. and Grossman, J. J. Appl. Physiol. 2, 217, 1949.
16. Schutz, H. G. and Pilgrim, F. J. J. Exptl. Psychol. 54, 41, 1957.

The development of sweet preference

James M. Weiffenbach, Ph.D.

Oral and Pharyngeal Development Section, Clinical Investigations Branch, National Institute of ·
Dental Research, Bethesda, Maryland 20014

SUMMARY

The link between sugar in the mouth and the attack of caries impels the
search for ways to reduce the consumption of cariogenic sugars. This challenge provides a pragmatic focus for investigating sweet preference in man.
Methodological and conceptual resources for the developmental study of sweet
preference have recently been assembled from a broad range of research areas[1]. Techniques for modifying food habits are needed to combat other health
problems such as obesity and hypertension. However, the relations between
these conditions and taste preference are known to be complex[2]. In contrast,
the relation between sweet preference and ingesting sweet sugar is, by definition, direct. Modification of the human "sweet tooth" provides a near-per-
fect opportunity for basic behavioral research to contribute to the amelior-
ation of a significant health problem, dental caries.

INTRODUCTION

Organisms undergo orderly change throughout their lives in the basic
phenomenon of development. The study of development involves describing
change, defining its mechanisms, and conceptualizing the factors that influence it. Anatomy, physiology and behavior can all be investigated develop-
mentally.

The developmental study of sweet preference is directly relevant to our
concern with sweetness and caries. It may provide the most reasonable base
for generating intervention strategies by which to modify the preference for
sweet and to reduce the ingestion of refined carbohydrates. Current attempts
to modify sugar-eating behavior are markedly unsuccessful. They are based
on an implicit developmental theory that views learning as the primary influence on behavior development. Such a theory may be inadequate. Current
developmental theory leads to a more comprehensive understanding of how
sweet preference develops and suggests alternative strategies to prevent
destructive food habits from dominating the lives of our children.

Advances within several research areas combine to make these times singularly auspicious for adopting a developmental approach to the study of human sweet perception. Recently, the study of taste in mature organisms provided new insights into the nature of sweet perception. Current manipulative studies of development in animals are moving beyond their traditional concern with the effects of early visual or auditory experience and examining the development of feeding behavior. Since the late 1960's, the study of taste in the human newborn has gone forward with a vigor unparalleled since the 1930's and now provides an adequate methodological base for prospective investigations that start at birth. Systematic observations are now being made that directly assess the preference of preschool children for food with added sugar.

Adopting a developmental perspective provides a stimulating challenge to our current understanding of the human preference for sweet. Existing theory, geared to account for adult taste, may need to be revised on the basis of new information about less mature subjects or expanded to account for how perception develops. Existing data must be examined for their relevance to developmental questions.

ADULTS AND INFANTS

Almost all information accumulated about the sense of taste comes from observations on mature organisms. Adults are the subjects of choice for anatomical, electrophysiological and psychological studies of taste. Relevant biophysical and biochemical investigations are made primarily with material derived from mature organisms. However, immature mammals are remarkably different from the mature members of their species in many ways. Cross-age comparisons of sensory capabilities are complicated by differences in response capabilities. Furthermore, infants are not "little adults" with respect to eating behavior. Newborn humans, for example, feed by a suckling maneuver that adults cannot even imitate, and the single fluid they ingest serves as both food and drink. Newborns swallow almost any fluid placed in their mouths and usually only reject it by regurgitation[3]. These substantial differences between infants and adults with respect to feeding make comparisons of taste perception particularly difficult. The newborn, and perhaps infants and toddlers too, might be viewed as members of an entirely different species, and the methodological safeguards of comparative investigations adopted.

Not all developmental studies compare younger and older subjects, and not all studies of young organisms are developmental. Furthermore, studies using only adult subjects could contribute to the understanding of development. A study showing differences in preferences between groups of adults exposed to different influences in infancy and childhood would, for example, demonstrate the extent of possible variation among adults and provide the first leads to factors underlying this variation[4].

FACTORS INFLUENCING DEVELOPMENT

Behavior is not caused by either heredity or environment, by nature or nurture. Behavior is, rather, a product of ongoing developmental processes involving multiple influences that can be various classified. A common response to this circumstance is to accept the simplification that behavior results form an interactive developmental process in which both hereditary and environmental influences participate[5]. As a practical matter, we cannot intervene to change the hereditary endowment of existing individuals. Thus, our attention must focus primarily on environmental influences[6]. However, as development is interactive, the same environmental manipulation may have different effects on groups of individuals who differ genetically.

Kuo provides a strong alternative to the generally-accepted solution to the problems posed by a nature-nurture dichotomy. Gottlieb summarized Kou's position as follows:

It was his life-long conviction that an explanation of an animal's behavior could be derived entirely from (a) its anatomy and physiology, (b) its current environmental setting and (c) its individual developmental history... All three must be taken into consideration in a comprehensive account of animal and human behavior .

For the present purposes, I adopt a classification that posits four factors influencing development: genetic, traumatic, chemical/metabolic and sensory[8]. Genetic influences arise from the physiological properties of the fertilized ovum and comprise the hereditary variable in behavior. Including a genetic factor among the influences on development cautions against an uncritical acceptance of a cultural explanation for differences between groups that differ both culturally and genetically. Traumatic influences arise from physical damage to nerves or other tissues and include birth injuries and postnatal accidents as well as lesions placed intentionally in the nervous systems of experimental animals. Traumatic influences, like genetic ones, are certainly potent and must be included for completeness. However, demonstrations of genetic or traumatic influences on taste are of limited usefulness in our concern with human responses to sweetness because they offer little opportunity for ethically acceptable intervention.

The chemical/metabolic factor includes all effects of chemical stimulation except those that are mediated by the sensory system and thus are considered sensory effects. Chemical/metabolic influences may be either nutritional or toxic and are the traditionally recognized determiners of postnatal physical growth. Malnutrition and therapeutic or abused drugs are increasingly recognized as determiners of learning and school behavior in children. Nizel reviewed nutritional influences on the growth and development of teeth[9]. Effects of maternal undernutrition upon anatomical development, as indexed by birth weight, have been demonstrated in humans under severe conditions such as those in Holland during World War II[10]. Paradoxically it appears that, except under extreme conditions, such prenatal chemical/metabolic effects are more evident in behavior than in measures of anatomical development[11]. Prenatal exposure to barbiturates is dramatically evident in the behavior of the newborn[12]. Chemical/metabolic influences in the uterine environment probably affect the functional or morphological development of the taste system.

The effects of postnatal exposure to chemical/metabolic influences are usually identified as such. However, when the exposure to chemical/metabolic influences is prenatal, the effect may be incorrectly attributed to the action of a genetic influence. Such misidentification has important practical consequences. While the exposure to a chemical/metabolic influence may be modified by changing maternal diet or avoiding specific toxins, genetic influences are not open to such modifications.

The sensory factor has been of great interest to developmentalists. The demonstration that early sensory experience is necessary for the normal development of perception spawned a strong research effort and a fascinating literature defining the critical aspects of stimulation and exploring their mechanisms of action, particularly with respect to vision[13]. Education, currently our most self-conscious attempt to engineer changes in behavior, can be considered a sensory influence on development. Here learning is a central concept. However, Gottlieb finds traditional concepts of learning inadequate to describe embryological phenomena and maintains that they are not adequate for describing how sensory stimulation maintains, facilitates, or induces behavior[14]. The development of human walking provides another instructive case. While skill in walking has been thought to be acquired by learning, the time at which walking begins is considered to be determined by maturation[15]. The time at which the infant walks is an amazingly stable human characteristic[16] and is resistant to manipulations generated from a developmental theory based on learning[15]. However, sensory elicitation of infantile reflexes produces early walking in human infants[17].

EXPERIENCE

Several distinctions are useful to make in thinking about the influence of the sensory factor on development. Since the sensory environment of the fetus is so dramatically different from that of the newborn, pre- and postnatal sensory experiences must be considered separately. Sensory influences may act before birth. Thus, the newborn's taste behavior may be determined not only genetically and by the prenatal action of the chemical/metabolic influences but also by prenatal sensory influences. The fetus exists in a potentially taste-rich fluid environment, has a morphologically mature peripheral taste apparatus, and responds to sweet tastants. The morphological and functional development of the taste system has been extensively studied in the sheep fetus. Knowledge derived from these investigations should contribute to our understanding of the development of taste preferences in humans. However, sheep undergo a greater portion of their cortical development before birth than do humans, and therefore, prenatal sensory experience may have a greater impact on development in this species than in man[18].

Sensory experiences that are normally inevitable for all members of a species can be usefully distinguished from sensory experiences that vary from individual to individual. Each normal infant grows up experiencing visual stimulation, but infants in different cultures see different sights. Similarly, all normal humans grow up experiencing the taste of their food, but the specific tastes to which they are exposed are culturally determined and, in addition, vary from individual to individual within the same culture.

Among those sensory experiences that are normally inevitable for each member of a species, one might expect to find experiences that are essential to normal perceptual development. Individuals deprived of patterned visual stimulation by congenital cataracts, for example, show gross deficits in visual perception when the cataracts are removed[19]. Children with esophageal atresia who are fed from birth by gastrostomy tube undergo what may be an analogous gross deprivation of taste experience. Incidental observations indicate that such deprivation of oral feeding experience does affect subsequent taste perception[20]. Clearly, systematic observations, such as those that have been made concerning visual, auditory, or tactual deprivation, are required.

Fructose intolerance is an inborn error of metabolism that provides a potentially rich source of data concerning the role of specific taste experience in development. Affected individuals vomit and become hypoglucosemic following exposure to the fructose molecule. Typically, this response is present at birth[21] and does not diminish with repeated elicitation[22]. Fructose intolerant individuals avoid not only fructose but all sweet tastes. Thus, they undergo a life-long deprivation of taste experience that is specific to the sweet taste. Studies of fructose intolerance from the perspective of the development of taste perception would contribute significantly to our understanding of the development of sweet preference.

While studies of normally inevitable taste experience will undoubtedly yield insights into the mechanisms of perceptual development, their relevance to the design of interventions may be only indirect. On the other hand, studies of developmental exposure to experiences that vary with culture or from individual to individual might be expected to suggest practical modifications of early sensory experience (e.g. changes in infant feeding practices) that would affect subsequent preferences for sweet. Jerome[23] recently surveyed early infant feeding practices transculturally and included data suggesting that parents readily change even long established patterns of infant feeding under some conditions.

The effect of a specific sensory experience upon development may depend on timing. Research on the role of timing as a determinant of the effects of sensory experience on development generated a large literature dealing with so-called "critical periods." Critical periods are more or less rigidly defined periods of time when certain stimuli have their greatest impact. Outside of the critical period, physically identical stimulation may be totally ineffective. Much of the literature on critical periods is concerned with imprinting in birds. However, the concept has been applied to the analysis of experiential effects occurring in mammals rather than birds and at periods of time somewhat further removed from birth[24]. If critical periods for dietary exposure to sweet could be defined, the degree to which feeding practices would have to be changed to effect a lasting modification of preference might be dramatically reduced. Perhaps short delays in introducing specific foods would be a useful technique for modifying later preference. Muto's survey of the ages at which Japanese and American parents introduce specific foods[25] can serve as a starting point for a line of research that is potentially of considerable practical use. Specific deprivations are imposed on growing children incident to the diagnosis and management of childhood food allergies. Viewing these dietary restrictions from the perspective of sensory experience could yield valuable information about the development of sweet preference.

Although critical periods are usually specified with reference to time, e.g. the first six hours or the 11th through 16th week, the events or processes going on in time, rather than time itself, are the functional entities. In the study of other mammals than man, weaning provides a clear marker for separating one period of time from another, for separating earlier from later experience. Exposures during pre- and post-weaning periods do, in fact, yield markedly different results even when the behavior under consideration is not directly related to food intake[26]. Weaning is a less definitive marker for the study of man. The transition of suckle feeding to fully independent feeding in man appears to be more extended in time and subject to considerable cultural and individual variation. Even within a single culture, some infants may be bottle or breast fed well into the second year while others begin to experience supplementary foods in their second month. A transcultural perspective upon the schedule with which various food items are introduced has recently been put forward[25].

Viewed broadly, the sensory experience associated with feeding in humans might be considered to include at least five distinct stages: (1) the fetus in utero, (2) the infant exclusively suckle feeding from the breast or the qualitatively different circumstance of suckle feeding from the bottle, (3) the infant passively fed semisolids or premasticated items from the adult dietary, (4) the toddler graduating to feeding as a coordinated performance of hand, eye and mouth, and (5) the child increasingly independent not only in the mechanics of ingestion but in the purposive seeking out of food.

CONTRIBUTIONS FROM THE STUDY OF MATURE ORGANISMS

Recordings of responses from single taste nerve fibers have recently been analyzed in a way that challenges the traditional notions of how taste quality is coded by the nervous system. Earlier analyses of single nerve fiber responses support the view that taste fibers are sensitive to many substances, i.e. they are broadly tuned. Each fiber responds to many substances and the fine discriminations between tastes are generated within the central nervous system from the pattern of excitation coming from multiple differently tuned, broad band receptors[27]. A new analysis groups fibers into classes based on which of the four qualities elicit the greatest response[28]. This procedure makes it clear that individual fibers are more narrowly tuned than previously believed. The fibers within the class that respond best to a specific quality (sweet, sour, bitter or salty) resemble one another in their profile of response more closely, and differ more markedly from fibers of other classes, than would be expected from the traditional view. Most importantly for our concern, this new way of conceptualizing the nervous system's coding of taste encourages the notion that the behavioral and subjective distinctiveness of the sweet taste quality is matched on the electrophysiological level.

Advances in adult psychophysics, particularly the development of direct scaling techniques, provide a significant contribution to the study of sweet perception. Direct scaling depends on the fact that adults can make responses (e.g. draw lines or assign numbers) that reflect the magnitude of their subjective experience. Direct scaling yields functions whose slope represents the growth of subjective experience with increasing stimulus concentration. Both the sensory intensity and subjective pleasantness of stimuli can be quantified by these techniques. While the full potential of direct scaling has not been realized, current studies clearly indicate that these techniques will make a major contribution to the understanding of the development of sweet preference. It has been shown, for example, that taste intensity and taste pleasantness can vary independently[29], suggesting, of course, that an intervention need neither enhance nor diminish the perceived intensity of a stimulus to change the degree to which it is preferred.

Direct scaling can free taste investigators from their dependence on the threshold as the single measure by which to characterize their subject's response to sweet stimuli. The stimulus concentrations employed for direct scaling are well above the threshold for detection (thus suprathreshold), and produce tastes that are more closely analogous to those experienced in everyday life. A clinical case reported by Bartoshuk[30] provides a striking example in which the effect on taste thresholds was clearly different from that upon the functions generated by direct scaling of suprathreshold stimuli. In this instance, x-irradiation initially raised thresholds, flattened direct scaling functions, and clinically disturbed the subjective experience of food. With time, thresholds returned to normal pre-irradiation levels. However, the direct scaling function remained depressed as did responsiveness to taste in the feeding situation.

Investigators of adult taste perception are beginning to contribute directly to the understanding of the development of taste perception as they include younger children in their study populations. Laboratory studies, both in this country and abroad, used subjects as young as six years of age[31]. Clinical investigations of taste sensitivity in cystic fibrosis and growth deficiency involved the testing of both affected and control children as young as five years of age[32]. These studies sometimes do not take sweet perception as their focus, and so far they have been limited to threshold measurement. However, they provide encouraging evidence of an expanding interest in the early roots of taste responding.

CONTRIBUTIONS FROM THE STUDY OF NEWBORN HUMANS

In a species, such as man, where prenatal measurement and manipulation

are restricted, the newborn provides the best available starting point for
developmental studies. However, the behavior of the newborn is the product
of considerable development under the influence of chemical/metabolic and
sensory factors, as well as the expression of genetic and perhaps traumatic
influences.

At birth, human infants show a preference for sweetened fluids. The
more concentrated a sugar solution, up to a point, the more of it they ingest[33]. They also seem to prefer those sugars which adults judge to be
sweetest[34]. In response to sweetened water, they display facial gestures
which resemble those that adults might make when presented with the same
fluids[35]. The newborns' heart rate and sucking pattern are related to the
sweetness of the fluids being ingested[36]. Stimulation of even restricted
portions of the newborn tongue with individual drops of sugar water elicits
a response that is dependent upon the taste of the stimulating fluid[37]. Extremely small samples of stimulus fluid delivered during pauses in non-nutritive sucking affect the pattern of sucking[38]. Significantly, the diversity
of neonatal taste related behaviors that are now being measured allows for
the separation of various taste parameters. Properly selected measures will
allow for the separate definition of age related changes in the subjective
intensity and pleasantness of taste stimuli.

In a notable extension of neonatal taste testing Desor, et al. recently
assessed the preference of infants 11 to 15 and 20 to 28 weeks of age[39].
Their results suggest that, while ingested volume increases with age, relative preference is stable over the first half-year of life.

CONTRIBUTIONS FROM THE STUDY OF FEEDING

Animal studies of the role of early auditory and visual experience in
the development of perception abound. In comparison, the effects of early
taste experience are virtually unexplored[40]. The initial failure to find
other than transient effects of early taste experience upon subsequent taste
preferences[41] may have discouraged further investigations of the role of
early experience in taste perception. However, the notion that food habits
and preferences are acquired through early experience persists. Recent investigations designed to study the acquisition of food preference are making
a direct contribution to understanding the development of taste based preferences. Galef uncovered effects of a pre-weaning factor (either sensory or
chemical/metabolic) on the later selection of solid food in rats. Specifically, he found a milk borne influence of maternal diet upon food selection
by rat pups at weaning[42]. Differential pre-weaning experience produced by

manipulating the match between a pup's postnatal age and postpartum age of its foster dam has been shown to affect the morphological development of the pup's peripheral taste apparatus[43], and thus is relevant indirectly to the development of taste preferences.

Systematic formal observation of the eating behavior of young children provide data that are directly relevant to understanding how the preferential eating of sweet foods develops. In her classic investigations, Davis observed infants from their first exposure to adult foods[44]. Infants selected their entire diet in a free choice or cafeteria situation and thrived under these conditions. Almost without exception, the infants sampled all the foods offered. Acceptance or rejection developed only after the foods had been tasted. Significantly for our purposes, the foods were all prepared without sugar and, although salt was supplied separately, sugar was not. Recently a direct comparison between sweetened and unsweetened versions of the same food items became available from a study by Filer[45]. Infants between two and six years of age were offered both plain and sugar-sweetened spaghetti in the context of a full meal. Not unexpectedly, most children ate more of the sweetened alternative. However, 24% of the subjects did not display a preference and of those that did, it was impossible to tell if preferences were equally strong for the youngest two-year-olds and the oldest six-year-olds. Perhaps sweet preference is in transition across the span of ages tested. Risley and co-workers already demonstrated that other food preferences change toward the adult form during the wide age range covered by this study[46]. They found that preference for cold over room temperature non-fat milk is absent in two- to three-year-olds, but that four- to five-year-olds react negatively to "warm" milk. A preference for salted over unsalted beef stew was demonstrated by the same procedure used to show the preference for sugared spaghetti. Evidence for age related change in salt preference comes from comparing this adult-like salt preference with the negative responses to salted fluids observed in the neonate[38] and the lack of salt preference among both four- and seven-month-old infants presented alternate forms of strained baby foods by their mothers[45].

Risley and co-workers recently described an extensive program investigating the eating behavior of one- to three-year-old children. Meals served at a day care facility are systematically varied and the children's intake measured. Under these conditions taste appears to be the primary determiner of acceptance with the visual characteristics, temperature and size of the serving not affecting acceptance[46].

The present state of research concerned with eating in children is encouraging. Increasing numbers of studies are focusing specifically on sweetness as a determiner of the choice among items in the adult dietary. In some studies measurements of preference are based on a child's choice between simultaneously presented alternatives and in others on a comparison between observations of intake on successive exposures. This diversity of method is healthy, and valuable normative data reflecting the state of sweet preference at successive ages will be obtained. However, to gain an understanding of the developmental process of evolving sweet preference, investigators have to muster the courage to restrict sweetness and expend the extra efforts required to begin their observations at weaning or, as is now possible, at birth.

REFERENCES

1. Weiffenbach, J.M., Ed. Taste and Development: The Genesis of Sweet Preference, Government Printing Office, Washington, D.C., 1977.
2. Grinker, J.A. in Taste and Development: The Genesis of Sweet Preference, edited by J.M. Weiffenbach, Government Printing Office, Washington, D.C., p. 309, 1977.
3. Desor, J.S. in Taste and Development: The Genesis of Sweet Preference, edited by J.M. Weiffenbach, Government Printing Office, Washington, D.C., p. 67, 1977.
4. Moskowitz, H.R., Kumaraiah, V., Sharma, K.N., Jacobs, H.L. and Sharma, S.D. Science 190, 1217, 1975.
5. Hainline, L. Contemp. Psychol. 22, 662, 1977.
6. DuBois, R. in Environmental Influences, edited by D.C. Glass, The Rockefeller University Press and Russell Sage Foundation, New York, N.Y., p. 138, 1968.
7. Kuo, Z-Y. The Dynamics of Behavior Development, enlarged edition, Plenum Press, New York, N.Y., pp. xiii-xiv, 1976.
8. Hebb, D.O. A Textbook of Psychology, W.B. Saunders, Philadelphia, Pa., p. 113, 1972.
9. Nizel, A.E. in The Pediatric Clinics of North America: Symposium on Nutrition in Pediatrics, C.G. Neumann and D.J. Jelliffe, Eds., W.B. Saunders, Philadelphia, Pa., p. 141, 1977.
10. Smith, C.A. J. Pediat. 30, 229, 1947.
11. Osofsky, H.J. in Intrauterine Asphyxia and the Developing Fetal Brain, L. Gluck, Ed., Year Book Medical Publishers, Chicago, Ill., 25, p. 395, 1977.
12. Kron, R.E., Litt, M. and Finnegan, L.P. Int. J. Clin. Pharmacol. Biopharm. 12, 63, 1975.
13. Berry, M. in Advances in Psychology, Volume 3, edited by A.H. Riesen and R.F. Thompson, John Wiley and Sons, New York, N.Y., p. 125, 1976.
14. Gottlieb, G. Psychol. Rev. 83, 215, 1976.
15. McGraw, M.B. The Neuromuscular Maturation of the Human Infant, Hafner, New York, N.Y., 1963.
16. Kessen, W., Haith, M.M. and Salapatek, P.H. in Carmichael's Manual of Child Psychology, edited by P.H. Mussen, John Wiley, New York, N.Y., p. 613, 1970.
17. Zelano, P.R. in Developmental Psychobiology: The Significance of Infancy, edited by L.P. Lipsitt, Lawrence Erlbaum, Hillside, N.J., p. 87, 1976.

18. Mistretta, C.M. and Bradley, R.M. in Taste and Development: The Genesis of Sweet Preference, edited by J.M. Weiffenbach, Government Printing Office, Washington, D.C., p. 51, 1977.

19. von Senden, M. Space and Sight: The Perception of Space and Shape in the Congenitally Blind before and after Operation, Methuen, 1932. Trans. of German, Free Press, Glencoe, Ill., 1960.

20. Dowling, S. in The Psychoanalytic Study of the Child, Yale University Press, New Haven, Conn., p. 215, 1977; Dowling, S. and DeMonterice, D. Pediatric Research 8, 343, 1974.

21. Schaeffer, A.J. and Avery, M.E. Diseases of the Newborn, W. B. Saunders, Philadelphia, Pa., p. 381, 1971.

22. Mühlemann, H. in Taste and Development: The Genesis of Sweet Preference, edited by J.M. Weiffenbach, Government Printing Office, Washington, D.C., p. 350, 1977.

23. Jerome, N.W. in Taste and Development: The Genesis of Sweet Preference, edited by J.M. Weiffenbach, Government Printing Office, Washington, D.C., 17, p. 235, 1977.

24. Scott, J.P. Science 138, 949, 1962.

25. Muto, S. in Taste and Development: The Genesis of Sweet Preference, edited by J.M. Weiffenbach, Government Printing Office, Washington, D.C., p. 249, 1977.

26. Forgays, D. and Read, J.M. J. Comp. Physiol. Psychol. 55, 816, 1962.

27. Erickson, R.P. in Olfaction and Taste I, edited by Y. Zotterman, Pergamon Press, Oxford, 1963.

28. Frank, M. and Pfaffmann, C. Science 164, 1183, 1969.

29. Moskowitz, H.R., this volume.

30. Bartoshuk, L.M. J. Am. Soc. of Clin. Nut., in press.

31. Murphy, C. Ph.D. dissertation, Dept. Psychology, University of Massachusetts, 1975; Yasaki, T., Miyashita, M., Ahiko, R., Hirano, Y., Kamata, M. and Iizuka, Y. Jap. J. Dent. Health 26, 200, 1976.

32. Desor, J.A. and Maller, O. J. Pediat. 87, 93, 1975; Hambidge, K.M., Hambidge, C., Jacobs, M. and Baum, J.D. Pediat. Res. 6, 868, 1972; Hertz, J., Cain, W.S., Bartoshuk, L.M. and Dolan, F.F. Physiol. Beh. 14, 89, 1975; Wotman, S., Mandel, I.D., Khotim, S., Thompson, R.H., Kutscher, A.H., Zegarelli, E.V. and Denning, C.R. Am. J. Dis. Child. 108, 372, 1964.

33. Desor, J.A., Maller, O. and Turner, R. J. Comp. Physiol. Psychol. 84, 496, 1973; Desor, J.A., Maller, O. and Greene, L.S. in Taste and Development: The Genesis of Sweet Preference, edited by J.M. Weiffenbach, Government Printing Office, Washington, D.C., p. 161, 1977; Houpt, K.A. and Houpt, T.R., Ibid., p. 86; Crook, C.K., Ibid., p. 146; Nowlis, Ibid., p. 190.

34. Nowlis, G.H. and Kessen, W. Science 191, 865, 1976; Nowlis, G.H. in Taste and Development: The Genesis of Sweet Preference, edited by J.M. Weiffenbach, Government Printing Office, Washington, D.C., p. 190, 1977; Desor, J.A., Maller, O. and Greene, L.S., Ibid., p. 161; Jacobs, H.L., Smutz, E.R. and DuBose, C.M. Ibid., p. 99; Desor, J.A., Maller, O. and Turner, R.S. J. Comp. Physiol. Psychol. 84, 496, 1973.

35. Steiner, J.E. in Taste and Development: The Genesis of Sweet Preference, edited by J.M. Weiffenbach, Government Printing Office, Washington, D.C., p. 173, 1977.

36. Lipsitt, L.P. in Taste and Development: The Genesis of Sweet Preference, edited by J.M. Weiffenbach, Government Printing Office, Washington, D.C., p. 125, 1977; Crook, C.K. in Taste and Development: The Genesis of Sweet Preference, p. 146, Ibid.

37. Weiffenbach, J.M. in Taste and Development: The Genesis of Sweet Preference, edited by J.M. Weiffenbach, Government Printing Office, Washington, D.C., p. 205, 1977.
38. Crook, C.K. Infant Behavior and Development, in press.
39. Desor, J.A., Maller, O. and Greene, L.S. in Taste and Development: The Genesis of Sweet Preference, edited by J.M. Weiffenbach, Government Printing Office, Washington, D.C., p. 161, 1977.
40. Hunt, J. McV. J. Abnorm. Soc. Psychol. 36, 338, 1941; Hess, E.H. in Roots of Behavior, edited by E.L. Bliss, Harper and Brothers, New York, N.Y., 1962.
41. Warren, R.P. and Pfaffmann, C. J. Comp. Physiol. Psychol. 52, 263, 1959.
42. Galef, B.G. in Taste and Development: The Genesis of Sweet Preference, edited by J.M. Weiffenbach, Government Printing Office, Washington, D.C., p. 217, 1977.
43. Henderson, P.W. and Smith, G.K. in Taste and Development: The Genesis of Sweet Preference, edited by J.M. Weiffenbach, Government Printing Office, Washington, D.C., p. 70, 1977.
44. Davis, C.M. Am. J. Dis. Child. 36, 651, 1928; Davis, C.M. Can. Med. Assoc. J. 41, 257, 1939.
45. Filer, L.J., Jr. Pediatric Basics 2, 6, 1975.
46. Twardosz, S., Cataldo, M.F. and Risley, T.R. Young Children 30, 129, 1975; Herbert-Jackson, E. and Risley, T.R. J. Applied Behav. Analysis, 10, 1407, 1977; Herbert-Jackson, E., Cross, M.Z. and Risley, T.R. J. Nutr. Ed. 9, 76, 1977.

DISCUSSION AFTER PRECEDING FOUR PAPERS

DR. KARE: As chairman, I would like to ask Dr. Moskowitz the first question. If insulin is administered to an animal, its preference for sugar goes up to select a higher concentration. Likewise, if you expend energy, your consumption of energy sources goes up. Would not the desire for greater sweetness in children be a reflection of physiological and metabolic needs?

DR. MOSKOWITZ: That is a very difficult question to answer experimentally. I am not sure anyone has done the proper experiment. Let me describe the paradigm that I would follow. One might have children either exercise or not exercise (two conditions) and then test their hedonic reaction to either a drink sweetened with sugar (A) or a drink sweetened with saccharin or some other artificial sweetener (B). One would follow each of the four conditions over several days, and measure how much of drink A or B was consumed. If you follow that design and equate for sweetness and for texture, then you ought to be able to determine how much of the drive for sweetness is motivated by taste and how much can be attributed to exercise.

DR. WEIFFENBACH: Offering a choice is only one way of measuring preference or acceptability. With children the data generated from offering a choice is considerably different from what is obtained in a more complicated design where a single item is offered.

DR. MOSKOWITZ: What happened in the case of children? Did they prefer the sweeter product?

DR. WEIFFENBACH: It is my understanding that the preferences are attenuated in the one bottle situation. That is, things that would not get in the way of acceptance in the two bottle situation become problems in the one bottle situation.

DR. MOSKOWITZ: If you watch children up to the age of seven or eight in a paired preference situation or in a rating situation, they always go for the extremes when they have to verbally state or otherwise indicate what they want. When you test older children, there is an understanding of intermediacy so that they may not like nor choose the sweeter product. In adults, paired preference data versus single stimulus data tend to agree with each other.

Grinker and Hirsch showed that if you give people two bottles, one of which is sweeter and the other less sweet, then the peak preference occurs at about 9% which also gets the most positive votes for acceptance.

DR. HERBERT: What do studies of quantity, strength, or hedonics of the satiety signal indicate in normal versus obese people?

DR. MOSKOWITZ: Data published by the French investigator, Cabanac, suggest that obese people are not sensitive to their own internal signals of satiety. What he does is have them sequentially taste and drink a certain volume of water containing sucrose. He then gets ratings of degree of acceptability of that sweetened water over time after the subjects have ingested the sample.

In theory they should become increasingly satiated because of the calories provided by the ingested material. However, ratings by obese subjects for sweetness and for liking do not systematically change as they continue to ingest the solution. In the control experiment with individuals of normal weight, an increasing disliking rating for that sweet material was given as the solution was continually presented. Initially the sweet material for the normal may have been acceptable by itself, but as they get filled up, the taste of sweet material became less and less acceptable and was eventually rejected.

DR. BOWEN: Dr. Moskowitz, is there any way in which you can change the perception or level of perception of sugar by feeding high concentrations of sugar continuously?

DR. MOSKOWITZ: I do not know if anybody has ever done that for sugar with humans. However, you can change the acceptability of a pure acid, such as citric, by rearing individuals in an environment where they are exposed to citric acid. We did this with people in India where we investigated laborers from the Karnatka region who were reared with a lot of tamarind flavored foods (an exceedingly sour material). Their ratings of acceptability of the sour taste increased. I do not know what would be parallel findings with sugar.

DR. NEWBRUN: Your data on hedonics were for solutions. The problems in solids are more complex by reason of the presence of other tastants. Since this conference is concerned with sweetness and caries and most foods are solids, our concern needs to be more with solids than with solutions. Is there any comparable work to show that there is a maximum sucrose concentration in solids for humans which is most pleasant beyond which a higher concentration is no longer as pleasant? If so, how does this relate to the fact that most candies and chocolates as well as many pre-sweetened breakfast cereals are in the range of 40 to 50% sugars by weight?

DR. MOSKOWITZ: There are two questions. First, are there data? Unfortunately, relatively little has been published in the open literature in terms of laboratory experiments. Much more information is in the proprietary files of companies where their products were evaluated with different levels of sugars. We studied puddings and cakes and, of course, sugar water and Kool Aid. In puddings the preferred sugar concentration was somewhat higher than for solutions. The break point for hedonic level vs. concentration was less noticeable than in solutions. When a solid food containing 40 to 50% sugar by weight is studied, a lot of the sugar provides functional properties other than sweetness. This means that the sugar never really was dissolved in saliva and evaluated as a tastant. Under these conditions, you may have 40% concentration in a solid food, but it may appear only as sweet as a 10% sugar solution. We do not know how much of that 40% is acting to influence and determine taste perception and how much of it is just there for bulk.

DR. MACKAY: I think the answer is that many of these high sugar solid foods end up giving about the same 10% sugar level in the saliva. In confections, the determining factor of the sugar content of the product may be the level perceived in saliva. As Dr. Moskowitz points out, 9 to 10% has the maximum liking. The value is affected by the amount and type of acidity and by the flavor, all of which alter salivation.

Soft drinks are what they are for very good reasons--the sugar concentration, the sugar/acid ratio, the level of carbonation, the nature of the acid, the nature of the flavor--all play specific roles in the design to obtain maximum consumer acceptance. In the use of confections, e.g. sucking a candy, the saliva may end up with about the same 10% preferred sugar level found in soft drinks. The rate of solution of the candy or extraction of sugar from chewing gum has to be a major factor, but other factors which influence salivation have to be considered to determine why one candy is preferred to another one.

DR. BOLLENBACK: Dr. Emily Wick did some work on sweetness detection at very high concentrations some years ago at the Massachusetts Institute of Technology. As I recall, the ability to reproduceably detect the differences in degree of sweetness was lost at or above 20% concentration.

I would like to ask Howard another question. With regard to the Grinker and Hirsch findings, obese people had a loss of preference for extremely sweet solutions compared with those of normal weight for the same degree of solution sweetness.

DR. MOSKOWITZ: You are referring to a publication in the early seventies where Grinker and Hirsch instructed obese people to do two things. First, for a graded series of sugar solutions in water, to rate on a fixed category point scale how much they liked them, and then by a pair comparison method to select which one from various pairs they preferred. Their data showed that, as sweetness (i.e. concentration) was increased, normal weight people had a break point for liking at around 9%, but obese people showed no break point. In fact, obese subjects disliked the sweetener starting from approximately 1% as the concentration was increased. In contrast to what I said earlier, these individuals found sweet solutions to be less pleasing.

In neither my study with obese subjects nor the Grinker and Hirsch study was there ever a break point. My conjecture is that we are dealing with the same phenomenon, viz., an inability to discern the difference between sweetness as information and the hedonic tone provided by a sweetener. Normal weight people can discern the difference, and, therefore, one sees two different functions, one with a monotonic increasing sweetness versus concentration, one with a peak point, of pleasantness vs. concentration. In the case of obese people, one either has individuals liking or disliking, and their ratings are directly related to sweetness. I think Grinker and Hirsch's subjects may somehow have been attuned to classifying sweetness as "bad" and, therefore, they said that they disliked it. But since they were unable to scale true "hedonics," they evaluated sweetness, but misreported it as unpleasantness.

In several studies we found that obese people tend not to know the difference between liking and hedonics, as Grinker and Hirsch showed. However, our subjects reported pleasantness rather than unpleasantness. Yet it's really the same thing.

DR. EMELE: Howard, with regard to optimal sweetness concentration, do any data indicate a difference in preference or liking between sweeteners? Where each sweetener would be optimally acceptable, can you discern differences between them as to liking and preference?

DR. MOSKOWITZ: There certainly are differences. They are not particularly marked in the case of the carbohydrate sweeteners, although glucose has a burning, bitter characteristic which makes it less pleasant. In the case of saccharin versus cyclamates versus sugar, there are noticeable differences in sweetening characteristics. The maximum preference that you can achieve with saccharin tends to be lower than the maximum preference level of sucrose where it reaches it maximum. The pleasantness-sweetness curve, for example, can be obtained for two different sweeteners, sugar and saccharin. Where each peaks (a) they are going to be different in the case of the achieved degree of sweetness, but also (b) the highest level of preference that is achievable is going to be higher with the sugar stimulus than with the saccharin stimulus. There are inherent differences in the potential liking score that each substance can achieve.

DR. BOWEN: Dr. Beidler, could you elaborate a little more on the mechanisms of action of taste modifiers and what future do you perceive that these have as sugar substitutes?

DR. BEIDLER: Gymnemic acid inhibits sweetness without affecting other taste qualities to any considerable degree. Such an inhibitor would be useful for those who crave sweets at certain times of the day. It would dull the sweet taste for up to an hour or more. We demonstrate it by incorporating the gymnemic acid into a lollipop.

Miracle fruit is a sweet protein that is only effective at low pH and is operable for up to an hour after high doses. The sweetness can be turned on and off by altering the pH. Thus lemon added to iced tea makes it sweet. A whole meal can be made desirable without addition of sugars or artificial sweeteners merely by eating a miracle fruit or its active principle before the meal and adding a normally sour ingredient like citric acid to those foods that taste good when sweetened. The glycoprotein of miracle fruit is a useful product for diabetics, obese people, and others who want sweetness without calories and without artificial sweeteners. Use is not limited to meals since popsicles also have been made with miracle fruit as well as certain candies, gums, fruity yogurt, etc. The possibilities are unlimited but industry must consider unconventional uses and development.

DR. KARE: Lloyd, could I ask if you are concerned when you use these non-carbohydrate sweeteners that they fail to initiate a release of insulin? When you use a substitute for carbohydrates, a change in the pattern of pancreas function may result.

DR. BEIDLER: You are illustrating a most important fact that is normally overlooked. Many physiological effects may be associated with sweeteners and some, such as those associated with the licorice sweetener, are well known. FDA emphasizes cancer and gives little attention to other roles. Insulin release is a good example.

DR. NAVIA: I would like to ask how important are the various aspects of the response to the consumption of a snack. The person eating a snack at a certain time not only receives pleasure from the sweet taste, but also receives side effects such as relief from a physiological demand for energy. How important are the various components in the overall picture of developing a sweet preference?

DR. KARE: Many things happen as a result of oral stimulation. Activity along the gut is stimulated and metabolic hormones are released. There is an immediate release of insulin, followed by another moments later, and then 20 minutes later the circulating hormone level rise parallels the hyperglycemia. In addition, there may be a substantial effect on the exocrine function of the pancreas. Taste can initiate a many-fold increase in enzyme release. For example, Fischer and Hommel used glucose as an oral stimulant and traced the release of immunoreactive insulin. They also administered the sugar intravenously, as opposed to oral administration, and measured the difference in effect on circulating hormone level. Also, Nicolaidis used oral stimulants and reported on their effect on respiratory quotient.

Sweetener economics: analysis and forecasts

Bruce J. Walter

CPC International, Inc., Englewood Cliffs, New Jersey 07632

INTRODUCTION -- THE SWEETENER MARKET

The so-called "sweetener market" is an international, multi-product, multi-industry complex. Among the products generally included in this market are the following:

> Sucrose sweeteners
>> Cane sugar
>> Beet sugar
>
> Starch sweeteners
>> Crystalline glucose (dextrose)
>> Glucose syrups
>> Isomerose syrups (HFCS)
>
> Other caloric sweeteners
>> Honey
>> Maple syrup and maple sugar
>> Molasses
>
> Noncaloric sweeteners
>> Saccharin
>> Other noncaloric sweeteners

Of course, this list merely indicates the major classifications and types of products included in this market. Each of these products is available in a wide spectrum of variations, each with very detailed specifications. In addition, there are a wide variety of other sweeteners, such as aspartame, miraculin, talin, and acetosulpham, which exist primarily as laboratory curiosities with little or no commercial application at the present time. Still others, such as xylitol and neohesperidin dihydrochalcone (neo-DHC), are just beginning to become commercially significant.

While all of the products listed above compete with one another as

*This manuscript was not prepared by Dr. Walter in an official capacity as an employee of CPC International Inc.; the opinions expressed are his personal ones rather than corporate opinions.

sweeteners, each has a unique set of characteristics--such as flavor, texture, calorie content, moisture retention, etc. -- which differentiates it from the others and makes it especially suitable for particular uses. Thus, the overall sweetener market is, in effect, a collection of overlapping "mini-markets" in which particular sub-sets of these products compete as substitutes or partial substitutes for particular uses.

My analysis is restricted to sucrose and the starch sweeteners, since they are by far the most important sweeteners from a commercial or economic standpoint. I am going to concentrate primarily on the United States sweetener market, but I will have to discuss the international situation in considerable detail due to the dependence of the U.S. on sugar imports.

INTERNATIONAL SUGAR MARKET

Sucrose, commonly referred to as sugar, is produced in most of the countries of the world. This broad geographical distribution of production is due, in part, to the fact that sucrose is commercially extracted from two significantly different plant sources--sugarcane, a tropical perennial grass, and sugarbeets, a temperate-zone annual root crop. The United States is one of several countries which produce both sugarcane and sugarbeets.

The United States normally ranks third or fourth (behind the USSR, Brazil, and sometimes Cuba) among the sugar-producing nations of the world, but is nevertheless the world's largest sugar importer. Domestic production accounts for roughly half of the sugar market in the United States, and the other half is supplied from abroad.

The pie-chart in Figure 1 represents world sugar consumption. Of the tremendous amount of sugar produced throughout the world, roughly 20 percent is traded in the so-called "world free market." The remainder is either consumed in the country of production or traded under preferential arrangements. These preferential arrangements include (a) Cuba's exports to the socialist countries, (b) the USSR's exports to other socialist countries, and (c) EEC imports from 17 less developed countries (former colonies) under the Lomé Agreement. The percentages presented on this chart are relatively stable from year to year. If we apply them to the September-August crop year of 1976-77 (the crop year which ended in September), during which an estimated 82.8 million metric tons of raw sugar were consumed worldwide, it indicates that 71 percent, or roughly 59 million tons, were consumed in the country of production. Of the remaining 29 percent (24 million tons) which entered world trade, over one-third (roughly 9 million tons) was exchanged under preferential arrangements. This left only 18 percent (roughly 15

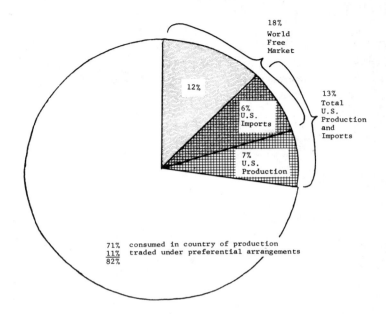

Figure 1. United States sugar production and imports in relation to world trade.

million tons) to be traded in the so-called "world free market." During crop year 1976-77, the United States consumed 10.6 million metric tons of sugar or roughly 13 percent of total world consumption. Of this, 4.5 million metric tons (or almost 6 percent of world consumption) were purchased on the world free market and imported, and 6.1 million metric tons (or 7 percent of world consumption) were produced in the U.S. Thus, the United States is dependent upon imports from the world market for roughly half of its sugar consumption, and those imports constitute a very significant portion (roughly one-third) of the world free market.

World production, consumption, and stocks for the past 23 years are shown in Figure 2. These lines illustrate the basis for what is commonly referred to as the sugar cycle. Note that consumption, which is determined primarily by (1) population and (2) per capita income, increases rather steadily--with the exception of 1974, which will be discussed later. Production, on the other hand, is considerably more erratic since it depends on such factors as (1) planted acreage, (2) weather, and (3) milling capacity.

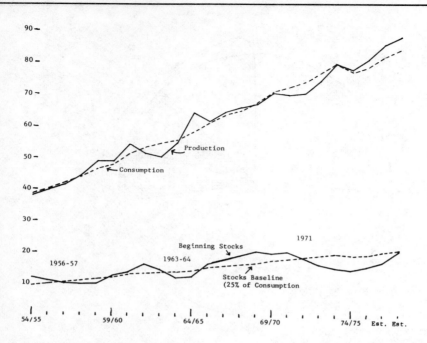

Figure 2. World production, consumption and stocks of sugar from 1954-55 to 1977-78 in millions of metric tons. Source: USDA.

The short-run inflexibilities inherent in both sugarcane and sugarbeet production and the psychology of the market lead to a cyclical production pattern for sugar: a period of short supplies (and high prices) followed by a period of expansion and then a period of surplus production (and extremely low world free market prices).

Stocks are shown here because they constitute an important part of total supply (or availability) and represent a "buffer" to smooth out small variations in year-to-year production. The medium term trend in world sugar stocks is the best leading indicator for world sugar prices. When stocks dip below 28 to 25 percent of consumption, world sugar prices begin to soar. Stocks above 25 to 28 percent of consumption can, therefore, be thought of as "surplus," while stocks below that level are generally considered to be inadequate. Actually, of course, the line of demarcation is not that clear cut-- the direction and duration of a stock buildup or decline are also of considerable importance in influencing price trends.

Please take note of the years 1956-57, 1963-64, and 1971-72, where the solid line representing stocks begins to dip below the dashed baseline representing 25 percent of consumption. Also note the relative size of the

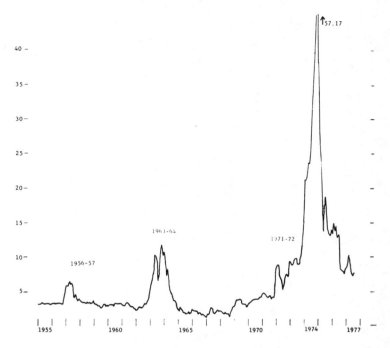

Figure 3. World raw sugar prices from 1955 to 1977 FOB Caribbean ports in cents per pound. Source: USDA.

stock "deficit" between the years 1971 and 1977.

The world free market price for sugar since 1955 is shown in Figure 3. Again, note the years 1956-57, 1963-64, and 1971-72--each with a very significant price peak. Also note the relative magnitude of the price rise beginning in 1971 and extending into late 1974, when stocks began to recover.

The magnitude and duration of the sugar price increase of 1971-74 caused two new variables to enter the sugar consumption equation--(1) the price of sugar and (2) the relative price of sugar substitutes. World sugar consumption declined in response to the fantastic price increase. This was a revolutionary development, since sugar demand had heretofore been considered highly "inelastic" or unresponsive to price.

UNITED STATES GOVERNMENT PRICE REGULATION

The world free market price for sugar as a dashed line and the United States sugar price as a solid line are shown in Figure 4. It illustrates the effect of U.S. government control of domestic sugar prices through 1974. That effect was a generally higher but generally more stable price than that

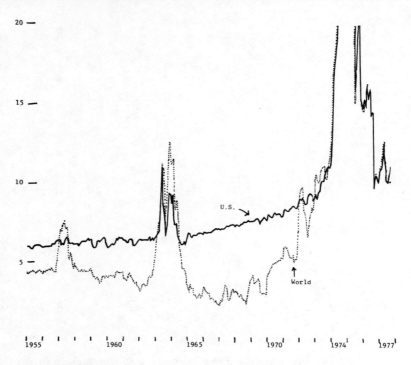

Figure 4. United States and world raw sugar prices 1955-1977. U.S. spot price versus world in cents per pound.

which existed in the volatile world free market.

Actions of the Federal Government have affected the U.S. sugar market since 1789, when a tariff was imposed on all sugar imports. A quota system of control was initiated in 1934 and was amended and extended periodically until, on June 5, 1974, the House of Representatives voted against extension. The Sugar Act formally expired on December 31, 1974, and the U.S. sugar market shifted from the preferential arrangement category to become a part of the world free market.

It should be noted at this point that most national governments through-out the world regulate sugar heavily. The major objective of this regulation is generally aimed at insulating the domestic price of sugar from the typically low but extremely volatile and cyclical world free market price. Thus, most exporting countries pool producers' returns from higher-priced (domestic and preferential) and lower-priced (world free market) sales. Importing countries, on the other hand, (a) pool higher cost sugar from domestic and preferential purchases with lower cost sugar bought in the world free market; (b) tax the difference between these higher and lower cost markets; and/or (c) pay stabilized and remunerative prices to their foreign suppliers in return for an implied assurance of adequate supply.

Among the factors which render the sugar industry universally susceptible to government control are:

1. Cyclical production--the short-term inflexibilities inherent in both sugarcane and sugarbeet production and the psychology of the market lead to a cyclical production pattern for sugar: a period of short supplies (and high prices) followed by a period of expansion and then a period of surplus production (and extremely low world free market prices). Due to the technology of sugar extraction and its associated economies of scale, both sugarcane and sugarbeets require relatively large and sophisticated processing plants. Furthermore, since these crops are both perishable and low in value per unit of weight and volume, they must be processed in or near the area in which they are grown. Thus, a given production area generally must either produce enough cane or beet to operate a processing plant or else not produce these crops at all. The output of any given plant (and, hence, the production of the beet or cane growers dependent upon that plant) is constrained by both the minimum economic and maximum physical capacities of the plant. The high investment cost of beet or cane processing facilities is a severe constraint on excess capacity; and the time required to plan, finance, and construct such facilities is an important constraint on short-run expansion. Sugarcane has the additional supply response inflexibility of being a perennial crop which is normally harvested annually but is replanted only once every three to five years.

2. Ease of regulation--due to the relatively few and well established channels through which refined sugar is distributed, the regulation of this flow is a rather simple administrative task.

3. Source of government revenue--due to its rather inelastic demand and low price, this "cheap food" is a natural target for income-producing taxes. Most importing countries impose tariffs on sugar imports, and exporting countries impose export taxes or employ other means of raising government revenue from the industry. Between 1789 and 1890, the major purpose of U.S. sugar regulations was, in fact, to raise revenue for the Federal Treasury.

4. Protectionism--the goal of most countries is to attain some degree of self-sufficiency in sugar production. Throughout the world a wide variety of techniques are utilized, where necessary, to protect sugar industries of varying degrees of efficiency. This was the policy of our own country from 1870 through 1974.

5. National security implications--dependence on overseas sources of supply and a proven strategic importance during prolonged conflict have been used to justify government regulation of sugar production and refining both in anticipation of and during war periods.

6. International relations--sugar has and continues to play an extremely important role in the commercial and political relationships between nations. Importation preferences of various types have traditionally been used on a quid pro quo basis by the temperate zone nations in dealing with their neighbors to the south.

7. Structure and conduct--throughout the world, the economies of scale inherent to sugar refining and distribution have led to the development of highly concentrated, oligopolistic industry structures which, in turn, have led to the formation of monopolies, cartels, or trusts. This seemingly inherent tendency has, in most countries, resulted in government intervention and control--either as a partner or as an antitrust policeman.

CURRENT U.S. SUGAR PRICE REGULATION

The 2½ years since the expiration of the U.S. Sugar Act happen to have coincided with a downswing in the world sugar price cycle. As prices fell, U.S. sugar growers and processors applied increasing pressure on the President and Congress to restore price protection. In response to this pressure, the U.S. International Trade Commission (USITC) conducted an investigation and, in March 1977, reported that there existed "threat of injury" to the U.S. sugar industry from increased (and projected increased) imports of raw sugar. The USITC recommended that the President establish a quota system to control sugar imports. The President refused to institute quotas but, as an alternative, proposed in May to establish a system of subsidy payments for U.S. sugar growers under existing legislation (a very liberal interpretation of the 1949 Agricultural Act). In August, before these subsidies could be established, the Justice Department declared the scheme illegal. However, through skillful rewording, an approved subsidy program, commonly called the "Bergland plan," was announced on September 15, 1977. Its provisions are as follows:

    Price objective:

            Sugar: Free market
            Sugarbeets and sugarcane: Equivalent to
              13.5¢ sugar

    Control mechanism:

            Variable subsidy paid to processors
            provided that they in turn pay growers
            equivalent to 13.5¢ raw sugar

Term: "Interim"

Effective date: September 15, 1977
                     Retroactive for 1977 crop

Meanwhile, as the Executive Branch was developing its sugar subsidy scheme during June and July, Congress was writing the omnibus Agricultural Act of 1977. On August 5, the House and Senate conferees agreed upon a bill which included a sugar price support program commonly referred to as the de la Garza amendment. The provisions of this program are as follows:

Price objective: 13.5¢ minimum
                      Range 52.5-65.0% parity

Control mechanisms:

          Nominal: Sugar loans and purchases
          Actual: Import duties and, possibly, quotas

Term: 2 years (1977 & 1978 crops)

Effective date:
          Legal: October 1, 1977 or before
          Administrative: ?--November, 1977

Other provisions:
          Purchased sugar not to be resold for less
          than 105% purchase price

          Minimum wage rates
          "Self-destruct" if ISA successful in raising
          U.S. price to 13.5¢

The President signed the 1977 Agricultural Act into law effective October 1, but his Administration did not implement the sugar provisions of that law immediately. Instead, they chose to implement the Bergland plan, despite strong objections from Congress (Senators Dole, Humphrey, etc.) and the corn wet milling industry. Finally, on October 20, Congress reached a compromise -- the de la Garza plan would replace the Bergland plan in early November.

INTERNATIONAL SUGAR AGREEMENT

In Geneva on October 5, after prolonged negotiations during last spring and summer, the representatives of roughly 72 nations tentatively acceded to all of the major elements of an international sugar agreement (ISA) designed to stabilize the world free market price of sugar. The provisions of this agreement are as follows:

Price objective: 11-21¢
                      Free market between 15-19 cents

Control mechanisms:
          Export quotas -- 15.9 mmt basic quota
                           Adjustable
          Buffer stocks -- 2.5 mmt
                           Held by exporters

Term: 5 years

Financed by import/export tax
(28¢/cwt)

Effective date:
          January 1, 1978 (if ratified)

Other provisions:
          Quotas renegotiated after second year

It should be noted that: (1) the 11 cents per pound minimum price objective of the ISA is roughly equivalent (after adding freight, insurance, etc.) to a U.S. price of 13.5 cents per pound; and (2) either the Bergland plan or the de la Garza amendment (whichever prevails) would become non-operative if this price level is achieved.

International agreements designed to stabilize the world free market price of sugar have existed in the past, and there are strong arguments both pro and con concerning their effectiveness. This agreement contains certain important provisions, such as buffer stocks, which have not been used in the past for sugar. Only time will tell if this agreement is workable and effective.

STARCH-BASED SWEETENERS

Having ignored the starch-based sweeteners up to this point, it should be noted that, in contrast to sucrose, starch sweeteners do not enter international trade in significant quantities. Instead, the source commodities--such as corn, wheat, potatoes, rice, and tapioca (also known as cassava, manioc, or yuca)--are traded internationally and enter domestic markets for a wide variety of agricultural and industrial uses. Only after they have been purchased in domestic markets and processed by starch sweetener producers are they identified with the sweetener industry. Only a very small percentage of the production of these commodities is used for the manufacture of sweeteners. For this and many other reasons, starch sweeteners have generally not been subjected to government economic controls to the same degree as sucrose. However, during the past year, pressure from European sugar producers has caused the EEC to impose harsh economic controls on the starch sweetener manufacturers in that market. These controls are aimed primarily at the recently developed isomerose (HFCS) sweeteners, which are much more highly competitive with sucrose than conventional glucose syrups.

U.S. SWEETENER CONSUMPTION

Before making a few forecasts or comments on the future, some brief comments on U.S. sweetener consumption are appropriate. Figure 5 indicates per capita U.S. sweetener consumption from 1910 to 1976. The term "consump-

Figure 5. PER CAPITA U.S. SWEETENER CONSUMPTION

(Pounds)

| Calendar | Refined cane and beet sugar | | | | | Corn sweeteners (a) | | | | Honey and edible sirups | | | Total caloric | Noncaloric sweeteners (b) | | | Total sweetener consumption |
| | U.S. grown | | | Imported | Total | Corn sirup | | Dextrose | Total | Honey | Edible sirups | Total | | Saccharin | Cyclamate | Total | |
	Beet sugar	Cane sugar	Total			High-fructose	Other										
1910....					75.4		5.4	1.1	6.5	1.2	6.8	8.0	89.9				89.9
1920....					85.5		10.1	0.5	10.6	1.1	7.1	8.2	104.3				104.3
1930....					109.6		7.4	5.8	13.2	1.4	4.5	5.9	128.7				128.7
1940....					95.7		7.9	2.9	10.8	1.6	4.0	5.6	112.1				112.1
1950....					100.6		9.2	4.5	13.7	1.6	1.9	3.5	117.8				117.8
1960....	25.2	28.1	53.3	44.3	97.6		8.2	3.4	11.6	1.2	0.8	2.0	111.2	1.9	0.3	2.2	113.4
1961....	26.1	28.7	54.8	43.0	97.8		8.6	3.4	12.0	1.1	0.8	1.9	111.7	2.1	0.4	2.5	114.2
1962....	23.9	28.0	51.9	45.4	97.3		9.3	3.6	12.9	1.1	0.9	2.0	112.2	2.5	0.4	2.9	115.1
1963....	27.2	27.8	55.0	41.7	96.7		9.9	4.3	14.2	1.1	0.7	1.8	112.7	3.0	0.7	3.7	116.4
1964....	28.5	30.3	58.8	37.9	96.7		10.9	4.1	15.0	1.0	0.7	1.7	113.4	3.5	1.3	4.8	118.2
1965....	29.4	30.3	59.7	37.1	96.8		11.0	4.1	15.1	1.0	0.7	1.8	113.7	4.0	1.7	5.7	119.4
1966....	28.3	28.6	56.9	40.3	97.2		11.2	4.2	15.4	1.0	0.7	1.7	114.3	4.5	1.9	6.4	120.7
1967....	26.6	29.9	56.5	41.8	98.3		11.9	4.2	16.1	0.9	0.5	1.4	115.8	4.8	2.1	6.9	122.7
1968....	27.8	26.5	54.3	44.7	99.0		12.6	4.3	16.9	0.9	0.7	1.6	117.5	5.0	2.2	7.2	124.7
1969....	30.1	25.2	55.3	45.4	100.7		13.2	4.5	17.7	1.0	0.6	1.6	120.0	5.3	1.6	6.9	126.9
1970....	31.4	25.0	56.4	45.5	101.9		14.0	4.6	18.6	1.0	0.5	1.5	122.0	6.2	(c)	6.2	128.2
1971....	31.1	22.8	53.9	48.5	102.4		15.0	5.0	20.0	0.9	0.5	1.5	123.8	5.7		5.7	129.5
1972....	30.4	25.4	55.8	47.0	102.8	0.9	15.6	4.4	20.9	1.0	0.5	1.5	125.2	5.7		5.7	130.7
1973....	30.4	24.9	55.3	46.2	101.5	1.4	16.7	4.8	22.9	0.9	0.5	1.4	125.8	5.7		5.7	131.5
1974....	26.1	21.0	47.1	49.5	96.6	2.3	17.4	4.9	24.6	0.8	0.4	1.2	122.4	7.0		7.0	129.4
1975....	30.5	24.9	55.4	34.8	90.2	4.7	17.7	5.1	27.5	0.9	0.4	1.3	119.0	7.0		7.0	129.4
1976(d)	32.5	22.7	55.2	39.5	94.7	7.1	17.7	5.1	29.9	1.0	0.4	1.4	126.0	8.0		8.0	134.0

(a)Dry basis. Recent corn sweetener consumption may be understated due to incomplete data. (b)Sugar sweetness equivalent--assumes saccharin is 300 times as sweet as sugar, and cyclamate is 30 times as sweet as sugar. (c)Cyclamate food use was banned by the Food and Drug Administration, effective in 1970. (d)Preliminary.

Source: 1910-1950: USDA, Economic Research Service, U.S. FOOD CONSUMPTION, Statistical Bulletin No. 364, June 1965, p. 53.
1960-1976: USDA, Economic Research Service, Sugar and Sweetener Report, Vol. 2, No. 9 (Sept. 1977), p. 28.

tion" is somewhat misleading when used to describe this type of data. In
reality, these data are based on sweetener deliveries or "disappearance."
Although disappearance tends to be equated with consumption, human consumption is actually somewhat lower due to non-food use, wastage, spoilage,
spillage, etc. Furthermore, this type of data is not adequately adjusted for
changes in user-consumer inventories. The latter may account for at least a
portion of the abrupt decline in apparent consumption during the price peak
of 1974-75.

From an economist's point of view, the most important features of this
table are (1) the recent decline in per capita sugar consumption and (2) the
larger and steadier gain in corn sweetener consumption. This means that corn
sweeteners are increasing their relative share of the U.S. sweetener market.
As you can see, this is primarily due to the phenomenal growth of isomerose
or high fructose corn syrup.

One important but often overlooked factor affecting the U.S. sweetener
market is the continuing increase in the proportion of total sweeteners distributed to industrial users and the resulting decrease in the proportion
distributed for direct home and industrial consumption. This shift in the
demand side of the market is extremely important, since it greatly increases
the potential for substitution between sweeteners.

Industrial food processors have become the largest users of sweeteners
in the United States. In the case of sucrose, the major U.S. sweetener and
the one for which household use has traditionally been a major market outlet,
the proportion of deliveries made directly to industrial users has increased
almost constantly since at least 1929, the earliest year for which data are
available.

Figure 6 provides the proportion of sucrose delivered to industrial
buyers and the proportion of sucrose delivered in consumer-size packages
(less than 50 pounds) from 1955 to 1976. The percentages do not sum to 100
because they are derived from different data series. The 1974-75 increase in
consumer-size packages and the concurrent decline in industrial deliveries
were due to hoarding and inventory depletions caused by severe price
volatility. The extremely low statistic for 1976 industrial deliveries
appears to be due to an aberration in the data (it is offset by an extremely
large increase in the "Other, non-industrial" category). The proportion of
total U.S. sweeteners accounted for by industrial users has actually increased even more than that indicated by these sucrose statistics, since
nearly all corn sweeteners sold in the U.S. are delivered to industrial
users and the market share of the corn sweeteners has (as discussed above)

Figure 6.   Proportion of sucrose deliveries to industrial buyers and in
consumer-size packages, 1955-1976

Year	Percent deliveries to industrial buyers	Percent deliveries in consumer-size packages*
1955 ..	48.2	NA
1956 ..	49.3	NA
1957 ..	51.2	31.0
1958 ..	49.8	35.3
1959 ..	52.2	34.2
1960 ..	53.6	34.0
1961 ..	53.8	34.2
1962 ..	55.1	31.6
1963 ..	56.1	30.2
1964 ..	57.2	30.7
1965 ..	59.6	29.0
1966 ..	61.4	27.7
1967 ..	62.1	26.9
1968 ..	64.2	25.3
1969 ..	65.1	24.8
1970 ..	65.1	24.0
1971 ..	65.5	24.6
1972 ..	66.5	23.9
1973 ..	66.8	23.5
1974 ..	66.2	24.5
1975 ..	61.6	26.0
1976 ..	60.2	24.3

* Deliveries in packages of less than 50 pounds.

Source: USDA

increased significantly.

The increase in the proportion of sweeteners delivered to industrial users and the associated decrease in the proportion delivered directly to households appear to be a reflection of (a) rising incomes and (b) changes in food habits. A number of studies have shown that direct purchases of sugar per capita are smaller for higher income families than for low income families[1]. Furthermore, the decline in home baking, canning, and similar activities and the rise in the use of prepared or convenience foods has obviously been an important factor in transferring an increasing proportion of the market for sweeteners from households to industrial food processors.

The importance of the increase in the proportion of sweeteners sold to industrial users is that industrial users are able and willing to substitute one sweetener for another to a much greater extent than household users. The extent of this substitution varies widely among food industries and even among segments of the same industry. The limits of substitution between given sweeteners (both in a given product and in aggregate) are determined by the unique technical characteristics of each sweetener and the desired characteristics of the food products in which they are used.

Sweeteners are a minor item in household expenditures and, hence, neither the general level of sweetener prices nor the relative prices of alternative sweeteners is of much interest or consequence to household users (with the extremely high prices of 1974 being an important exception). On the other hand, the cost of sweeteners is much more significant to industrial users. General increases in sweetener prices may cause food processors who use significant quantities of sweeteners to make changes in the prices of the products they manufacture. In turn, this may affect their volume of sales and, ultimately, the quantity of sweeteners which they purchase. Changes in the relative prices of sweeteners, on the other hand, encourage substitution between sweeteners. Surveys indicate that industrial users are much less concerned with (a) the general level of sweetener prices than with (b) price stability and (c) purchasing sweeteners at prices no higher than those paid by their competitors[2].

In general, the price elasticity of demand for sweeteners in the United States is very low (i.e., consumption is not very responsive to price changes). Of course, since individual sweeteners can be substituted for one another, the elasticity of demand for a given sweetener is somewhat higher than that for all sweeteners as a group. Furthermore, due to their greater willingness and ability to substitute sweeteners, the elasticity of demand for sweeteners, the elasticity of demand for sweeteners delivered to industrial users is somewhat higher than that for sweeteners sold to household users. While these views concerning the price elasticity of demand for sweeteners are generally held by most economists, it is not possible to adequately measure the response of consumption to changes in price for sweeteners due to the effectiveness of the Sugar Act in preventing significant fluctuations in the price of sugar and, indirectly, the prices of other sweeteners from 1934 through 1974. In other words, for any given sweetener, the quantity demanded by industrial users is more responsive to changes in the relative price of that sweetener than the quantity demanded by household users. Thus, the most important effect of the indrease in the proportion of sweeteners sold to industrial users is that it is accompanied by an increase in demand elasticity or responsiveness to price.

Sucrose deliveries in 1976 by product or business of buyer are shown in Figure 7. It indicates, in a crude way, the relative importance of various user industries in the total sucrose market and in the liquid sucrose segment

Figure 7. Sucrose deliveries, calendar year 1976

Percent

Product or business of buyer	All sucrose	Liquid sucrose
Industrial		
Bakery, cereal and allied products.................	12.2	4.3
Confectionery and related products................	8.6	5.0
Ice cream and dairy products	5.2	13.7
Beverages...................	21.5	54.3
Canned, bottled, frozen foods, jams, jellies and preserves................	6.8	13.6
Multiple and all other food uses.....................	4.9	3.8
Non-food products..........	1.0	1.5
Sub-total.........	60.2	96.2
Non-industrial		
Hotels, restaurants, institutions..............	.6	.1
Wholesale grocers, jobbers, sugar dealers............	20.3	2.5
Retail grocers, chain stores, supermarkets..............	12.6	.7
All other deliveries, including deliveries to government agencies.......	1.3	.5
Sub-total.........	34.8	3.8
Unspecified...................	5.0	
TOTAL DELIVERIES..............	100.0	100.0
Included in totals		
Deliveries in consumer-size packages (less than 50 lbs)...................	24.3	
Deliveries in packages of 50 lbs or more...........	43.6	
Deliveries in bulk (unpackaged)..............	32.1	

Source: USDA, AMS, Sugar and Sweetener Report, Vol. II, No. 4 (April 1977), p. 15.

of that market. It is primarily in the liquid sucrose segment of the market that HFCS is making its greatest penetration.

In calendar year 1977, total corn sweetener shipments for food use are expected to total around 3.4 million short tons (dry basis). Of that total, dextrose shipments are likely to remain near one-half million tons. Conventional corn syrup deliveries could approach two million tons (up slightly from last year), and HFCS shipments are likely to total around a million tons (up from three-quarter million tons in 1976).

FORECASTS

The single most important element in forecasting the future of the various sweetener markets is, of course, the price of sucrose. If sugar prices increase and/or become more volatile, the incentive to (1) replace sucrose with existing substitutes and (2) discover new substitutes, will be increased.

As noted earlier in Figure 3, the past sucrose price cycle was much more severe in both magnitude and duration than those which preceded it in the post-war period. This makes forecasting even more difficult.

Due to the recent development of both U.S. and world market control mechanisms, it is necessary to predict prices and production patterns both with and without these mechanisms, and then evaluate the likelihood or degree to which they will succeed and the period for which they will endure.

Concentrating first on the world free market, please recall that world sucrose stocks are on a strong upswing--indicating weaker world free market prices. Without an ISA, world sucrose prices could have continued downward to the 3-5 cent level and would have probably remained below 10 cents through 1979. World sugar prices then would have begun to recover and would have reached a peak of 25 cents or more by 1982-83. Under the ISA, the price trough will be moderated or postponed, but the price peak will also be delayed.

During the next month or two, prices will continue to be unsettled and depressed as world buyers and sellers maneuver to position supplies in anticipation of both the ISA and U.S. sugar programs. During the next two years prices will struggle upwards toward the ISA objectives as the International Sugar Organization (ISO) purchases sugar for its buffer stock. However, I feel that the proposed buffer stock of 2.5 mmt will not absorb enough surplus sugar, and therefore, excess supply will overhang the market.

Anticipation and reliance on ISA buffer stock purchases and the ISA free market price objective of 15-19 cents will, when coupled with the current drive for sugar self-sufficiency in certain less developed countries, overstimulate expansion of world sucrose production capacity. Eventually this excess production will undermine the ISA export quota system and flood the world free market in the traditional cyclical pattern. In other words, I feel that we may have "bought time" with the ISA, but we have not solved the long run problem of cyclical surpluses. I hope that I am wrong.

Turning now to the U.S. sweetener market, Congress has clearly indicated that it intends to protect the domestic sugar industry and that it desires to do so through market prices rather than through grower subsidies. The de la Garza plan will apparently be instituted before the end of 1977, but will probably be used only through the first half of calendar year 1978. By mid-1978 the ISA will probably have raised world market prices to the equivalent of 13.5 cents in the U.S. and the de la Garza plan will automatically terminate. It should be noted that the de la Garza plan is only authorized for the 1977 and 1978 crop years and, therefore, will not be automatically available to protect the U.S. price if the ISA fails in three or four years as I have predicted.

I forecast that starch-based sweeteners will continue to expand their market share in the U.S., EEC, and Japan, particularly during periods of high sucrose prices. Further refinements in isomerose (HFCS) production processes will lead to greater user acceptance/satisfaction and further market penetration. Starch-based, crystallized fructose will probably be commercially available in the 1980's and will further expand the importance of starch-based sweeteners.

REFERENCES

1. Rockwell, G.R., Income and Household Size: Their Effects on Food Consumption, USDA Marketing Research Report No. 340, 1959.
2. Ballinger, R.A. and Larkin, L.C., Sweeteners Used by Food Processing Industries: Their Competitive Position in the United States, USDA-ERS Agricultural Economic Report No. 48, 1964.

Patterns of use of sweeteners

Sidney M. Cantor

Sidney M. Cantor Associates, Haverford, Pennsylvania 19041

INTRODUCTION

The most profound changes which have occurred in the American diet in the past century relate to the consumption of carbohydrates and fats. It is apparent from data available to us that the average fat consumption has increased while the average total carbohydrate has decreased, but within the decreased total carbohydrate the simple sugars or sweeteners component has increased substantially. Statistical data also suggest an epidemiological relationship between these dietary changes and certain degenerative diseases. These diseases include obesity, cardiovascular disorders, diabetes mellitus and dental caries. However, unlike specific nutrition associated deficiency diseases, degenerative diseases are clearly multifactorial. Moreover, they are associated with overconsumption and also with food behavior changes and their accompanying socioeconomic stresses. Of the various diseases classified as degenerative, only in the case of dental caries has a causal relationship been established between sweetener consumption and the disease. But such relationships may be expected to be clarified as we learn more about the realities of food consumption and changing consumption of food components-- fat and sugar as an important example in our case. Surely this combination is closely related to richness of diet and therefore the multifaceted subject of oral stimulation and indeed the entire complex of overconsumption, affluence and degenerative diseases.

Accordingly, the purpose of this paper is to consider the patterns of use of sweeteners (nutritive and non-nutritive) in the context of a dynamic food system. Additionally, factors will be examined which impinge on the system and which appear to have influenced consumption. Such an examination in the context of a workshop on "Sweeteners and Dental Caries" should assist in isolating promising points of intervention for an intensified program of caries reduction.

CONSUMPTION DATA: CHANGE AND SIGNIFICANCE

Changes in the consumption of carbohydrates in the United States[1] over approximately the last 60 years (data beyond 1972 have been added) are clearly evident in Figures 1 and 2 which present trends in consumption of

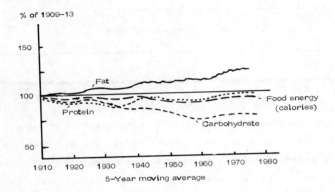

Figure 1. Per capita civilian consumption of food energy (calories), protein, fat, and carbohydrate, 1909-1913 to 1972[1].

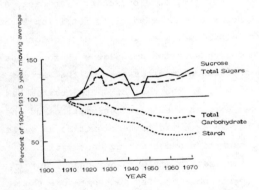

Figure 2. Per capita consumption of total sugars, refined sugar, starch and total carbohydrate. Agricultural Research Service, U.S. Department of Agriculture, 1972.

food elements and of carbohydrates in terms of 5-year moving averages respectively. The comparison base (100) is the period 1909-1913 when these data were first gathered. Total carbohydrate on an average per capita daily basis was about 500 grams then of which 156 grams was sugars. Today total carbohydrate is about 380 g. of which 200 g. is sugars which represents an increase in the 60-year period from 31.5% of sugars to 52.6% of sugars in the total carbohydrate consumed daily. In this same period, dietary fat increased by one-fourth and currently represents about 155 grams per capita per day. Protein for comparison is nearly 100 grams daily which is no great change from 1909-1913 and total calories reflect only an approximate 5% change in 60 years.

The apparent simultaneous relationship between increase in fat and sugar consumption has been noted by several investigators[2], but Wretlind[3] has shown the particularly interesting interrelationship in some European countries to which has been added the comparable U.S. fat and sugar values (Figure 3).

Several indicators show that greater affluence, increasingly sedentary occupations and in general the changes in lifestyle, including food behavior resulting from urbanization and industrialization, account for this relationship. Before examining this relationship more closely, it is important to establish some qualifications about what have been called consumption data.

First, the figures do not represent true consumption, but rather, <u>average</u> <u>daily</u> <u>disappearance</u>. They are derived by dividing the amount available in the market to consume (what disappears) by a population figure.

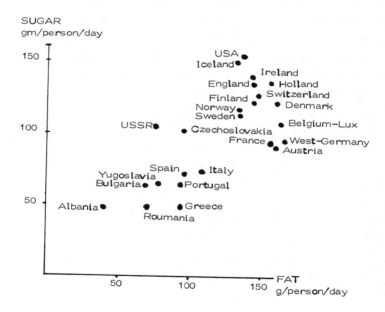

Figure 3. Interrelation between sugar and fat consumption in some European countries. The values have been obtained from Production Yearbook (1972) and are for the period 1964-1966 for Albania, Bulgaria, Czechoslovakia, Denmark, Greece, Ireland, Romania, Yugoslavia and the USSR. All other values are for the period 1969-1970.

Actual consumption data are almost non-existent for the U.S., although occasional "one-day" surveys have provided some information related to pre-aggregated population segments, such as age groups, sex, geographic areas, etc. Even these figures, however, are subject to question since they are also averages, and the limits of individual consumption as a measure of real-life experience have not been established. So-called average consumption is really the average of the sum of true consumption and waste. The waste factor includes the waste of industrial manufacture, household waste such as spillage and discarded leftovers. Second, the "sugars" portion of the diet cited above includes estimates of intrinsic sugars, that is, those contained naturally in foods as well as those added in preparation which are principally the sum of sucrose and corn-derived sweeteners. The latter figures (sweeteners added), however, are those used in most calculations of average sweetener consumption because industrial distribution data are most easily obtained. Third, industrial uses such as brewing for alcohol production and baking for carbon dioxide production are included in the sugars added data, and these figures are not insignificant. One estimate puts the brewing and baking figure at over eight pounds per capita annually. Actual consumption may therefore be as much as 10 to 20% lower although no reliable figures will be available until the household survey currently in progress by the USDA is concluded two to three years from now. This survey is the first to use a 24-hour recall procedure as well as computerized accumulation and analysis procedures to provide actual consumption data.

One may inquire why there is such a paucity of true consumption data and why has it taken so long to develop an acceptable methodology for accumulating such data. The reasons reach to the roots of our economic system, but here is one possible answer. The USDA, which is responsible for collection of such data is a production-oriented agency[4]. Ever-increasing crop production and acceptable prices are in the best interests of the U.S. farmer, U.S. agribusiness, and therefore the U.S. economy. Limits to growth are not a part of our collective experience. Marketing policy, following regular production increases, must accordingly be concerned with increased consumption, and much of our product distribution data--both food and industrial--are supplied by the industries which make or distribute the products. How the consumption actually takes place is of lesser interest so long as it takes place. Disappearance statistics thus become a natural part of this consumption dominated philosophy and its operating policies. The social consequences of increased consumption are of limited concern in such an economy and in its associated food system. Overconsumption then may be regarded as equivalent to air pollution, water pollution and comparable ills. These are problems which in the beginning are secondary to industrial development and which emerge for consideration as a society becomes more concerned about its public ills.

The pollution analogy is useful but far from perfect because the relationship of man and food consumption in the ideal sense of regular and frequent repetition is unique. Its special nature is particularly exemplified in the relationship between man and sweeteners. We see it featured in the biblical literature, i.e., "land of milk and honey" and "manna from heaven" and in such instances as the association of sugar with royalty, the acceptance of sugar as pay during the Crusades and the aphorisms about an endless supply of sugar being equated with wealth. Perhaps the best evidence of the human attitude toward sweetness and its active pursuit is the association in most languages of only very pleasant words and ideas with sugar and sweetness. In short, the association between sugars and sweeteners and the "good life" has been well established and repeatedly reinforced over the centuries.

DEVELOPMENT OF THE SWEETENER SYSTEM

The progressive increases in average sweetener disappearance are reasonably well documented. A brief examination of the growth and development of the sweetener system over the last century is instructive since it provides further insights into the factors which have contributed to increased sweetener consumption. In Figure 4 is shown the per capita sweetener consumption for the hundred year period 1875-1975[5]. In 1875, the development of the French beet sugar industry from its inception during the Napoleonic Wars to domination of the European sugar markets by beet sugar was well underway. New markets for colonial cane sugar were needed. While the institution of

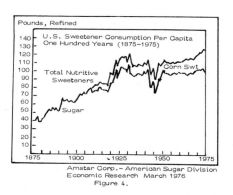

Pounds, Refined

U.S. Sweetener Consumption Per Capita
One Hundred Years (1875-1975)

Total Nutritive
Sweeteners

Corn Swt

Sugar

140 130 120 110 100 90 80 70 60 50 40 30 20 10

1875   1900   1925   1950   1975

Amstar Corp.- American Sugar Division
Economic Research March 1976
Figure 4.

115

slavery, which made possible the rapid expansion of tropical raw sugar production as well as the coastal refining industry, had been abolished, its
cheap labor heritage was at its peak. The era of overproduction and of
dumping low cost sugar on the U.S. market--where beet sugar development was
just getting started--was already visible. By World War I in less than 50
years, per capita annual sucrose consumption was near the 100 pound mark.
This familiar number remained reasonably constant (Figure 5) until 1975.
Low priced sugar and a climate of rapid industrial expansion including the
start of convenience food manufacture (confections, jams, jellies) were con-
tributors to the rapid consumption increase. The development of an indigenous sugar supply (i.e., the French sugar beet experience) and the ensuing
competition for markets was a valuable lesson and is of considerable importance in understanding the current sweetener situation and in emphasizing
the importance of sweeteners in maintaining a contented population. In the
light of this historical precedent, the alleged Chinese proverb which
reportedly says, "In times of stress, sweeten the tea" becomes quite under-
standable. Low cost and a favorable industrial development climate catalyzed
man's pursuit of sweetness. Sugar, originally identified with royalty,
became everyman's food.

The large increase in the price of sugar following World War I reinforced
the desirability of sugar price control that was enacted in the thirties.
This legislation followed another period of increased dumping of offshore
supplies of sugar and preferential tariff grants. The first U.S. Sugar Act,
taking previous experience into account, imposed an import tariff designed to
protect indigenous production, provided a carefully controlled supply situation by assigning quotas to preferred raw sugar sources, and also controlled

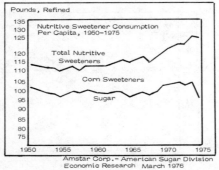

Figure 5.

the amount of sugar reaching the market at any time so as to protect the consumer. Sucrose consumption as a result appeared to stabilize around a figure of 100 pounds annually.

Corn sweeteners began to be significant around 1910 (Figure 5), even though the discovery of starch hydrolysis dates from 1811. The lesser sweetness of corn starch-derived sweeteners was responsible for characterizing them as substitute sweeteners. This identity also succeeded in establishing restrictions on the amounts of corn sweeteners which could be used in such standardized foods as jams and jellies and also canned fruits. Such restrictions were not lifted until the last decade when high fructose corn syrup (HFCS), a corn sweetener having about the same sweetness as an equimolecular mixture of glucose and fructose, became available in commercial quantities. Meanwhile, industrial consumers learned how to use sweetener blends in which regular corn syrups, which were being developed in increasing variety, contributed functional improvements. And because these syrups were less costly than sucrose, they also diluted the rising price of sucrose. As can be seen in Table 1, the calculated average daily per capita consumption of various sugars has changed significantly from the base period 1910-1913[6]. This table, which consolidates data from a number of sources, incorporates the intrinsic sugars of the average diet. However, it is misleading since it does not reflect the effect of food processing nor the use of sweetener blends by the food industry on what the consumer actually eats and therefore the mix of sugars which enter the mouth. In Table 2 the major sugar identities in the corn sweetener component have been redistributed. Also the effect of processing on sucrose has been estimated as well as the decrease in

Table 1. Calculated Daily Average Consumption of Various Carbohydrates[a] (g/day)

Year	1910-1913	1960	1974
Starch	342	188	179
Sugars			
Sucrose	101	121	123
Corn sweeteners	8	19	33
Lactose	21	25	23
Glucose	6	11	12
Fructose	6	4	3
Maltose	2	4	4
Others	12	5	3
TOTAL SUGARS	156	189	200
TOTAL CARBOHYDRATE	498	377	379
PERCENT OF SUGARS	31.5	50.0	52.6

[a]Compiled from USDA/ARS 1972 data and sugar statistics (USDA).

Table 2. Calculated Daily Average Consumption of Various Carbohydrates*
(grams per day)

	1910-1913	1960	1974 Disappearance	1974 Adjustment	1976 Disappearance	1976 Adjustment
Sucrose	101	121	123	83	118	59
Corn Sweeteners (oligosacch.)	8	19	33	17	38	21
Lactose	21	25	23	23	18	18
Glucose	6	11	12	40	12	48
Fructose	6	4	3	25	3	44
Maltose	2	4	4	12	4	10
Others	12	5	3	--	3	--
TOTAL SUGARS	156	189	200		200	

*Compiled by S.M. Cantor from USDA/ARS 1972 data and sugar statistics (USDA)

the use of dairy products and the increase in distribution of high fructose corn syrup. Some impact of change in the average diet of new sweeteners--of the dynamics of this segment of the food system--are clearly evident. It must be emphasized that this is a conservative redistribution and the limits of the various components that might contribute to caries incidence are probably extremely wide.

CURRENT SWEETENER SITUATION--INDEX TO SOCIAL AND TECHNOLOGICAL CHANGE

The significance of the revolution in food preparation--the move from kitchen to factory, the increased use of convenience foods, the increase in eating away from home was not clearly apparent until after World War II. These changes reflected in the food system were the result of social changes--more women in the working force, the rise of the nuclear family and the ultimate flowering of our consumer society. The sweetener situation, however, is a useful index of change in the food system. Changes in the use of sweeteners are clearly visible in Figure 6[7] which reveals the rapid rise in the industrial market for sweeteners in contrast to the sharp decrease in direct consumption or home market. Currently about 75% of sweeteners, sucrose and corn-derived, are delivered to industrial users. In the 1910-1913 base period, industrial use was about 25% of total sugar distributed. The use lines crossed about 1950.

The significance of the rise in the industrial use of sugar[8] is shown in Figure 7. Here use is plotted against selected products and processed foods while total sweetener and sucrose curves are also shown since sucrose is the principal household sweetener and virtually all corn sweeteners are industrial products. The uses showing most rapid growth rate are those-- snack food or fast foods--that best exemplify our changing food behavior as

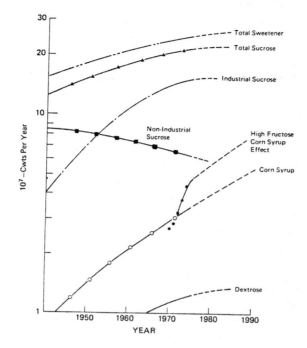

Figure 6. Trends in sweetener use basis: Disappearance from stocks[7].

well as our growing appetite for processed foods which, along with other components, combine sweeteners with fats into many products best described as rich.

Along with the growing appetite for richness, however, is a growing anxiety about calories, another post World War II phenomenon. Judging by one quantitative measure of this anxiety, that is, the disappearance of non-nutritive sweeteners, it is a fairly modest anxiety and expressed largely in the consumption of "diet" soft drinks. The sucrose equivalent of the average annual per capita in food of saccharin--our only currently acceptable non-nutritive sweetener--is 7 pounds. This is about 7.4% of the present level of sucrose consumption or about 5.5% of total sweetener consumption. Just over 5 pounds of this 7 pounds, about 75%, is used in soft drinks. The remainder is used in diet foods. On an average daily basis saccharin usage in diet beverages is equal to about one and one-half teaspoons (6 g.) of sucrose per person. Before the FDA's proposed ban on saccharin was delayed by Congress, it was estimated by the USDA that about one-half of its equivalent sugar market or 375,000 short tons would be lost by consumers simply cutting sweetener usage. The remainder was expected to be about equally divided between

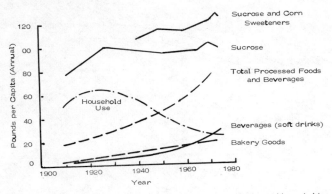

Figure 7. Use of sugar in selected products, processed foods, and households.

sugar and high fructose corn syrup (187,500 short tons each). The expected
equal impact of high fructose corn syrup is a significant observation.

The effect of the availability of a sweetener roughly equivalent in
sweetness to sucrose but derived from corn, the major American grain crop,
has had an impact on the sucrose market comparable in many respects to the
development of beet sugar by the French. The estimated distribution of this
new product in 1976 was 7% of the total sweetener market and near 25% of corn
sweetener deliveries. The reaction of the American beet sugar industry has
been defensive as well as one of joining the new industry. The cane sugar
refining segment of the sugar industry has seen most of the displaced sucrose
business disappear from its markets and largely the liquid sugar market; in
other words, it has displaced some raw sugar imports. This has affected
world production and our relationship with friendly producing countries.
In the European Economic Community, dominated by French agricultural policy,
a different reaction appeared. A tax has been established on the new sweetener making it prohibitive to manufacture there. Here in the absence of
control afforded by the U.S. Sugar Act which expired in 1975, there have been
calls for new controls, for quotas assigned to all segments, for a new support price for sugar (which would increase the profitability of the new
product also). In any case, a revolution has occurred, a new and less costly
source of an industrial sweetener has been established which can compete
favorably with sucrose in certain uses on a world basis, and the sweetener
situation which right now is in a somewhat chaotic state will never be the
same again. This revolution could not have occurred without the following:
1) Rapid industrialization of the U.S. food system characterized by the fact
that the sugar market became almost three-fourths an industrial market in
which sweeteners are used primarily for incorporation in other foods.

2) Growth in convenience foods and eating away from home.

3) A technological achievement of very considerable significance featured by a commercial process for isomerizing glucose to its sweeter counterpart fructose and acceptance by the food industry of the new product.

Translated into changes, we can recognize:

1) Social changes - changes in the family structure and working practices.

2) Changes in food habits - reduction in grain consumption which translates into reduction in starch consumption.

3) Changes in the food system - industrialization of food preparation.

4) Response of the sweetener system to industrial requirements by the development of new technology - changes in sweetener production and consumption.

All of these changes taken together have resulted in increased sweetener consumption along with other food components, notably fat, and have also reduced the discretionary limits of the consumer. The interrelationships of the variables at work in the system are emphasized. It seems clear at this time that the impact of the new technology was not nearly understood by its developers.

The readjustment of the system was displayed most dramatically in 1974-1975 when the price of sugar increased fourfold. High fructose corn syrup which had been enjoying a very good growth rate suddenly was in short supply, and plant capacity was increased rapidly. Currently there has been some slowing down, but the sweetener industry agrees generally that the product will continue to grow substantially in the United States as food uses for it continue to be explored.

The effect of high fructose syrup availability as well as sugar price rise is shown in Table 3, in terms of the changes in industrial sucrose consumption in the five major categories of food processing use. These data provide an updated and detailed extension of Figure 7. While the sugar price rise did reveal a price elasticity of demand for sweet products, its exact size is somewhat obscured by the availability of high fructose corn syrup. However, it appears that sweetener consumption can be reduced if the price is high enough.

What appeared to be the beginning of a decline in sweetener use has now, however, turned upward again as all of the data indicate. Estimates for 1977 show both the reduction in sucrose now generally regarded as stabilized at a level below 100 pounds annually and the increase in corn sweeteners which has been steady since World War II and which became more rapid with the introduction of high fructose syrups (Figure 8.) Table 4 presents details of consumption since 1960 and projections into the next decade[9].

Table 3. Industrial Sucrose Consumption for 1973-1976 and Estimates for 1977 in Five Major Food Processing Categories

	Annual Pounds Per Capita				
	1973	1974	1975	1976	1977 (Est.)
1. Bakery products, cereals, etc.	13.8	13.6	11.0	11.6	(12.5)
2. Confectionery, chewing gum, etc.	9.8	9.6	7.1	8.1	(8.3)
3. Ice cream, dairy products	5.7	5.4	4.5	4.9	(5.5)
4. Beverages	23.5	22.1	18.7	20.5	(23.8)
5. Canned, bottled, frozen foods, jams, jellies	9.7	8.9	6.5	6.3	(6.3)
Total Five Categories	62.5	59.6	47.8	51.4	(56.4)
Total All Industrial Food	69	65	52	56	(62)
Total Sucrose	103.5	96.6	90.2	94.7	(94.7)
Percent Industrial Food of Total Sucrose	66.7%	67.3%	57.6%	59.1%	(65.5%)
HFCS Deliveries	2.7	3.8	6.8	8.9	(11.6)
Total Corn Sweetener	23.3	26.8	28.4	31.0	(34.4)
Total Sweetener	126.8	123.4	118.6	125.7	(129.1)

Data Sources: S. Kolodny; Sugar Assn.; Private Estimates

Sweeteners as Food Components - The sweetener industry has always regarded its products as foods and components of formulated foods. Present day attitudes which argue that sweeteners are additives and therefore subject to careful control are vigorously disputed. When consumers were able to exert greater discretionary control over sugar consumption prior to the sharp increase in industrial usage and the change in food preparation practices, there was some reason to believe that what are referred to as built-in satiety controls would prevail. The presentation of a large fraction of sweeteners as food components in which sweetness is not separated from flavor enhancement or flavor integration or richness is a different situation. The apparent close connection between fat and sugar consumption which is equated to richness needs careful examination. It is logical to ask whether sweetness per se is the major determinant of levels of consumption of sweeteners, or whether sweetness associated with other food ingredients as well as lifestyle invites increased consumption.

The consumption of some major food commodities in selected years is shown in Table 5. The fat content of the diet continues to rise. Butter consumption is declining but margarine is increasing. The consumption of cheeses and their incorporated fats is increasing sharply. Ice cream is on the rise again after a decline. Does this substantial but not obvious fat consumption increase signal a further increase in sweeteners? Is sweetness addictive and does the reduction in discretionary control add up to a drop in

Figure 3.

Corn Sweetener Shipments

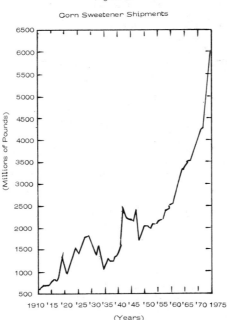

(Years)

Source: USDA 1910-1972

Kolodny 1973-1975 (estimates)

Saul Kolodny —"Economic Factors Affecting Future Sweetener
Consumption", American Chemical Society
Centennial Meeting, Sweetener Symposium,
April 8, 1976

individual resistance to sweetener consumption? If we wish to control sweetener intake, must we also control other factors such as fat intake? There is
reason to believe that proposing a major reduction in sweetener consumption
without careful systematic consideration is overly simplistic and like most
simplistic interventions can be expected to fail.

I was asked in these remarks to consider the reactions of industry to a
major reduction--perhaps as much as 40%--in sweetener consumption. I think
it hardly needs any elaboration to report that such a proposal would be
rejected immediately for many reasons most of which any of us could list.
Not the least of these reasons is the matter of personal choice or freedom to
eat what one wishes. And who is to say in view of mounting evidence, at least
in some age groups, that there is not a substantial physiological requirement
for high sweetener consumption. But the close connection of apparent sweetener consumption with social changes, with changes in food habits or, more
broadly, food behavior and with technological changes strongly suggests that

Table 4. U.S. Population, Corn Sweeteners, Sucrose and Nutritive Sweetener Consumption Data, Selected Years and Projections[9]

YEAR	U.S. Population Mid-Year (millions)	Corn Sweetener Disappearance Per Capita				Sucrose Disappearance Per Capita (lbs.refined)	Total Nutritive Sweetener (lbs.)
		Dextrose	Reg. Corn Syrup (lbs. dry basis)	HFCS	Total		
1960	180.0	4.5	9.4	0	13.9	97.4	111.3
1965	193.9	5.3	12.3	0	17.6	96.8	114.4
1970	204.6	5.4	13.1	0	18.5	102.3	120.8
1971	206.3	5.6	12.8	0	19.4	102.7	122.3
1972	208.0	5.2	15.4	0	20.6	102.1	122.7
1973	209.6	5.7	16.2	1.4	23.3	103.5	126.8
1974	211.1	5.9	17.1	3.8	26.8	96.6	123.4
1975	212.5	5.4	16.2	6.8	28.4	90.2	118.6
1976	213.8	5.5	16.6	8.9	31.0	94.7	125.7
1977	215.1	5.6	17.2	11.6	34.4	94.7	129.1
1978	216.5	5.5	17.6	13.9	37.0	91.5	128.5
1979	218.0	5.5	17.9	14.7	38.1	91.4	129.5
1980	219.5	5.5	18.2	15.5	39.2	91.3	130.5
1981	221.2	5.4	18.5	15.8	39.7	91.8	131.5
1982	222.5	5.4	16.9	15.7	40.0	92.5	132.5

S. Kolodny "Economic Factors Affecting Future Sweetener Consumption"
American Chemical Society Centennial Meeting., NYC, April 1976
USDA "National Food Situation" September 1977

any intervention which proposes to reduce sweetener consumption must be multi-factorial.

Can we be expected to reverse our changes in food behavior? To return to more elaborate and careful food preparation, for example, if indeed we were ever there? There doesn't seem to be much evidence that this will happen until some sense of limits is conveyed to both producers and consumers in our society. The current energy situation is very much to the point. Concerns about the national nutrition status are those of a relatively small group. While degenerative diseases or what have been called by epidemiologists "man-made diseases" appear to be on the rise, so is life expectancy. The current drive for delaying retirement reflects concern about degenerative disease only in the sense that stopping some absorbing activity too early encourages such disease. I do not wish to be interpreted as recommending that we ignore fat, sugar and richness relationships or that quality of life is not a criti-cal consideration. Rather, I think we need to know more about the complexi-ties and interrelationships of our food system with our social and economic systems. Should food be merchandised like any other consumer item?

In the case of sweeteners, there is obvious need for an acceptable non-nutritive sweetener. Acceptable means pleasant tasting like sucrose which is the standard, adequately sweet and safe to use. We also need acceptable

Table 5. Civilian Per Capita Consumption of Major Food Commodities (Retail Weight) and Civilian Population, Selected Years*

Commodity	1960	1967	1974	1975	1976	1977 as percentage of 1976
Meats:	134.2	145.1	152.2	145.2	154.6	100
Beef	64.3	78.8	86.4	88.9	95.4	97
Pork (excluding lard)	60.4	59.6	61.9	51.0	54.1	105
Fish (edible weight)	10.3	10.6	12.2	12.2	12.9	98
Poultry products:						
Eggs	42.4	40.7	36.6	35.4	35.0	98
Chicken (ready-to-cook)	27.8	36.5	41.1	40.3	43.3	103
Dairy products:						
Cheese	8.3	10.1	14.6	14.5	15.9	106
Fluid milk and cream (prod. weight)	321.0	303.0	288.0	291.1	292.0	99
Ice cream (product weight)	18.3	17.8	17.5	18.7	18.1	98
Fats & oils-total, fat content	45.3	49.4	53.2	53.3	56.0	99
Butter (actual weight)	7.5	5.5	4.6	4.8	4.4	98
Margarine (actual weight)	9.4	10.5	11.3	11.2	12.5	95
Fruits:						
Fresh	89.6	79.1	76.7	81.9	84.6	96
Processed:						
Canned juice	13.0	11.7	14.7	14.8	15.0	98
Vegetables:						
Fresh	96.0	90.8	93.6	93.9	94.4	99
Potatoes, (including fresh equivalent of processed)	105.0	105.5	112.3	120.3	113.5	100
Grains:						
Wheat flour	118	112	106	107	111	100
Other:						
Coffee	11.6	11.1	9.5	9.0	9.4	93
Peanuts (shelled)	4.9	5.7	6.4	6.6	6.3	106
Sugar (refined)	97.3	98.3	96.6	90.2	94.7	100

*USDA National Food Situation September 1977

simulated systems for conveying the sense of richness and other complex oral stimuli. For these sensations are a part of our food behavior as the histories of both shortages and excesses in various cultures clearly demonstrate. Calling for voluntary changes in food behavior or applying restrictive measures or taxing beyond sensible levels doesn't work effectively. Adam

Smith's two-hundred-year-old declaration[10] that rum, tobacco and sugar were legitimate taxables because they were not necessities of life has not succeeded in controlling the use of either alcohol or tobacco. Direct taxing of sugar has not been practiced in recent times, a situation which I believe is very significant. There is an implication here that the use of alcohol and tobacco represent a larger sin. But common word associations with each of Adam Smith's taxables strongly indicate that sweeteners are not at all sinful. This situation can probably be changed, however, by appropriate public statements from authoritative sources and perhaps this is already underway.

The increasing plurality of our sweetener system is worth noting despite the difficulties involved in finding acceptable non-nutritive sweeteners. The polyols[11], principally sorbitol and mannitol, are used alone and in blends at allowable levels with both nutritive and non-nutritive sweeteners to improve flavor, retard crystallization, provide sweeteners in dietary foods and also sweeteners in sugarless confections such as special chewing gums. Sorbitol resists fermentation, and therefore, the caries activity that has occurred in some animal experiments was not expected. However, substantial usage has not developed. In 1974, 30 million pounds of 70% aqueous sorbitol was used in foods out of a reported production of 156 million pounds. On average, this is a per capita use annually of about 45 grams which is negligible in sugar terms.

Xylitol, because of its greater sweetness than sorbitol and its resistance to fermentation, is of particular and current interest as a dietary sweetener. The possibility, however of low cost xylitol (by low cost is meant a price near sorbitol) seems remote until the market reaches a significant figure--estimated to be about 70 million pounds--and projected uses of xylitol are approved. The kraft paper pulp industry is the most likely economic source of xylose for extensive xylitol production, since I have been told by a paper manufacturer that "a typical large kraft pulp mill could be a source of more than 30 million pounds of impure xylose per year." Yet the approximately 100 such pulp mills in the United States would produce less than 1.5 million short tons of xylitol in a food system which uses 10.3 million short tons of sucrose. Thus, the substitution of large amounts of special sweeteners seems hardly likely.

But the development of special nutritive sweeteners for special uses can be expected to increase and also broaden the choice for blends by industrial food formulators. Higher fructose corn syrups--that is, fructose concentration over the 42% originally introduced--have already been made available and are being tested. Crystalline fructose also can be expected to be produced in significant quantity. Both the syrups and the crystalline product are expected to be used principally for the reduction of calorie density while maintaining desirable sweetness. New corn syrups or syrup blends featuring particular glucose polymers can be expected on the market to supply body, adjusted sweetness and provide other functional properties.

It is to be expected also that flavor and sweetness enhancers will be examined as a means of reducing sweetener consumption. The use of flavor enhancers in cigarettes with concomitant "tar" reduction is a parallel concept and the current experiment now underway in the market will bear close watching as to public acceptance of other means of reducing the consumption of products regarded as "health hazards."

As the number of sweeter products increases, we can expect market segments to develop which, because they do not share mass market characteristics, will tend to raise the price of the sweetener component. This may ultimately have a deterrent effect on sweetener consumption, but the profitability necessary for industrial survival in our system may well make the difference.

Finally, the appearance of real consumption data, the collection of which is now in progress, should supply a rational basis for a more educated approach to marketing sweeteners and other food components as well. It is time, I believe, for us to recognize that our food, in respect of merchandising techniques, cannot be treated like any old consumer product. Our food and its consumers deserve considerably more respect. The public health costs which relate to the nutrition status of specific target groups--the aged, the obese, the economically disadvantaged, a growing and seriously needy group--unwed, teenage mothers--need careful attention. These target groups can become special markets for the food industry not just in sweetener terms but in terms of specific food requirements relating to systematic assessments of their nutrition problems. In the process of approaching food problems in a rational manner, in other words, in a nutrition surveillance system, sweetener consumption should be adjustable in a socially and economically acceptable way.

REFERENCES

1. Page, Louise and Friend, Berta. Level of Use of Sugars in the United States in Sugars in Nutrition, edited by H.L. Sipple and K.W. McNutt, Academic Press, New York, 94, 1974.
2. McGandy, R.B., Hegsted, D.M. and Stare, F.J. N.E. J. Med., 277, 186,1967. Perissé,J., Sizaret, F. and Francois, P. Nutr. Newsletter 7, 1, 1969.
3. Wretlind, A. in Sugars in Nutrition, edited by H.L. Sipple and K.W. McNutt, Academic Press, New York, 87, 1974.

4. Cantor, S.M. Associates Preliminary Technology Assessment of U.S. Food
    Nutrition and Agricultural Information Systems: Food Consumption and
    Nutrition Status, Off. of Tech. Assess., Washington, D.C., 1975.
5. Kolodny, S. Economic Factors Affecting Future Sweetener Consumption,
    Am. Chem. Soc. Mtg., Sweetener Symposium, NYC, 1976.
6. Cantor, S.M. in Sweeteners: Issues and Answers, Natl. Acad. Sci. Fourth
    Forum, Washington, D.C., 21, 1975.
7. Cantor, S.M. and Shaffer, G.S., Jr. Cereal Sci. Today, 19, 266, 1974.
8. Cantor, S.M. in Sweeteners: Issues and Answers, Natl. Acad. Sci. Fourth
    Forum, Washington, D.C., 27, 1975.
9. Kolodny, S. Economic Factors Affecting Future Sweetener Consumption,
    loc. cit. and USDA National Food Situation, September 1977.
10. Smith, Adam, 1776, Cited in History of Sugar, Vol. II, N. Deerr,
    Chapman and Hall, London, 420, 1950.
11. Benson, F.R. Sugar Alcohols in Encycl. Chem. Technology 35d Edition.

DISCUSSION AFTER PAPERS BY DRS. WALTER AND CANTOR

DR. MOSKOWITZ: I wonder how much of the increased consumption of sugar
is due to (1) a changing taste of the population for more sweetened products,
or (2) the increased use of sweeteners by food producers due to its functional properties, such as in baked goods.

DR. CANTOR: It's very likely both. I tend to accept the more
complicated responses to any situation of this kind. Your point about the
quality of sweetness is extremely important. As I noted, and this is
probably an oversimplification, I think of the fat-sugar combination, the
texture of the particular product as very significant and not enough re-
spected in terms of why people consume numerous foods and confectionery
items. The perception of a sweet emulsion being dispersed at a constant
rate from the chemical mass is extremely important and the amount of fat and
sweetener must be just right for the individual taste.

I want to make another point in response to the economic aspect of
sweeteners that no one has touched on. Since we live in a consumer society,
why don't we collect real consumption data? The answer appears to be that
numerous people think that only information about what disappears into the
market is needed. Such data initiate new products, new merchandising techniques and distribution methods. What disappears into the market, whether
food or automobiles, has been the determinant of what will disappear into
the market. However, the USDA now is collecting actual 24-hour recall
food consumption information that will provide a picture of the limits of
individual consumption.

DR. COYKENDALL: I wonder if you would comment on the impact that cyclamate made when it was being used on the total consumption of sweeteners and
sucrose.

DR. CANTOR: I can't, except to give you a percentage in sucrose equivalent at the time cyclamate was banned—about 6% for cyclamate and saccharin.
The current annual use of saccharin is equivalent to seven pounds of sugar
per capita.

Of that seven pounds, 5.2 pounds goes into the beverage industry and
most of the remainder into diet foods. I don't think we know the full significance of that amount. I have a feeling that fundamentally we are talking about the sin of overconsumption. The whole idea of consuming a
carbonated beverage that is sweetened with a non-nutritive sweetener repre-

sents a kind of release from anxiety, the anxiety created by the sin of overconsumption.

DR. WALTER: I would like to return to Dr. Moskowitz' question about the reasons for increased sugar use. A lot of sugar and other sweeteners are used for bulking agents and other non-sweetening purposes. The reason that they are used is because they are so cheap. Even under the artificially high prices derived from the "tax" effect of the Sugar Act, sugar was very very inexpensive relative to other ingredients.

The current world sugar price (New York basis) is something like $6.80 per hundredweight. On a real or deflated dollar basis, $6.80 per hundredweight in 1977 dollars is equivalent to only $1.30 in 1934 dollars. In January 1967, the price was $2.30, or the equivalent of only 81 cents per hundredweight in 1934 dollars. In 1934 the world sugar price (New York basis) averaged $2.56 per hundredweight. Clearly the price of sugar has fallen in real purchasing terms.

This is the economic motivation for using sugar in these products—it's simply very cheap compared to just about anything else that you want to compare it to. Sucrose is a very underpriced product. Of course, part of the reason that sugar prices have remained low and, in fact, declined in relative terms is that the fantastic increases in the productivity of both labor and capital in sugar production. In fact, the tremendous productivity gains that have come about in the industry since sugar was cut by hand and crushed by oxen driven stone mills, etc. are probably worthy of considerably more analysis than they have received.

I don't think that many other products have the same relationship between their current price and their price back in, say, the 1930's or the 1900's or whenever you want to choose.

DR. KARE: Would the banning of saccharin increase the amount of per capita consumption of sucrose?

DR. WALTER: I'm certainly not qualified to say that. But I think that studies done by Ballinger and others in the 1950's indicate that a certain percentage of non-caloric sweetener sales were substitutions for sucrose. Presumably, this substitution effect is at least partially reversible. Sugars would be substituted for some portion of saccharin consumption if the latter were removed from the market. Of course, diabetics and people like that can't substitute sucrose for saccharin, but the secretary who sits and drinks a bottle of low calorie or dietetic soft drink at 10 A.M. and 3 P.M. will probably switch to a caloric sweetened beverage if saccharin is removed from the market.

I think that many are hooked on the "soft drink break" habit. Many would prefer to drink a low calorie beverage, but if that's not available many would switch to a caloric beverage. But that's not really in the realm of economics, per se; that's more in terms of the needs, tastes and preferences of individuals.

Regulatory constraints on sweetener use

Richard J. Ronk

Division of Food and Color Additives, Food and Drug Administration, 200 C Street, S.W. Washington, DC 20204

Consideration of the regulatory constraints on sweeteners flows in two directions: safety questions and economics. To examine these considerations rationally, one must be familiar with the Federal Food, Drug and Cosmetic Act as amended. Further, one should specifically be concerned with Section 409 of this act, which defines certain statutory requirements for introducing new food ingredients into our food supply. In the case of those ingredients which also have drug uses, one must also be familiar with Section 505 which outlines provisions for "new drugs."

Economic considerations relate to our food standards authority contained in Section 401. In addition, the misbranding provision of both the food and drug portions of this basic law must not be forgotten.

In addition, our seeker of status--regulatory status, that is--must also be completely familiar with all of the regulations which have been promulgated by FDA down through the years--truly a task of formidable proportions. In order to save you a trying period of study in some dusty law library, I have consolidated the important features of this law and its accompanying regulations.

It is important to understand that prior to the Food Additives Amendment of 1958, just 20 years ago, no regulatory constraints were placed upon the users of food additives, other than an individual's conscience or FDA court action in cases of demonstrated hazard. Pretesting of food additives was not required, and prior to 1958 it was necessary for the FDA to prove in court that a substance added to food was poisonous or deleterious to health before regulatory action could be taken.

The 1958 Food Additives Amendment changed this state of affairs, however, and required food additive users to obtain premarketing clearance of food additives from the FDA, i.e. clearance for use prior to marketing. The Food Additives Amendment further defined a food additive requiring such clearance as,

> "any substance the intended use of which results or may reason-
> ably be expected to result, directly or indirectly, in its be-
> coming a component or otherwise affecting the characteristics
> of any food (including any substance intended for use in pro-
> ducing, manufacturing, packing, processing, preparing, treating,
> packaging, transporting, or holding food; and including any
> source of radiation intended for any such use) . . . ."

Although this definition would appear to be all-inclusive, the definition exempted from this terminology substances that are

> "generally recognized, among experts qualified by scientific
> training and experience to evaluate its safety, as having been
> adequately shown through scientific procedures (or, in the
> case of a substance used in food prior to January 1, 1958,
> through either scientific procedures or experience based on
> common use in food) to be safe under the conditions of its
> intended use . . . ."

Thus, while the definition of a food additive, and the law pertaining to food
additives, appears to require all food ingredients to have premarketing
clearance, substances could continue to be used without premarketing clearance if they were exempted from the definition of a food additive because of
a history of long use (i.e., they were used in food prior to 1958) or because
of a judgment of safety by experts, through scientific procedures. Because
the law did not attempt to define what constituted an expert, or what was to
be included in an evaluation through scientific procedures, the Food Additives Amendment still had little effect on persons who did not wish to comply
with the spirit of the law, or were not forced to petition the FDA for a food
additive regulation through competitive sales pressure. The FDA's only re-
course in these cases has been to go to court. To add a postscript to this
matter, however, the FDA has recently defined, through regulation, what it
regards as GRAS ("generally regarded as safe") through scientific procedures.
This definition requires establishment of the same quality and quantity of
data as is required for a food additive petition, and further requires common knowledge and judgment of safety by scientists knowledgeable about the
safety of food ingredients. Common knowledge must also be generally established by the existence of publications of applicable safety data.

The 1958 Food Additives Amendment provides that a party seeking to have
a food additive recognized by regulation must submit a food additive petition to the FDA. The petition must: (1) identify the new additive; (2) give
its chemical composition; (3) describe how it is manufactured; and (4) give
an analytical method for the additive, capable of detecting and measuring its
presence in the food supply at levels of expected use. The petition also

must: (5)establish that the submitted method is of sufficient sensitivity to determine compliance with the regulations; (6) provide sufficient data to establish that the additive will accomplish its intended effect in the food, and that the level sought for approval is no greater than the level necessary to accomplish the intended effect. Finally, (7) data must be provided to establish that the additive is safe for its intended use. This usually requires scientific evidence derived from feeding studies of the additive, fed at various levels in diets of two or more species of test animals.

One can not fail to take into consideration these substantial process constraints in the developmental planning for a new sweetening material. It must also be recognized that it would not make a substantial difference whether the proposed new sweetening agent was derived from natural ingredients, or whether it was totally synthetic product.

Once the FDA agrees that a new ingredient is safe for a particular use, then anyone can formulate a food product using that ingredient within prescribed limits, as long as the food is not one which has a standard of identity. Standardized foods are those for which standards of quality and identity have been established. Sweeteners used in foods have generally been covered by standards and thus amendment of these pose a problem since these standards have generally been changed to allow for any "safe and suitable" nutritive sweetener. One must remember, however, that the term "sugar" means sucrose.

What about products already on the market? As many of you are aware, we have also been recently required to re-evaluate many existing regulations because of new information. Thus, we may all correctly assume that a food additive regulation that was permissible yesterday, based upon the scientific evidence of yesterday, may have to be re-evaluated tomorrow, based upon tomorrow's scientific information.

The GRAS list review is an example of the FDA's efforts to update safety data for food ingredients. I should explain here that, because GRAS ingredients have been specifically exempted from the definition of food additives in the law, we have a natural tendency to identify GRAS items as food ingredients, to distinguish them from legally defined food additives. At any rate, when the food additives amendment was passed, the FDA identified a number of food ingredients as GRAS and listed such substances in Section 121.101. Our regulations have recently been recodified and the GRAS regulations are now in Parts 182, 184 and 185 of the Code of Federal Regulations. Although never attempting to accumulate an all-inclusive GRAS list, and continuing to issue opinion letters on the GRAS status of food ingredients, the

FDA listed 675 substances in Section 121.101, based generally upon scientific expert opinion of safety and a history of use prior to 1958.

The FDA in 1970 undertook a systematic review of the GRAS list. The review is, of course, still in progress, but has been organized along the following lines:

(1) A 50 year literature search, called a scientific literature review, has been prepared for each substance.

(2) Particularly where there is an absence of safety data, or some reason to suspect the integrity of a substance, mutagenic and teratogenic animal testing have been conducted for GRAS ingredients.

(3) The FDA contracted with the National Academy of Sciences to survey food manufacturers regarding their use of GRAS ingredients, and attempt to establish from such information, the consumer consumption of such GRAS items.

(4) After the literature search, animal testing and consumption data had been accumulated, the FDA contracted with the Federation of American Societies for Experimental Biology (FASEB) to have this body establish the scientific expertise required to make safety judgments for each food ingredient, based upon the available data.

The FASEB committee makes recommendations to the FDA on the safety of each substance, and the FDA interprets each evaluation and publishes a proposed and final regulation for each substance. If the FASEB committee believes that sufficient safety data are available for present and projected levels of use for an ingredient, the FDA will generally affirm the GRAS status of the substance. If only sufficient safety data are available to substantiate present levels of use, however, we have established the means to affirm the GRAS status but limit levels of use to present uses until further safety data are established. On the other hand, depending upon the degree of uncertainty or alarm expressed by the FASEB committee on the safety of other ingredients, we have established the means to either interim regulate or prohibit the use of a previously listed GRAS ingredient, pending further study. It is also clear that for those substances requiring further study, the burden of establishing the required safety data will lie with either the user or the manufacturer of the ingredient, and not with the FDA.

Procedures have also been developed for public input into the GRAS review. When scientific literature review contracts have been established, the public has been so advised, and urged to submit unpublished data and information to the contractor. Upon completion of the scientific literature reviews, their public availability has also been announced through the Federal Register, and the public once again is urged to submit any additional data or information to the FASEB committee. Again, when the FASEB committee has reached a tentative conclusion on an ingredient, their tentative conclusion is published, and the public urged to either submit written or oral arguments to the committee in the form of a hearing. Finally, public comments are solicited when proposed regulations are published for each ingredient.

Now, as I am sure you appreciate, this process is not quite as expeditious as may be accomplished by locking up a dozen experts for a week or so. It is, however, the public way of doing business and permitting everyone to have some input into the decisions of government. General recognition of safety cannot, after all, be based upon the opinions of a dozen experts. It must be based upon the scientific judgment of all experts capable of making such judgments and in the light of all available data. Our recent regulations have in fact stated that all safety data and information used to make a judgment of GRAS status, must be available to the public generally. This has a particular impact in distinguishing whether a substance should be petitioned for GRAS affirmation or food additive status.

Accompanying the GRAS safety review, the FDA has also established procedures for submission of GRAS affirmation petitions to the FDA. Although the format of such petitions is very similar to that required for a food additive, as mentioned above, no trade secrets can be contained in such a petition, if that information is required to make a judgment of safety on the ingredient. Unless the petitioner can also establish that his ingredient was in common use in food in the U.S. prior to 1958, his petition can only be considered for GRAS affirmation after publication of the safety data upon which his petition relies. If a petition is also erroneously submitted for GRAS affirmation, when it properly should have been submitted for a food additive regulation, the FDA also has the authority to treat the petition as a food additive petition.

Clearly, then, one of the restraints on sweetener development is a regulatory restraint. This restraint is a necessary one, however, as the public has every right to expect that a particular product presents no hidden or unreasonable risks. Safety is after all an essential part of doing business.

The status of some of those sweeteners mentioned by other speakers is presented below:

Status of Aspartame A petition proposing the use of aspartame as a food additive was received from G. D. Searle and Company on February 12, 1973. The petition was regulated on July 24, 1974, to provide for the limited use of aspartame in foods and beverages. However, preliminary results of an Agency audit of the records of some of the animal studies presented by the petitioner raised doubts about the authenticity of the data which had been used to establish the safety of the additive. The questions were serious enough that an order announcing a stay of effectiveness of the aspartame regulation was published in the Federal Register of December 5, 1975. Subsequently, efforts by the Agency to act as a "third party" participant in a contract between Searle and an outside group to authenticate the studies were not successful. This contract would have involved the examination for authenticity of 15 "pivotal" (i.e., integral to the approval decision) and related studies on aspartame.

In lieu of the contract approach a decision was made to implement a direct inspection of certain non-clinical studies submitted to FDA in support of food additive petition No. 3A2885. This investigation began on April 25, 1977, and encompassed the authentication of all raw data and summary data relating to several studies chosen for review by the FDA. The authentication review is continuing. If data from the studies are found to be not authentic, the food additive regulation for aspartame will be revoked. If the data from the studies are found to be in order and correct, an announcement will be made to convene a Board of Inquiry promptly, to resolve issues raised by objectors.

Status of Cyclamate   Abbott Laboratories submitted a petition on November 16, 1973, proposing the use of cyclamic acid, calcium cyclamate and sodium cyclamate as non-nutritive sweetening agents. This petition contained over 300 toxicological reports including assessment of carcinogenicity, metabolism and mutagenicity. Following a preliminary review, FDA concluded that the petition did not contain sufficient evidence to refute earlier studies on cyclamate safety. FDA suggested the need for a well-designed lifetime study in rats to determine carcinogenicity of cyclamate. In a meeting held November 13, 1974, additional data were submitted by Abbott. FDA, on March 14, 1975, requested that the National Cancer Institute conduct an evaluation of the carcinogenic properties of cyclamate and its degradation products. This report, presented to us on March 15, 1976, did not reach a conclusion regarding the question of cyclamate's potential carcinogenicity in humans. Since this, as well as questions concerning cyclamate's potential contribution to genetic defects and effects on the growth and reproductive capabilities were unresolved, Abbott was notified on May 11, 1976 that the petition could not be approved. Abbott Laboratories, on June 16, 1976, advised us that they would not withdraw their petition; therefore, on October 4, 1976, an order denying the petition was published. Abbott Laboratories and the Calorie

Control Council requested a formal evidentiary hearing. On March 4, 1977, the Notice of Hearing granting this hearing and the pre-hearing conference date of April 20, 1977 were published. Oral cross-examination and redirect of testimony took place on July 13, 14, 18 through 21, and August 25 and 26. At the present time the hearing is ongoing. Upon the issuance of a decision by the evidentiary judge, the Commissioner will announce a final decision concerning the safety of cyclamate as a food additive.

Status of Neohesperidin Dihydrochalcone   Nutrilite Products, Inc., Lakeview, California petitioned FDA on August 25, 1975 for use of neohesperidin dihydrochalcone (NDHC) as a sweetening agent and flavor enhancer in chewing gum, mouthwash and toothpaste. A letter was sent to the firm May 18, 1976 requesting additional chemistry and technology data. The petition included a long-term, two-generation reproduction, teratology study in rats. The petitioner emphasized in a conference on March 29, 1976 that the product would not be considered for use in soft drinks at this time. Nutrilite has requested a conference on November 2, 1977 to present additional chemical and technological data.

Another petition received from California Aromatics and Flavors, Inc., on March 18, 1977 requested use of the compound in foods as a naturally derived sweetener. However, at a subsequent conference, held on April 5, 1977, representatives of the firm indicated that they were interested in having the compound regulated for use in chewing gum, mouthwash, toothpaste, cosmetics, drugs, and as a table-top sweetener which would be formulated with sucrose and/or another nutritive sweetener. At that conference, the firm was also made aware that additional data were required to support the toxicological, chemical and technological aspects of the petition. A letter was sent on June 16, 1977 requesting additional chemical, technological and toxicological data. The toxicological data outlined in the letter included (1) chronic feeding in utero, 3-generation reproduction study in rodents, (2) lifetime carcinogenicity study in another rodent species, (3) subchronic (one-year) study in a non-rodent species, (4) mutagenic study and (5) metabolic studies for information on the safety of the metabolites.

A report on the two-year feeding study in dogs was received on August 31, 1977. Before any use other than the flavor enhancer in chewing gum can be considered, the toxicity data mentioned above, as a minimum, will need to be submitted by an interested firm. Also, technological data to support all uses of this compound are required. To date, the FDA has not received any technological data to demonstrate that the compound is effective for the uses requested.

Status of Saccharin   The issue is concerned with whether saccharin and its salts should be banned as additives in human and animal foods and beverages, cosmetics, all animal drugs and most human drugs, and its use permitted as a "single ingredient over the counter drug."

Background   Saccharin and its ammonium, calcium and sodium salts were considered generally recognized as safe (GRAS) until February, 1972 when they were removed from the GRAS list and regulated under an interim food additive regulation.   Three long-term animal toxicity studies with saccharin conducted by the Wisconsin Alumni Research Foundation (1971), the Food and Drug Administration (1973), and the Health Protection Branch of the Government of Canada (1977) indicated that saccharin causes cancer of the bladder in laboratory animals.   The Canadian study confirmed that the increase in cancerous tumors was not due to the impurity orthotoluenesulfonamide.

History of Actions   In April 1977, a proposal was published to revoke the interim food additive regulation permitting use of saccharin in certain foods and beverages, all cosmetics and most drugs.   Consideration was given to the sale of saccharin as an over-the-counter drug for medical uses pending resolution of its medical effectiveness.

In July 1977, the comment period to allow comments on this proposal which was to expire on June 14, 1977 was extended to August 31, 1977 to permit review by the public and the scientific community of two epidemiological studies on the safety of saccharin for human consumption.   One study was coordinated by the National Cancer Institute of Canada.   The other, Epidemiology of Bladder Cancer: A Second Look, is by E. L. Wynder and R. Goldsmith. The Canadian study indicates a positive dose- and duration-related correlation between saccharin use and cancer of the bladder in human males.   The second study found no association between the use of saccharin and bladder cancer.   In September 1977, the comment period was extended again until October 3, 1977 to provide time for the determination of the disposition of freedom of information requests filed by the Calorie Control Council and to allow further comments from interested persons on the saccharin data.

Current Status   The voluminous comments received in response to the April 15, 1977 proposal are being reviewed and evaluated.   FDA is also awaiting the comments that are being submitted during the present extension period.

Future Consideration   Congress has appropriate $1,000,000 for FDA to conduct research on saccharin and related artificial sweeteners from the

standpoint of testing dosage levels equivalent to not more than several times
the probable cumulative human intake. Plans to implement such studies on
artificial sweeteners by FDA personnel will have to consider the logistics
of breeding, raising, feeding, housing, sacrificing and examining an extraordinary number of test animals. $1,000,000 may only fund the start-up of
such a project.

   The Saccharin Working Group (SWG) composed of scientists from FDA and
the National Cancer Institute (NCI) are recommending additional epidemiological studies for the determination of possible association between the consumption of saccharin and the incidence of human bladder cancer. The SWG
recommends: (1) that the NCI's SEER program consisting of cancer reporting
from nine states would be a good study resource for a case-control study.
(2) that a case-control study carried out with the New Jersey State Health
Department would have several advantages. This state includes one of the
highest bladder cancer rates in the country. The reporting of bladder cancer is required by the state which provides not only for the reporting of
cases (1300/year) but provides the Health Department with authority to access hospital and other health records. (3) that a potential study population exists in the Los Angeles area where a rapid reporting system for bladder cancer already exists.

   Because of the present limitations of NCI's SEER program and the New
Jersey Health Department programs there will be need to develop a rapid reporting system so that follow-up of cases may take place as soon as possible
and additional personnel are needed to provide the resources for follow-up
in the various programs.

   Members of the SWG plan to investigate further the use of these three
programs. Since contracts betwen NCI and these three programs already
exist, additional studies could be carried out without a great deal of delay
in contract development.

   The SWG has estimated the costs ($1,375,000) and time to carry out such
studies (a minimum of 24 months). FDA would carry out such studies in cooperation with NCI if Congress provides the necessary resources.

DISCUSSION OF DR. RONK'S PAPER

   DR. BEIDLER: I would like to know how any product can be brought into
the market from the private sector in the future in view of the cost and the
uncertainty about the minimum kinds of tests for safety required. Some
people say that even a figure of a million dollars is too low for adequate
testing. In order to raise venture capital, the producer must be able to
promote the product for some desirable property.

DR. RONK: The Bureau of Drugs has laid out very carefully the requirements for marketing a new drug in the United States.

In terms of a food additive, the requirement is on the producer to demonstrate the safety, not on the government to figure out how to demonstrate the safety of that product. For instance, I am at a loss to know how to demonstrate that radiation preservation of food is in fact safe. I would be equally at a loss to figure out how to demonstrate that the cooking of meat is safe.

There are inherent difficulties in trying to figure out in these indirect ways that a food ingredient is safe. That requirement falls on the producer to do the intellectual kinds of things that are necessary for him to decide for himself that the product is safe and appropriate for the proposed use, and to convince the government of that.

This year the FDA will publish protocol guidelines that outline minimum criteria, but most scientists have not favored that approach. They would like the opposite; they would like the situation to stay as fluid as it is so each individual problem can be looked on by an individual approach.

DR. GUMBMANN: Based on the FDA philosophy that you intimated here, is it right to assume that FDA may not approve any new non-nutritive sweetener for general food additive use or do you see any prospect of a changing attitude?

DR. RONK: Well, what I am saying is that all the special dietary food regulations would have to be amended to allow for that.

DR. GUMBMANN: Since the FDA is unhappy with the use of saccharin in soft drinks, this would eliminate any replacement for saccharin.

DR. RONK: The FDA is not happy or unhappy with the use of saccharin in soft drinks. It is a matter of enforcing the law. If this group today should disagree with the saccharin decision and say that the animal tests are inappropriate models for evaluating safety, then this group should also agree to approve of nothing until a more appropriate system to evaluate the safety of food ingredients has been developed. Either the animal systems are appropriate models to evaluate human health or they are not.

Considering the defect of the systems for evaluating safety that we have, and the animal system is far from perfect, I would have a great deal of pause before changing a significant dietary habit of the people of the United States away from a cultural pattern.

To answer your question directly, though, special dietary food regulations would have to be changed. In fact, if you consider the totality of materials that are being proposed for introduction into the food supply in general, and the new kinds of food additives that are at a developmental stage, you will find a tremendous emphasis on low calorie types of foods. Indeed, one wonders if we won't be presented certain kinds of foods with such a low caloric density that people will be able to get themselves into serious problems if they are able to freely select their diet from such items.

DR. GUMBMANN: In finding a substitute for saccharin, then, we would be looking for a lot more than just safety studies; we would be trying to actually change this regulation.

DR. RONK: On about July 17th we published a regulation related to low calorie foods that sets the scene for how these products can be offered to the public.

But, clearly, if we were able to entirely substitute for sugar in our diet, I think that there would be a great deal of concern on the part of biologically trained scientists and medical people to ascertain what the true health impact of that substitute might be.

DR. CANTOR: There is some concern in the sweetener industry about the identification of sweeteners as additives rather than ingredients.

DR. RONK: Well, the basic statute defines food in terms of itself; I think that the act says in the definition section food is food or drink for man or animal. Then it says that a food additive is also a food. So you start to run into some great difficulties here. I think we could all agree that meat and potatoes are food. When we get to sugar, I think I could say that's food also, but others might not agree. When we get to coffee and tea, we would have less agreement about whether or not they are truly foods or food ingredients. And when we get to monosodiumglutamate, there would be hardly any discussion at all that that in fact is a food additive.

Sugar is generally recognized as a safe ingredient in terms of its legal status and is being considered along those lines. If the participants all could agree today, for instance, as experts on carbohydrate sweeteners, that some serious health hazards are associated with sugar, sugar would no longer be generally recognized as safe. In other words, a very significant body of people would no longer recognize sugar as safe for its intended use.

That would in fact throw sucrose into a food additive category and at that point it would have to qualify as a food additive. Cyclamate originally was generally recognized as safe. If it was to have a status in the United States now it would have to be as a food additive because its recognition as safe came under question.

I hate to use the term food additive or GRAS material any more, because what we really are talking about is food ingredients and the regulations that relate to those food ingredients.

DR. HERBERT: If I understand your presentation correctly, the definition of an ingredient is something added prior to 1958 and the definition of an additive is something added after 1958.

DR. RONK: That's right. The basis for that really was that Congress at that point in time said we are willing to accept the risk only from traditionally used food ingredients. However, in 1970 we started to change that by saying that we are in fact going to look at even the generally recognized safe ingredients themselves.

DR. SWANGO: You used the term special dietary foods. If your physician asked you not to use sucrose because you are diabetic and you used a dietetic food that did not contain sucrose, for example, a can of peaches. That is a special dietary food by your definition.

DR. RONK: That's right.

DR. SWANGO: If your dentist asked you not to use sucrose in candy and you wanted to buy candy sweetened by a sucrose substitute, why does that not qualify for a non-standard food or a special dietary purpose?

DR. RONK: Well, I think it primarily doesn't because traditionally dental caries is not considered as a disease.

DR. SWANGO: But now that we know it is a disease--

DR. RONK: Well, I'm not convinced that people really consider that dental caries is a disease. Some of the problems that we have had with the regulation of sugar and these kinds of products would in fact tell me that we really haven't established that concept. I think Bob Choate and the dental community would agree with that. However, as regulators, that thought has has really not penetrated absolutely, and we do not regulate on that basis.

DR. SWANGO: Would you accept periodontal disease as a disease? Suppose periodontal disease is shown to be related to sucrose consumption. Would this influence regulation of sucrose use?

DR. RONK: You are pointing up a problem. I don't think the food additive amendment that we are talking about here covers all kinds of things. The caloric density that we are talking about clearly is an important feature in the use of these materials.

It's like trying to consider whether this particular confectionery item as it sits on my plate is safe for its intended use. If someone is going to consume 50 of them, then there is a problem. I doubt very much that we could ever approve, as was asked earlier, a table top sweetener, because of the potential for abuse of that product. We traditionally have used 100-fold as our margin for safety. Can you actually send that product home and put it in the sugar bowl because you know it could be abused by someone using 50 packets or more per day?

Over the next few years I think that a number of things are going to happen and sweetener development is going to take place in this context. First, we will have a careful reevaluation of the food additive amendment itself. Congressman Rogers indicated that this is high on his priority list. Second, we are going to have to consider how to handle all these other kinds of situations in terms of the Food and Drug Act. I think that labeling considerations and informed consent must be considered. If I know what the percentage of sweetener is in a product, I can make the kind of decisions about whether I want to use that material or not.

As we saw in earlier presentations, the thing that is characteristic about sweetener use is that less and less of it is discretionary. I want to be able to choose to eat these products or not on a knowledgeable kind of take it or leave it basis. If I am going to have to be associated with processed foods because I don't care to cook or don't have time to cook, I need to know more about the kinds of things that are put in most products.

Clearly, the government is responsive to those kinds of requests and I think we will work out some sort of system to do that. I don't think the food additive amendment is currently the vehicle to do that and I think we have to have some changes.

DR. NOZNICK: What do you mean by the term approved?

DR. RONK: By approved, I mean that we have published a regulation that prescribes the safe conditions for use for that product. If a request is made to use aspartame as a sweetener, we review the data and decide that the data indicated whether or not the material was safe for its intended use.

Its intended use was essentially in dry systems, beverage bases, puddings, anything that didn't have a high temperature that would cause the product to break down to the diketopiperazine. We published a regulation and at that point we had <u>approved</u> that product. At that point in time the government was a party to the decision. When objections were filed so that a Board of Inquiry became involved, then the Bureau of Foods of the Food and Drug Administration was responsible for presenting the scientific case for that material.

So once we approve of something, we are a party to it. Therefore, as most scientists would, when they become a party to a particular action, they are very careful that the action that they are approving of is appropriate.

DR. SHAW: I wonder   if Dr. Talbot would comment on the FASEB report on the relation of sugar to health because he was involved in its preparation. How did your working group determine whether dental caries is serious enough to be concerned about in the whole evaluation of sugar consumption and health?

DR. TALBOT: This subject was debated by the FASEB Select Committee on GRAS substances fairly extensively. The evaluation began with a literature review, and then involved expert outside consultants, people who were not currently members of the committee, who were consultants in dentistry and dental research.

DR. SHAW: As I recall that report, you put the dental caries-sugar relationship in a different category than you did coronary heart disease and diabetes. In the latter diseases, you didn't see the same direct tie that you did between caries and sugar consumption.

DR. TALBOT: A special section of the report was devoted to sugar and dental caries. In the final evaluation, this subject fell into place as another potential hazard of sucrose used in excess and without due care for dental hygiene.

DR. RONK: That recommendation, of course, was sent to the Food and Drug Administration, and we have to act on that in terms of a regulatory response. I think the FASEB report says that other than contributing to obesity and the problem of dental caries, they saw no hazard to public health if sugar and corn sweeteners continue to be used at current levels.

Now, that says a couple of things to us. First, it says at the level they are used now; if consumption increased by some unknown factor, one then could take into question whether or not the material was still generally recognized as safe.

The second thing you would have to consider is: if you are going to hold consumption at current levels, how does one do that when tolerances are generally set on a particular product? Would we have to say that each sugar containing food or candy can contain only this much sucrose? If so, you start to run into some immense kinds of regulatory problems. As Dr. Cantor pointed out, you are really talking about food ingredients here; it really doesn't fit the kinds of constraints that the food additive amendment addresses. That was set up in concern for minor constituents of food.

MR. CHOATE: The curve of industrial use of sugar and sweeteners has been rising while the per person purchase of sugar and sweeteners for use in the home has been dropping. Are there any studies on the increase of sucrose in individual prepared prepackaged products over the last 20 years?

DR. RONK: Well, as Dr. Cantor mentioned earlier, the only thing that you can be sure about up to this point is disappearance data. We are going to try to get, by cooperation between the Food and Drug Administration and the Department of Agriculture, surveys on portion size through recall data to find out what amounts people actually eat. We should be able to finally catalogue exactly how much dietary exposure a particular person has from any particular category of foods, for instance baked products.

MR. CHOATE: Does the FDA have any knowledge about the sugar or sucrose content of a pudding or a cake mix or a bread mix and whether it has changed over the last 10 or 15 years?

DR. RONK: No, because we are not entitled to the formulation itself.

MR. CHOATE: Have any studies been published on this subject?

DR. RONK: Not that I am aware of. The Rogers bill that relates to food use in the United States has a provision that would give the Food and Drug Administration recipes used in the manufacture of foods. If we had that information, a number of things could be done in terms of the intake of sugar, food additives, and contaminants from packaging and other kinds of things.

One of the things that scientific groups like this can and should contribute is a very thorough look at how the Food and Drug Administration or any agency attempts to decide on the safety of food ingredients. I think that the procedure of food safety assessment requires a good overhaul. We need to develop in the next 10 years in this country a science of food safety assessment.

Some of the things being done in FDA, in the Food and Safety Council and in industry will go a long way toward developing the state of art of the assessment of food safety. I think that the university community itself must become more involved in these kinds of projects, and to date there has not been that kind of involvement. We will be looking for mechanisms to involve people such as yourself in the process and if you are called upon, I hope that you will serve.

SESSION II.

DENTAL CARIES : ROLE OF SWEETENERS

Moderator:

Victor Herbert

Hematology and Nutrition Laboratory
Veterans Administration Hospital
130 West Kingsbridge Road
Bronx, New York 10468

Role of carbohydrates in dental caries

William H. Bowen

Caries Prevention and Research Branch, National Caries Program, National Institute of Dental Research, Bethesda, Maryland 20014

SUMMARY

It is apparent that carbohydrates and sugars in particular have a profound effect on the development of dental caries. Sugar ingestion may enhance the susceptibility of teeth to caries, promote the formation of a cariogenic flora, and result in the formation of pathogenic plaque which has the ability to form acid from a wide variety of carbohydrates.

The evidence which shows that carbohydrate is essential in the etiology of dental caries is unequivocal. The results of epidemiological investigations[1] and experiments carried out in animals[2-4] clearly show that both the prevalence and incidence of dental caries are related to the frequency of ingestion of readily fermentable carbohydrate.

In an extensive investigation carried out in Sweden by Gustafson et al, subjects who consumed 94 kg per year of sugar with meals were clearly shown to have developed substantially fewer lesions than a group of subjects who consumed 85 kg, 15 of which were ingested between meals[5]. It was also apparent from the results of this study that the level of dental caries is directly related to the duration for which sugar is present in the mouth.

Similar observations were made by Harris on the children in Hopewood House, Australia. Children who resided in this institution and had restricted access to candy, cookies, honey and related foodstuffs Monday through Friday developed substantially fewer lesions than cohorts who had unrestricted access to sugar containing products[6].

Further evidence to support the importance of sugar in the pathogenesis of dental caries may be gleaned from the observation of Marthaler and Froesch[7] who reported that persons who are fructose-intolerant and who, therefore, must restrict their intake of sucrose and sweet-tasting substances are virtually caries-free.

A number of investigators have reported they were unable to find any correlation between the total intake of sugar and the incidence of dental caries[8,9]. A similar observation was made by Campbell and Zinner[10] who, however, additionally reported a correlation between the prevalence of caries and the frequency of ingestion of such items as soft drinks and chewing gum.

Indirect evidence to support the concept that sugar is cariogenic may be found in human subjects who received their entire diet by gastric intubation. Although these patients form dental plaque, it has little ability to lower the pH of sugar solutions[11,12].

Results of experiments carried out in animals also clearly show that dental caries do not develop in the absence of carbohydrate, and sucrose in particular, from the diet. Rats which received their entire cariogenic diet by gastric intubation were shown by Kite et al[13] to remain caries-free. However, if the same diet was fed per os, the animals developed dental caries.

The influence of specific sugars and proteins on the bacteriological and chemical composition of plaque in primates has been determined[14,15]. Although animals which received their entire diet by gastric intubation developed plaque, it differed considerably from that formed in monkeys which received sucrose, glucose and fructose or casein alone by mouth. Plaque formed in the absence of sugar lacked the ability to lower the pH of sugar solutions. In addition, it contained substantially fewer microorganisms and less carbohydrate, both extracellular and intracellular, than that formed in the presence of sucrose or glucose and fructose. Plaque formed in the presence of casein had little acid producing ability; in addition, considerable amounts of calculus were formed in these animals. It is clear, therefore, that sugar has a considerable influence on the pathogenic potential of plaque.

Controversy exists over the relative cariogenicity of different carbohydrates[16]. In general, it appears from the results of animal experiments that monosaccharides and disaccharides are substantially more cariogenic than starch[17,18]. Sucrose is usually regarded as the most cariogenic of sugars, probably because it is the sugar most frequently ingested. However, results of animal investigations show considerable variation and appear to indicate that there is little difference in the cariogenicity of sucrose, glucose and fructose[19,20]. Some investigators have observed that sucrose appears to promote the development of smooth surface caries in animals to a greater extent than other sugars, although this observation may be a reflection of the composition of the oral flora of the animals[21,22]. In general, it appears that little benefit would accrue, even were it possible, if sucrose in the human diet were replaced by glucose or fructose.

The physical form in which sugar is ingested also influences its cariogenicity[23]. Results from animal experiments have shown that sugar solutions are substantially less cariogenic than sugar ingested in solid form[24]. Similar observations have been made in humans; for example, Gustafson et al[5] reported that subjects who chewed sticky toffees developed more caries than subjects who ingested a comparable amount of sugar in a non-sticky form. Infants who suck for prolonged periods on bottles filled with syrup or other sugary solutions develop rampant dental caries, particularly on the palatal surfaces of the upper molars, and are frequently edentulous before the age of 5 years[25-27].

The ingestion of sucrose may have a profound effect on the composition of plaque flora. Unequivocal evidence shows that eating sugar enhances the ability of Streptococcus mutans to colonize tooth[28] surfaces even though S. mutans can become established in the absence of sugar, even in fructose intolerant patients. S. mutans, S. sanguis and other oral microorganisms have the ability to synthesize glucan and fructan from sucrose through the enzymes glucosyltransferase and fructosyltransferase[29]. The glucan formed was formerly termed dextran in the erroneous belief that it was composed primarily of α 1-6 linked glucose units. However, this material is now termed "mutan" because it is characteristic of S. mutans and is composed mainly of α 1-3 linked glucose units[30]. S. mutans produces both water soluble and insoluble glucans which because of their adhesive properties appear to enhance the ability of S. mutans to adhere to the tooth surface[31]. In addition, the polysaccharide produced by S. mutans carries a charge, probably a phosphate group in the form of lipoteichoic acid[32]. It appears probable that this charge is also implicated in the colonization process.

The precise role that extracellular polysaccharide plays in the pathogenesis of dental caries is uncertain. However, it can act as a reserve source of carbohydrate when extraneous sources are lacking[33,34]. It can protect microorganisms from inimical influences and prevent the diffusion of charged substances into and out of plaque[35].

Many microorganisms in plaque have the capacity to store polysaccharide intracellularly which as judged by its tinctorial properties is glycogen of the amylopectin type[36]. This material is readily catabolized. Results from at least one investigation show a positive correlation between the numbers of intracellular polysaccharide containing organisms and the number of decayed, missing and filled tooth surfaces[37].

The mere physical presence of plaque on the tooth surface will not of itself cause dental caries. Although a wide variety of biochemical activities occur in plaque, the available evidence appears to suggest that its ability to produce acid rapidly from sugar is the property most associated with the development of dental caries[38,39]. Each ingestion of sugar is followed by a rapid production of acid by plaque and values as low as 4 have frequently been recorded[40]. It is generally believed that demineralization of enamel occurs below pH 5.5[41]. Lactic is the predominant acid formed in dental plaque, even though less than half the titratable activity in plaque can be accounted for by the amount of lactate present[42]. A variety of acids, including valeric and propionic, have been identified in dental plaque obtained from either humans or primates[43]. Acetic and formic acid were the major products in 78% and 14% of plaques studied by Gilmour and Poole[44]. However, the microbial composition of the plaques was not established nor was it clearly established that these plaques were indeed cariogenic. Plaque which had not been exposed to carbohydrate for several hours was found by Geddes to contain 3×10^{-5} moles of acid/mg wet weight, and that approximately 5 minutes after exposure to sugar the concentration increased to 5×10^{-5} moles. The major change was 5-fold increase in D(-)lactate and 8-fold increase in L(+)lactate[45].

Available evidence also suggests that carbohydrate and sugar in particular may directly affect the susceptibility of enamel to caries. It has been observed that the teeth of rats which were exposed to high sugar diets showed delayed maturation, a process which leads to enhanced mineral uptake post-eruptively[46]. Presumably teeth so affected are more susceptible to decay. Rats bred on a diet conducive to the formation of severe protein-caloric imbalance appear to be particularly susceptible to dental caries. The enhanced susceptibility was ascribed to altered tooth size and to alterations in salivary composition[47,48].

At least three sugar alcohols are frequently used as alternative sweetening agents for sucrose. Mannitol, sorbitol and xylitol are either non-cariogenic or substantially less cariogenic than other sugars[49-51]. Xylitol apparently is not metabolized by plaque microorganisms, and no significant adaptation by the plaque flora was observed in clinical studies extending over two years[51]. Sorbitol and mannitol are slowly metabolized by oral microorganisms[52]. Dallmeier et al[53], using pure cultures of streptococci have observed that formic acid and ethanol are the main end products when sorbitol is fermented, which may explain in part at least its lack of cariogenicity. In a study carried out over two years in monkeys, sorbitol was found to be minimally cariogenic and enhanced ability of plaque microorganisms to metabolize sorbitol was not observed[54]. Similar observations were made by Clark, et al[55] in humans which are in contrast to those reported by Frostell[56] who also reported substantial increase in the number of plaque microorganisms capable of fermenting sorbitol following prolonged ingestion of sorbitol by humans.

Claims that xylitol is cariostatic remain to be substantiated, even though results of investigations carried out in humans and primates appear to suggest that the ingestion of this pentitol affects the levels of lactoperoxidase and amylase in saliva. It has not been precluded that other sugar alcohols have the same effect.

REFERENCES

1. Read, T. and Knowles, E. Brit. Dent. J., 64, 185, 1938.
2. Frostell, G. and Baer, P.N. Acta Odont. Scand. 29, 253, 1971.
3. Green, R.M. and Hartles, R.L. Arch. Oral Biol. 14, 235, 1969.
4. König, K.G., Larson, R.H. and Guggenheim, B. Arch. Oral Biol. 14, 91, 1969.
5. Gustafson, B.E., Quensel, C.E., Swenander, L., Lundquist, C., Grahnén,H., Bonow, B. and Krasse, B. Acta Odont. Scand. 11, 232, 1954.
6. Harris, R.M. J. Dent. Res. 42, 1387, 1963.
7. Marthaler, T.M. and Froesch, E.R. Brit. Dent. J. 124, 597, 1967.
8. Zita, A., McDonald, R.E. and Andrews, A.L. J. Dent. Res., 38, 860,1959.
9. Weiss, R.L. and Trithart, A.H. Amer. J. Pub. Health, 50, 1097, 1960.
10. Campbell, R. and Zinner, D.D. J. Nutr. 100, 11, 1970.
11. Littleton, N.W., Carter, C.H. and Kelley, R.T. J. Amer. Dent. Assoc. 74, 119, 1967.
12. Carlsson, J. and Egelberg, J. Odont. Rev. 16, 112, 1965.
13. Kite, O.W., Shaw, J.H. and Sognnaes, R.F. J. Nutr. 42, 89, 1950.
14. Bowen, W.H. and Cornick, D.E.R. Int. Dent. J. 20, 382, 1970.
15. Bowen, W.H. Arch. Oral Biol. 19, 231, 1974.
16. Frostell, G., Keyes, P.H. and Larson, R.H. J. Nutr. 93, 65, 1967.
17. Shaw, J., Krumins, I. and Gibbons, R.J. Arch. Oral Biol. 12, 755, 1967.
18. Green, R.M. and Hartles, R.L. Brit. J. Nutr. 21, 921, 1967.
19. Colman, G., Bowen, W.H. and Cole, M.F. Brit. Dent. J. 142, 217, 1977.
20. Guggenheim, B., König, K.G., Herzog, E. and Mühlemann, H.R. Helv. Odont. Acta 10, 101, 1966.
21. Larson, R.H., Theilade, E. and Fitzgerald, R.J. Arch. Oral Biol. 12, 663, 1967.
22. Larson, R.H. and Goss, B.J. Arch. Oral Biol. 12, 1085, 1967.
23. Caldwell, R.C. J. Dent. Res. 49, 1293, 1970.
24. Sognnaes, R.F. J. Nutr. 36, 1, 1948.
25. Winter, G., Hamilton, M.C. and James, P.M. Arch. Dis. Child, 41, 216, 1966.
26. Syrrist, A. and Selanda, P. Odont. Revy. 61, 237, 1953.
27. Goose, D.H. Caries Res. 1, 167, 1967.
28. Krasse, B. Arch. Oral. Biol. 11, 429, 1966.

29. Wood, J.M. and Critchley, P. Arch. Oral Biol. 11, 1039, 1966.
30. Hotz, P., Guggenheim, B. and Schmid, R. Caries Res. 6, 103, 1972.
31. Gibbons, R.J. and van Houte, J. Ann. Rev. Micro. 29, 19, 1975.
32. Rölla, G. and Mathiesen, P. in Dental Plaque, edited by McHugh, W.D., E and S Livingstone, Ltd., Edinburgh, 129, 1970.
33. van Houte, J. and Jansen, H.M. Arch. Oral Biol. 13, 827, 1968.
34. Wood, J.M. Arch. Oral. Biol. 14, 161, 1969.
35. Rorem, E.S. J. Bacteriol 70, 691, 1955.
36. van Houte, J., Winkler, K.C. and Jansen, H.M. Arch. Oral Biol. 14, 45, 1969.
37. Loesche, W.J. and Henry, C.A. Arch. Oral Biol. 12, 189, 1967.
38. Stephan, R.M. J. Dent. Res. 23, 257, 1944.
39. Kleinberg, I. Int. Dent. J. 20, 451, 1970.
40. Graf, H. Int. Dent. J. 20, 426, 1970.
41. Mühlemann, H.R. Int. Dent. J. 21, 456, 1971.
42. Gilmour, M.N., Green, G.C., Zahn, L.M., Sparmann, C.D. and Pearlman, J. IADR Abstr. 299, 1975.
43. Cole, M.F., Bowden, G., Bowen, W.H. Caries Res. in press, 1977.
44. Gilmour, M.N. and Poole, A.E. Caries Res. 1, 247, 1967.
45. Geddes, D.A.M. Caries Res. 9, 98, 1975.
46. Nikiforuk, G. J. Dent. Res. 49, 1252, 1970.
47. Menaker, L. and Navia, J. J. Dent. Res. 52, 680, 1973.
48. DiOrio, L.P., Miller, S.A. and Navia, J.M. J. Nutr. 103, 856, 1973.
49. Larje, O. and Larson, R.H. Arch. Oral Biol. 15, 805, 1970.
50. Shaw, J.H. J. Dent. Res. 55, 376, 1976.
51. Mäkinen, K.K. Int. Dent. J. 26, 14, 1976.
52. Guggenheim, B. Caries Res. 2, 147, 1968.
53. Dallmeier, E., Bestmann, H. and Kronke, A. Deut. Zahn Z. 25, 887, 1967.
54. Cornick, D. and Bowen, W.H. Arch. Oral Biol. 17, 1637, 1972.
55. Clark, R., Hay, D.I., Schram, C.J. and Wagg, B.J. Brit. Dent. J. 111, 244, 1957.
56. Frostell, G. in Nutrition and Caries Prevention, Almquist and Wiksell, Stockholm.

DISCUSSION OF DR. BOWEN'S PAPER

DR. TURNER: As a dentist responsible for the care of children, I want to add my support to those who recommend the reduction of sucrose and other sugars in the diet, and in particular, in those items used between meals. The dentist who assumes the responsibility to provide exemplary dentistry for children knows that he does not have an easy task. Daily I see children with advanced dental caries. The need for prevention and early detection cannot be underestimated. Inadequate or unsatisfactory dental treatment too often results in premature loss of the primary and permanent teeth, leaving the child with many of the dental problems that are so commonly seen in our adult population. Prevention must be emphasized with increasing vigor, and a keystone in a preventive program is the reduction of sugar intake, especially between meals.

DR. BOLLENBACK: Dr. Bowen, did I understand you to say that fructose and glucose are equally cariogenic compared with sucrose?

DR. BOWEN: Yes. A number of studies have been carried out on this subject. In the instances where the flora has been controlled by only having Streptococcus mutans present or predominant, sucrose is in fact more cariogenic than fructose or glucose. However, in animals "with a mixed flora" containing S. mutans and other microorganisms, there is no difference in the cariogenicity of fructose, sucrose and glucose. We have just completed a study in primates, again with a mixed flora, which contained S. mutans, and found no difference in the cariogenicity of glucose and fructose, a mixture of glucose and fructose, and sucrose (Colman, G., Bowen, W.H., Cole, M.F., Brit. Dent. J. 142, 217, 1977).

DR. AMOS: Since Streptococcus mutans has the capacity to use the energy released from hydrolysis of sucrose to make the extracellular polysaccharide, glucan or mutan, how do you account for the fact that fructose or glucose independently or as a mixture of fructose and glucose can be as cariogenic as sucrose?

DR. BOWEN: In our gastric intubation studies we looked at the amount of extracellular polysaccharides formed from glucose and fructose, and we were able to demonstrate quite clearly that significant amounts formed, even in the complete absence of sucrose (Bowen, W.H., Arch. Oral Biol. 19, 231, 1976).

Many other microorganisms are present in dental plaque in addition to S. mutans, some of which have the capacity to form extracellular polysaccharides of one kind or another. Unfortunately, much of our attention has focused only on the extracellular polysaccharides produced by S. mutans. To the best of my knowledge, only one study has been carried out on the type of polysaccharide formed in human dental plaque (Hotz, P., Guggenheim, B. and Schmid, R., Caries Res. 6, 103, 1972). We know very little in detail about the type of extracellular polysaccharides in human dental plaque.

DR. NEWBRUN: A lot of people seem to identify only sucrose as cariogenic. Until 1972 with the very sharp price rise in sucrose in the U.S.A. consumption figures for dietary carbohydrates were largely sucrose and starches. The consumption of glucose and fructose prepared from corn starch and of corn syrups did not become a major factor until very recently. Therefore, the previous publications where people identified sucrose as the main problem in regard to human caries have to be taken in their historical context. In practical terms I noticed in Dr. Bowen's paper--and indeed many people in discussions about caries talk about the role of "fermentable carbohydrates"--he pointed out that starches per se are not particularly cariogenic except in some unique animal model. In humans, for example fructose-intolerant individuals, starches are eaten abundantly in rice, potatoes and bread, but these people do not eat foods containing fructose, sucrose or syrups and they are virtually caries free. It is high time that we stopped using the term "fermentable carbohydrates" because starches are fermentable under suitable circumstances but do not appear to be cariogenic in man.

Finally a comment in regard to sorbitol fermentation. The only study that Dr. Bowen mentioned where adaptation was claimed was in an unreferred paper at a symposium (Frostell, G. in Symposium of Swedish Nutrition Foundation, G. Blix, ed., 1965). Theoretically adaptation would not be expected because the NAD oxidoreductases, which are required to required to ferment sorbitol, are inducable enzymes only if sorbitol is the only available carbohydrate (Brown, A.T., Wittenberger, C.L., Arch. Oral Biol. 18, 117, 1973). However, that would never be the situation in vivo in dental plaque. I do not think the problem of adaptation is serious because the use of sorbitol, mannitol or xylitol in food items is not likely to result in any one of them becoming the only carbon source for organisms in the dental plaque.

DR. CANTOR: Referring to the sweetener disappearance before 1972, it should be made clear that much of the sucrose in the diet undergoes processing before consumption and has been hydrolyzed before entering the oral cavity. The sucrose used in canned fruit and in carbonated beverages to a very large extent has become a mixture of fructose and glucose before consumption. For this reason substantial amounts of glucose and fructose were present in the human diet long before 1972, even though sugars from corn did not have a prominent place in the market until recently.

High fructose corn syrup (HFCS) is the principal product that is displacing sucrose now. This product contains 42 percent fructose, 50 percent glucose, and the balance glucose polysaccharides--a mixture very close to invert sugar. The impact of HFCS is somewhat limited in terms of changing the character of the sugars in the oral cavity. It is only when we begin to use new sweeteners such as fructose and syrups containing more than 50 percent fructose that the carbohydrate sweeteners will provide a somewhat different oral environment than has been true historically.

This point must be remembered in relation to sugar consumption. Many people talk about sucrose consumption as though all the sucrose placed in food during processing was consumed as sucrose. This is simply not the case because of its hydrolysis in several categories.

DR. BIBBY: I want to make a point in this discussion of sucrose and fructose and other mono- and disaccharides. In our present society very little of these sugars are used as such but are present in mixtures with other food components. The point of whether sucrose is more cariogenic than starch is really unimportant because they generally are consumed in complex mixtures in foods rather than in pure form. We must focus on the cariogenicity of mixtures as used in foods rather than the pure carbohydrates as such.

DR. GEY: In view of Dr. Bowen's comments about pH decreases after sugar consumption, I should like to comment on some studies on saliva conducted with five volunteers who chewed gum, unsweetened or sweetened with xylitol or sucrose, or sucked boiled candy sweetened with sucrose or xylitol. The results are shown in the following figure. All five products stimulated the flow of saliva with the two sweetened gums being the most effective and the unsweetened gum base and two candies being somewhat less effective but not strikingly different from each other. The two sweetened chewing gums resulted in almost identical increases in calcium concentration in the saliva. The other three products resulted in lower increases in salivary calcium but did not differ from each other appreciably. The results with respect to pH were more striking. The unsweetened gum base caused the largest increase in pH from 7.0 to 7.5 within 5 minutes after its use. Both xylitol-sweetened chewing gum and candy caused increases in pH at 5 minutes to about 7.4 with the pH still above 7.0 after 25 minutes. The gum sweetened with sucrose resulted in an increase in pH to 7.3 at 5 minutes but at 25 minutes the pH had decreased to 6.8. Candy sweetened with sucrose resulted in an increase in pH to only 7.1 at 5 minutes with a decrease to 6.8 by 25 minutes. This failure of pure sucrose candy to increase the pH is obviously masked in the sucrose chewing gum by mastication since the chewing of the unsweetened chewing gum base caused the same immediate rise of the pH as xylitol-sweetened gum.

In consequence, in total saliva, any conceivable pH dependent benefits of xylitol, for instance, as so-called remineralization of white spots might become more obvious by the comparison of sucrose and xylitol in candy than in chewing gum. This effect might be even more true if sites are taken into account where sucrose would be metabolized into lactic acid as in dental plaque. Sucking candy results in markedly less mastication and thus less mechanical stimulation of salivation than does chewing gum.

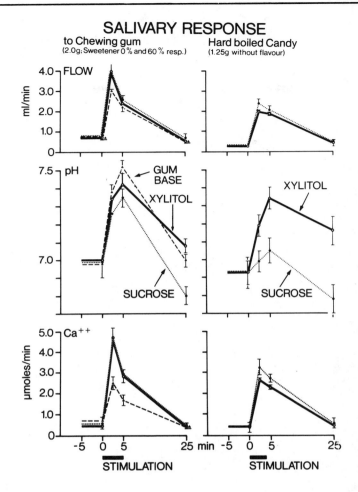

SALIVARY RESPONSE

to Chewing gum
(2.0g; Sweetener 0% and 60% resp.)

Hard boiled Candy
(1.25g without flavour)

FLOW
ml/min

pH

GUM BASE

XYLITOL

SUCROSE

XYLITOL

SUCROSE

Ca⁺⁺
μmoles/min

-5 0 5 25 min -5 0 5 25

STIMULATION   STIMULATION

FIGURE. Comparison of sweetened chewing gum and candy with sucrose or with xylitol in regard to salivary composition in man. The values represent means ± S.E. for five volunteers.

Thus, studies with xylitol candy instead of chewing gum may also be helpful to elucidate the question whether the clinical effect of xylitol containing chewing gum in Turku was primarily due to xylitol as the major ingredient or to mastication.

The metabolism of the polyols and their potential for greater use as sweetening agents in foods and confections

James H. Shaw, Ph.D.

Harvard School of Dental Medicine, 188 Longwood Avenue, Boston, Massachusetts 02115

SUMMARY

The ability of the human body to metabolize polyols and the potential for their increased use in human diets in replacement of sugars as a means to reduce dental caries incidence are sufficiently great to justify additional investigation. Xylitol currently appears to be the polyol of choice for further studies because of the human body's greater ability to utilize it without adverse influences and the possibility of its ability to exert an anti-caries influence that is disproportionate to the amount consumed. Future studies should be conducted in highly caries-susceptible populations and emphasize the search for the most suitable vehicle or vehicles and the optimal frequency of usage to obtain benefit.

INTRODUCTION

The polyols are not sugars in the strict sense, but are, as the term indicates, polyalcohols with each carbon of the compound bearing an alcohol group. Hence, unlike the sugars there can be no reactive group such as the aldehyde group on carbon 1 in the aldosugars (aldoses), of which glucose is the best example among the monosaccharides, and no keto group on carbon 2 in the ketosugars (ketoses), of which fructose is a well-known simple sugar. Therefore, the pentitols, such as xylitol, and hexitols, such as sorbitol and mannitol, are not reducing agents. Similarly, maltitol and lactitol are 12-carbon polyols formed commercially by the hydrogenation of the single aldehyde group of maltose and lactose, respectively. The absence of any reactive group in the polyols results in no Maillard reaction when the polyols are sterilized or heated less rigorously with amino acids or with proteins. Thus browning or caramelization in food preparation, when the carbonyl group of glucose or sucrose reacts with free amino groups in proteins such as the ε-amino group of lysine, cannot occur when polyols are used as the sweetening agent in the absence of sugars.

Although the polyols are not sugars, their physiological and biochemical relationships may be discussed legitimately along with the sugars for the good reason that the polyols are formed by the reduction of sugars and can be converted to sugars by oxidation. Furthermore, they are metabolized physiologically in the same pathways as the sugars and indeed, xylitol, in particular, is an intermediary in an important pathway. For the above reasons, polyols such as xylitol, sorbitol, and mannitol cannot be classified as sugars; products sweetened by them rather than by sugars are correctly described as sugar-free.

Touster draws attention to the fact that the polyols occur widely in nature and have been isolated from a variety of plant and animal tissues which are normal components of the human diet, with the result that they routinely occur in small amounts in the urine[1]. His summary of the occurrence of the polyols in animal tissues and urine is presented in Table 1.

Polyols tend to have a higher endothermic heat of solution than the sugars so that xylitol, in particular, and sorbitol, to a lesser extent, exert a cooling effect in the mouth as with menthol in mints. This fact has to be taken into account in the use of xylitol in food and candy formulations. Since no carbon groups in the polyols are oxidized beyond alcohol, their caloric value after absorption is slightly higher per gram than for the comparable sugars.

POLYOL METABOLISM

Metabolism of the polyols in mammals takes place as a result of oxida-

Table 1. Polyols in animal metabolism [1]

Polyol	Occurrence in Tissues
sorbitol	fetal blood, seminal vesicles and plasma, erythrocytes, brain, nerve, kidney, aorta, lens of alloxan-diabetic rats or rats given cataractogenic dose of D-xylose
xylitol	lens of rats given cataractogenic dose of D-xylose
dulcitol	various tissues after cataractogenic dose of D-galactose
	Occurrence in Urine
erythritol D- and L-arabitol sorbitol D-mannitol	

tion to either a ketose or an aldose, with the entry of the resulting sugar
into a metabolic pathway.   Since the polyols of special interest are
sorbitol, mannitol, and xylitol, because of their availability for use in
the human dietary and the potential to reduce the dental caries experience,
this discussion will be limited to the metabolism of these compounds.

The reaction to oxidize sorbitol to D-fructose is catalyzed by a polyol
dehydrogenase, sorbitol dehydrogenase, or more accurately L-iditol
dehydrogenase, which has NAD as its coenzyme.   The glucose to sorbitol
reaction is catalyzed by the NADP-dependent enzyme, aldose reductase.
Touster points out that numerous investigators have shown that sorbitol is a
normal metabolic intermediary in mammals[1].   For example, sorbitol is the
precursor of fructose in fetal blood and is especially high in fetal blood
of hoofed mammals.   In addition, sorbitol produced from glucose is the
precursor of fructose in male accessory organs.   Less is known about the
intermediary metabolism of mannitol, but like sorbitol it is oxidized to
D-fructose by a polyol dehydrogenase.   Fructose enters the metabolic scheme
by being phosphorylated to fructose 1-phosphate by fructokinase.

The intermediary metabolism of xylitol is shown in Figure 1 entering
the pentose phosphate pathway in its general relationship to carbohydrate
metabolism[1].   Recognition of xylitol as a normal metabolite originated in
studies on the route of formation and utilization of the pentose, L-xylulose,
which is excreted in unusually large amounts in the urine of patients with
essential pentosuria, a genetic abnormality.   These individuals excrete
L-xylulose in gram amounts daily; in contrast, normal animals and non-
pentosuric people excrete this pentose in milligram amounts daily.
Recognition of the source of L-xylulose metabolically and of the presence of
xylitol in a metabolic pathway came with the observation of Hollmann and
Touster[2] and Touster et al.[3] that L-xylulose was reduced to xylitol when
this pentose was incubated in tissue preparations.   This reaction was
catalyzed by a specific NADP-linked polyol dehydrogenase which was
originally found in liver mitochondria[2] but was later shown to be largely a
soluble cytoplasmic enzyme[4].   This polyol dehydrogenase catalyzed reaction
is evidently deficient in pentosuria, permitting the accumulation of L-
xylulose and its urinary excretion rather than expediting its reduction to
xylitol.   L-arabitol is the other possible reduction product of L-xylulose
and is produced in small amounts by pentosuric individuals but is not known
to be present in appreciable amounts in normal human beings.

Xylitol is then oxidized by the removal of hydrogen in a reaction catalyzed by an NAD-linked enzyme, L-iditol dehydrogenase, to produce D-xylulose. The latter compound is then phosphorylated to D-xylulose 5-phosphate for metabolism in the pentose phosphate pathway to hexose phosphate for utilization as needed in the overall metabolic pool of the body[5]. Hollmann and Touster estimated that 5 to 15 gm of carbohydrate per day pass through this pathway in man, even when no supplement of xylitol is included in the diet. In other words, several percent of the carbohydrate metabolized daily may be metabolized in this pathway with xylitol as an intermediate[6].

The metabolism of xylitol has often been described as insulin-independent, in contrast to the dependence of glucose upon insulin. For this reason, xylitol has been evaluated frequently for use by diabetic persons. Indeed, much of the investigation of the use of xylitol intravenously or orally in man has been to evaluate its safety as a carbohydrate

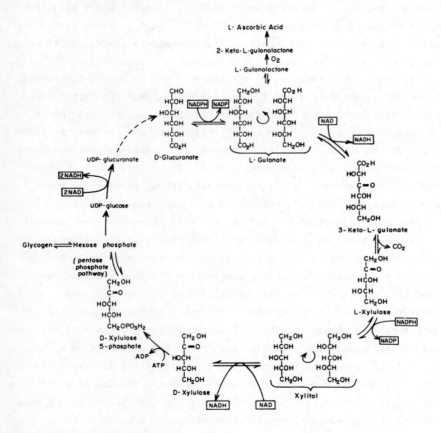

Figure 1. Summary of carbohydrate metabolism to indicate the relationship of xylitol.

source for diabetics. Interest in its possible use to reduce or prevent occurrence of dental caries came later and profited from substantial investigation in earlier studies. General agreement exists that the absorption of xylitol from the gastrointestinal tract and by individual cells, the metabolic steps from xylitol to D-xylulose to D-xylulose 5-phosphate and those in the pentose phosphate pathway to hexose phosphate, are independent of insulin. However, the hexose phosphate may be catabolized for energy in the citric acid cycle or glucose may be polymer-ized for storage as glycogen. Whatever glucose is realized from xylitol metabolism directly or by the later hydrolysis of glycogen is dependent upon insulin. The slow absorption of xylitol from the gastrointestinal tract results, when metabolized, in a much slower release of glucose and a lesser effect on the elevation of blood glucose than a similar amount of glucose and is more similar to the effect of a complex carbohydrate which is hydrolyzed in the gut to glucose.

With the exception of D-ribose, the pentoses are not efficiently metabolized by animals. When tracer doses of D-ribose tagged with ^{14}C were injected, 48% was metabolized to be expired as $^{14}CO_2$ while only 10% was excreted in the urine[7]. In contrast, only 0.8% of the tracer in L-arabinose was expired as $^{14}CO_2$ and 85% was excreted in the urine. D-arabinose, D-lyxose, and D-xylose were intermediate with 14 to 19% expired and 35 to 72% excreted. D-arabitol and L-arabitol were not metabolized well by the rat with 14 and 13%, respectively, expired as $^{14}CO_2$, 36 and 38% excreted in the urine and 0.2 to 0.3 and 0.9 to 3.2% detected in liver glycogen, respectively. D-ribitol was more efficiently metabolized with 31% of the tracer expired and up to 22% stored in liver glycogen, while xylitol was oxidized to a still larger extent.

Bässler et al. conducted kinetic studies on the decrease of the xylitol level in the blood after infusion of 0.5 gm of xylitol per kg body weight in premature infants, mature infants, and children four to eight years old and 20 gm to adults between 30 and 41 years of age[8]. The early decrease in blood xylitol was rather rapid, which the investigators attributed to distribution throughout the extracellular space with a temporary accumula-tion, especially in those tissues that were not capable of metabolizing xylitol. The second phase of the decrease was much slower and was attributed to a distribution through the total intracellular space of the body and to the metabolism and disappearance of xylitol. No distribution or elimination differences were observed between the groups of subjects with

even the premature infants possessing the full capacity to metabolize
xylitol. These investigators reported that 11.3 mg/min/kg body wt of xylitol was metabolized by the individuals in their adult group with the urinary
excretion of only 6.1% (0.67 mg/min/kg) of the administered dose. Similar
results were observed in diabetic persons after the infusion of xylitol by
Bässler and co-workers[9,10].

Mehnert et al. reported that infused xylitol was metabolized equally
well by normal persons and by individuals with diabetes or moderate liver
damage[11]. After the infusion of 0.5 gm/kg body weight over 90 minutes,
xylitol levels decreased rapidly in the blood. Urinary excretion varied
between 8 and 12% and no increases in blood and urinary glucose were
observed. Similar results were observed in 15 young individuals infused
with 1.0 gm xylitol per kg body weight during 30 minutes. Some evidence
of delayed utilization was observed in aged patients and ones with severe
liver damage.

Liver is the main organ for the metabolism of xylitol, to which liver
cells are readily permeable. The results of in vitro studies of xylitol in
the rat are shown in Table 2[12]. The amount of xylitol dehydrogenated by
the liver was estimated to be about 10 times that of all other tissues.
Likewise a comparison in liver perfusion studies and perfusion of intact
normal and alloxan-diabetic rabbits indicated that 70% of the xylitol was
metabolized by the liver, 15% by the non-hepatic tissues, and 15% was
excreted in the urine[13]. Hepatic glucose production and plasma glucose
concentration were not influenced during xylitol infusion at 0.5 gm/kg/body

Table 2. The ability of various rat organ homogenates to dehydrogenate
xylitol in vitro[12]

| Organs | Nanomoles of Xylitol Dehydrogenated at 30°C Per Minute Per- | | |
	Mg Protein	Gm Fresh Weight	Total Organ
liver	24.7	2,670	27,000
kidney	6.5	855	2,138
testes	1.5	77	238
small intestine	0.26	7	76
heart	0.40	23	24
lung	0.13	13	20
epididymal fat pad	0.56	7	16
brain	0.10	3	10

weight/hr in normal rabbits, but increased slightly in diabetic ones. The greater amount of xylitol metabolized by the liver cells may be attributable to their higher degree of permeability to xylitol, a characteristic which is typical for liver cells and many other compounds.

ADAPTATION TO POLYOL INGESTION

Bässler studied the adaptive processes associated with the absorption and metabolism of xylitol[14]. One group of rats was adapted to xylitol by feeding a standard laboratory diet containing 10% xylitol for 10 to 14 days. Then these adapted rats and controls not previously exposed to dietary xylitol were intubated with 200 mg xylitol. Rats were killed at hourly intervals from 1.5 to 10.5 hours after intubation. The gastrointestinal tract was removed quickly and the contents analyzed for xylitol. After 4.5 hours significant differences were observed; in the complete gastrointestinal tract, 54% of the xylitol was still present in the non-adapted rats in comparison with 40% of the adapted rats. After 10.5 hours, 24% of the xylitol was still present in the non-adapted rats, but none was detectable in the gastrointestinal tract of the adapted rats. The half-time of absorption was 6.5 hours in non-adapted rats and 4.5 hours in adapted rats. Bässler acknowledged the lack of evidence of any active transport of xylitol in the absorptive process. He postulated that absorption of xylitol was controlled by the xylitol level of the blood and, therefore, that adaptive processes should be suspected at the level of cellular metabolism. He studied the half-life of xylitol in the blood of non-adapted and adapted rats, and concluded that the values were 31 and 19 minutes, respectively. He proposed that the cytoplasmic polyol dehydrogenase of the liver limited the turnover of xylitol. In tests to evaluate this hypothesis, he concluded that prolonged xylitol feeding increased the activity of polyol dehydrogenase. In other studies, he determined the rate of disappearance of xylitol in healthy adult men to be 23 minutes, with a total capacity for xylitol metabolism approaching 80% that of glucose or fructose.

Various other investigators have studied the adaptation of animals to increasing amounts of xylitol in the diet. For example, Hosoya and Iitoyo reported that rats readily accommodated to 20% xylitol in the diet by successive 5% increases at weekly intervals[15]. Under these conditions, the liver of rats adapted to the 20% xylitol diet and then maintained at that level contained 15.08 units of NAD-linked xylitol dehydrogenase per mg protein. This value was approximately twice that observed in control rats whose diet did not contain added xylitol.   In contrast, no increase was observed in cytoplasmic NADP-linked xylitol dehydrogenase in liver.

POLYOL ABSORPTION

Absorption of the polyols from the intestinal tract clearly appears to be by simple diffusion without facilitation by any transport system.   Dehmel, Förster, and Mehnert compared the absorption of the monosaccharides (glucose, galactose, and fructose), the disaccharides (maltose, sucrose, and lactose), and the polyols (sorbitol and xylitol) from the human duodenum and from the rat intestine[16].   In man the lumen of the duodenum was isolated by means of double balloons.   Samples were injected into the lumen between the balloons with 5 gm of the test substance in 100 ml of solution.   Samples were withdrawn after 15, 30, and 60 minutes.   Galactose and glucose were absorbed most rapidly and sorbitol and xylitol least rapidly.   The results were similar in the rat where the contents of the entire intestinal tract were analyzed after the subjects were sacrificed.   In both species, xylitol was absorbed somewhat more rapidly than sorbitol.   Since absorption is believed to be due to simple diffusion, the larger molecular weight of sorbitol (182) and its less symmetrical molecule than xylitol, whose molecular weight is 152, would result in the slower absorption of the sorbitol molecule across the intestinal wall.

BLOOD AND URINE POLYOL CONCENTRATIONS

The amounts of the polyols in the blood of individuals who are not being provided with supplemental quantities beyond the amounts present in foods are very low, to the point of being difficult to measure.   The normal serum concentration of any polyol does not exceed a few tenths or hundredths milligrams per 100 ml.   Pitkänen and Sahlström stated that the xylitol value in blood was 0.03 to 0.08 mg/100 ml[17].   When 10 gm of glucuronate in solution was ingested during a one-hour period, the xylitol concentration in the blood of five subjects varied from 0.14 to 0.23 mg/100 ml 30 minutes after the end of the glucuronate ingestion period.   Likewise, oral ingestion or infusion of ethanol at levels of 60 ml in 1.5 hours and 50 ml over 3 hours, respectively, resulted in increased blood xylitol levels. Urinary excretion of xylitol increased after either glucuronate or ethanol administration.   For example, in the two hours before the oral ingestion of 60 ml of ethanol, 1 to 4 mg xylitol were excreted in five subjects.   In the first three hours after ingestion, 18 to 40 mg xylitol were excreted and an additional 18 to 35 mg in the second three hours.

Pitkänen observed that 8 ± 5.5 mg (52.8 ± 36 μmoles) of xylitol were excreted daily in the urine of normoglycemic patients[18]. The comparable values for mannitol and sorbitol were 23.7 ± 21.2 mg (156 ± 139.2 μmoles) and 8 ± 5.1 mg (52.8 ± 33.6 μmoles). The amounts of xylitol excreted in diabetic persons were not significantly different from the normal, while individuals with chronic renal failure excreted only about half as much xylitol. In contrast, mannitol and sorbitol excretion were both increased in individuals with diabetes or with chronic renal failure. The concentration of xylitol in the blood is in the neighborhood of 3 to 5 mg/100 ml during ingestion of substantial amounts in the diet. Pitkänen gave the mannitol and xylitol concentrations in serum of individuals with normal renal function as 34 ± 18 and 18 ± 7 μmoles/liter, respectively. In seven individuals with chronic renal failure, he observed levels of 55 ± 24 and 37 ± 12 μmoles/liter for mannitol and sorbitol.

HUMAN TOLERANCE TO POLYOL INGESTION

Numerous studies on the parenteral use of sorbitol and xylitol in comparison with glucose and/or fructose have been conducted over the past 15 years, especially in Europe and Japan and to a lesser extent in the United States and Australia. These studies were conducted largely to find sources of energy that would not be subject to the problems inherent in the use of glucose parenterally and that would also be more suitable for use in individuals with diabetes. Most investigators have been of the opinion that sorbitol and xylitol could be used safely for parenteral nutrition as long as the usual precautions in intravenous feeding with glucose solutions were taken. These polyols at reasonable levels of infusion did not cause increases in blood sugar or insulin in man. Concern was expressed by Donahoe and Powers about the use of xylitol[19]; a single two-hour infusion of 25 gm xylitol caused an elevation of serum uric acid, while a 100 gm infusion in two hours caused an even greater elevation. In both situations, despite the elevations, the serum uric acid levels were not beyond the normal range. Likewise, Australian workers expressed concern about adverse effects of xylitol infusion which included elevated serum uric acid, liver and cerebral disturbances, diuresis and calcium oxalate deposition[20]. The level of xylitol administration was not stated for most of the 22 subjects. In one case (#19), the patient received 200 gm of xylitol in a period of 2.5 hours, which is well beyond the level used by most investigators. In addition, several of the patients had terminal illnesses which complicated

the evaluation of the findings.   Three of the Australian workers reviewed
their continuing concerns about xylitol in parenteral usage[21].   Apart from
the above, the consensus is that levels of these polyols, sorbitol and
xylitol, not in excess of 0.25 g/kg body weight/hour can be given intravenously for periods of at least 48 hours as long as such solutions are
fortified with nutrients in the same fashion as is practiced for glucose
solutions.   Parenteral usage at rates of 0.25 g/kg/body weight/day would
mean the administration of 105 gm per day for a 70 kg man.

Brin and Miller recently reviewed the tolerance of humans to oral
ingestion of xylitol[22].   Three studies are of particular interest.   Two of
these, conducted in Switzerland and Germany, are described below[23,24]; the
third and the most extensive in time and number of subjects was that from
the Turku sugar studies described by Mäkinen[25].

Dubach et al. studied the oral tolerance for xylitol in three trials, in
the last of which xylitol and sorbitol were compared[23].   In 19 non-diabetic
students, an initial dose of 5 gm/day was increased in 5 gm increments to
the 14th day and then continued for an additional seven days at the 75 gm/
day level.   Intolerance was not observed except for frequent diarrhea in
two cases.   In the same group one month after termination of the first
trial, xylitol ingestion was started at 40 gm xylitol in one dose.   Then in
10-gm increments, the amount was increased as the individual was able to
tolerate the xylitol intake; by the end of 14 days, 12 were ingesting 150 gm
per day without adverse influence.   During the following week some individuals were able to tolerate higher amounts with two ingesting 200 gm and
one 220 gm/day.   Diarrhea was fairly common above 130 gm/day.   Amounts in
excess of 40 gm at one time were not tolerated well; xylitol consumption at
this level more frequently than every three to four hours gave unpleasant
side effects.   In the comparison of xylitol with sorbitol, 26 students
ingested the two polyols in consecutive intervals beginning with a 5 gm dose
per day, increasing to 75 gm by 5-gm increments over a 14-day period.
Twenty-one of the 26 preferred xylitol to sorbitol because of the much
higher frequence of diarrhea and flatulence with sorbitol at the same levels
of ingestion.

Förster et al.[24] studied the tolerance of xylitol among diabetic
children in the desire to find a sweetening agent which was safe and which
would obviate the necessity for the prohibition of all sweets as is
necessary when dietary sweetness is solely dependent upon sucrose, glucose
and fructose, or foods containing these sugars.   They sought a more

desirable sweet substance than sorbitol, which is only one-half as sweet as sucrose and also causes osmotic diarrhea at relatively small levels of intake. For a period of four weeks, 30 g xylitol was used daily in food with medical examination before use was begun and two and four weeks thereafter. Twenty-four children five to 15 years of age were studied. Only one child had a problem with diarrhea and was dropped from the experiment. Three children considered the xylitol too sweet. Data were presented for the 18 children for whom complete data were available. Their body weights varied from 21 to 63 kg. Insulin requirement and glucose excretion were not altered during the trial. Significant increases in uric acid, protein, and phosphate concentration in the serum were observed ($p <$ 0.05). No change in urinary excretion of uric acid, urea, or creatinine were observed. The authors pointed out that these children had been on a low sucrose diet because of their diabetic condition prior to this clinical trial. Under similar conditions in other trials, the daily administration of fructose or sucrose caused comparable increases in uric acid concentration. Indeed, elimination of sucrose from the diet of healthy volunteers has been shown to result in reductions in serum uric acid concentration.

The third study[25] on the oral tolerance will not be dealt with here in detail because of Mäkinen's participation in this conference, other than to note that the average daily consumption of xylitol over a two-year period was about 50 gm in comparison to 73 gm for sucrose and 70 gm for fructose. More than 200 gm per day of xylitol was consumed in 64 of the experimental days with a high of 430 gm; more than 200 gm of fructose was consumed on 117 days.

In the overall consideration of oral use of the polyols, the regulation of their absorption from the gastrointestinal tract reduces the likelihood of the introduction of excess amounts into the circulatory system, unlike the situation in intravenous administration.

POTENTIAL FOR GREATER USE OF POLYOLS IN HUMAN CONSUMPTION

The question of whether minor or major replacement of sucrose and other dietary sugars by polyols in selected food and confectionery items or in the majority of sugar-containing foods will reduce or prevent dental caries in highly susceptible populations of children has yet to be answered. On the basis of our present knowledge, the likelihood of substantial benefit seems to be reasonably good. On the basis of microbiological data, sorbitol and mannitol are metabolized at very low rates by oral microorganisms in comparison to the sugars. Xylitol appears to be less well-utilized by cariogenic microorganisms, and those microorganisms which do metabolize this pentitol produce less acid than the metabolism of comparable amounts of sugars. The information from animal experiments in general suggests that the polyols are not as cariogenic as the sugars, although some data raise questions about this generalization. Completely satisfactory testing of the polyols in rodent experiments is difficult because assays are designed to maximize caries production. The low levels of polyols which can be tolerated by experimental animals in comparison with the 60 to 70% sucrose customarily used in cariogenic diets, coupled with the need for adaptation to polyol ingestion and the short caries assay period, reduce the ability to obtain a definitive assay on the relative cariogenicity of the polyols in rats or hamsters in comparison to the sugars.

If one or more of the polyols prove to be appreciably less cariogenic than sucrose and the other sugars which are commonly used in human nutrition today, or if any of the polyols are unable to support the carious process or even are anti-cariogenic, the potential for increased use of such polyols in various human foods and confections appears to be considerably greater than is currently true. In my opinion, conclusive proof is not yet available about the exact relationship of any polyol to the initiation and progression of carious lesions in man, the minimum amount and frequency needed to achieve the maximal result, or the optimal system for delivery of the agent. These subjects lend themselves to clinical investigation, albeit arduous and expensive. However, the potential for significant contributions to the control of dental caries is sufficiently great that several additional clinical investigations on the polyols are certainly appropriate beyond those which have been completed or are currently underway. In the absence of answers to the above uncertainties, the following thoughts and specula-tions about the potential for greater use of one or more polyols in human foods and confections have to be based on the amounts that the human body can metabolize safely, the taste characteristics, and the cost relative to other sweeteners.

First, it is necessary to state clearly what the polyols cannot be expected to do as sweeteners. Since the polyols enter metabolic pathways with the production of energy at similar levels to the sugars, they are not useful in situations where sweetness is desired with a simultaneous reduction in calories. Also, the polyols do not qualify as intensely sweet compounds; the level of sweetness for xylitol is essentially the same as sucrose, while sorbitol and mannitol are substantially less sweet than sucrose. Thus to attain the same degree of sweetness in a product, polyol replacement for sucrose would be by an equal weight in the case of xylitol or by a greater weight in the case of sorbitol or mannitol. Consumer acceptability evidently was not a problem for the xylitol-sweetened products in the two Turku investigations. However, some products sweetened with sorbitol, or with sorbitol and mannitol mixtures, have allegedly not received adequate consumer acceptability because the characteristics of the sweetness were not considered by consumers to be suitable enough to pur-chase instead of the comparable sucrose-sweetened products. Customarily replacement of sucrose by sorbitol, or by mixtures of sorbitol and mannitol, has been accompanied by the use of some saccharin to make up for the lower sweetness levels of these two hexitols and the need to give sweetness characteristics that are more acceptable to the consumer. Since polyols yield about the same amount of calories as the sugars and are not intensely sweet, their use does not require the search for and addition of a suitable bulking agent to maintain the customary size and weight of a product. The polyols in this regard are unlike the intensely sweet compounds where their use will require the introduction of some material to increase the bulk of solid products or the viscosity of fluids.

Our considerations about the potential use of polyols in human consump-tion need to include the range from the low levels to be expected with their use in a confection, such as chewing gum, to the much higher levels of consumption if the polyols are used as sweetening agents in a wide variety of foods and confectionery items.

Low Levels of Polyol Ingestion

In the case of chewing gum, clinical trials with products sweetened by sorbitol indicate that these products do not increase the caries incidence above levels in control subjects, but neither do they decrease the caries incidence below levels in the controls. In contrast, the results of the Turku chewing gum study indicate that the xylitol-sweetened product may not have been just non-cariogenic but anti-cariogenic when chewed 4.5 times daily. In any case, the polyols used in these studies were at the level of about 1.5 to 2.0 gm per stick or chicle. No attempt was made to alter the amount and type of food intake of the experimental subjects from their usual dietary customs. If five sticks of polyol-sweetened gum were chewed daily, about 7.5 to 10.0 gm of polyol would be ingested. This level for xylitol, sorbitol, and mannitol is well below the amount readily absorbable and metabolizable daily by the human body. There would be no reason to expect any adverse influence on the human body at this level of consumption. Indeed, no adverse effect would be expected in children who chewed three or four times as many sticks of polyol-sweetened gum daily.

Additional studies on xylitol consumption in limited amounts, as in this chewing gum experiment, are needed to determine if the Turku results on dental caries can be reproduced during different circumstances by other investigators. In further studies, particular care should be given to the selection of populations of a younger age when the caries susceptibility is at its peak and a larger number of tooth surfaces per individual are available to decay. If these studies yield results which are even reasonably similar to the striking reductions in the Turku studies, strong impetus would be given to the introduction of xylitol into chewing gum and its widespread recommendation of use in our caries-susceptible population. It should be mentioned that the level of dental caries reduction reported in the Turku chewing gum experiment does not appear to be explainable in any way on the basis of our current understanding of the relationship of sugar consumption to the caries process. However, this inability to explain why xylitol appears to be so effective in the presence of all the sugar supplied from dietary sources other than chewing gum should not be a deterrent to further investigation and widespread usage if positive and confirmatory results are obtained.

Higher Levels of Polyol Ingestion

Sorbitol and mannitol have not been used to any extent as sweeteners in products in common use, apart from chewing gum. Apart from the Turku dietary study, evaluations of the relation of polyols in foods and confections other than chewing gum have not been conducted. In the case of the two-year Turku dietary study, the sucrose in about 100 widely different products was replaced by either xylitol or fructose, without undesirable changes in taste or appearance. The average sucrose consumption was 2.2 kg/month (73 gm/day) in comparison with 1.5 kg/month of xylitol (50 gm/day). This experiment was a monumental endeavor, not only because of the many products that were available for the participants but also because of the numerous biochemical evaluations which were made during the study. The striking reduction in dental caries in the xylitol group in comparison to the fructose or sucrose groups during the two-year period fits in well with the current widely-held consensus: namely, that major reductions in the fermentable carbohydrate consumption cause major reductions in dental caries experience, especially if reductions occur in the frequency of use of those foodstuffs which are retained in the oral cavity. Clearly this dietary study represents the clinical evaluation of a whole family of sweetened products and was not designed to and cannot give specific information about a specific product or type of product.

Clinical trials need to be conducted on candy and other confectionery items to evaluate what level of xylitol in a line of products will give the maximal benefit in the presence of sucrose and other sugars. Such trials will be difficult to conduct and could not be limited to a single candy or other product. Instead, the appropriate procedure seems to be to prepare a series of different confectionery products sweetened completely or partially with xylitol and make these available as the sole sources of confection to be used by the experimental population. The control subjects would not be influenced to alter the choice, quantity or time of consumption of products that they ordinarily use.

No physiological obstacles seem to prohibit substantially greater use of polyols in human foods until levels of intake are reached where osmotic diarrhea is produced due to the slow absorption of the polyols from the intestinal tract. Over the long term, intake of 30 to 50 gm/day of sorbitol or mannitol or 50 to 70 gm/day of xylitol or some combination appears to be feasible for human teenagers and adults. Some adaptation to these levels probably would be needed by most individuals. In view of the likelihood that polyol-sweetened foods and confections would be introduced to the marketplace slowly, the adaptation from the current small to such substantially increased intakes would occur automatically over periods of weeks or months. Most individuals probably could consume greater than the above amounts of xylitol, based on the results in the Turku dietary study and other tolerance trials. In view of the current average levels of usage in the United States at around 125 gm/day of sucrose (100 lbs/year) and 25 gm/day of other sugars, theoretically one-half or more of current sugar usage could be replaced by polyols without abnormal reaction by normal individuals.

Cost of Sugar Replacement by Polyols

If the replacement of small amounts of sucrose or other currently used dietary sugars by xylitol, and/or the replacement of sugars completely or partially by xylitol in particularly cariogenic items, prove to be effective anti-caries measures, the cost of xylitol does not appear to be prohibitive in view of the major benefit to be derived. However, if major replacements of sucrose by xylitol, as in the two-year dietary study in Finland, are necessary for the prevention of dental caries, the current and projected costs of xylitol appear to be beyond what all but the most concerned and affluent consumer can be expected to bear. If the less-expensive polyols, sorbitol and mannitol, prove with suitable testing to be non-cariogenic sweeteners, their lower cost would make major substitution for sugars feasible, providing that consumer acceptability can be obtained with regard to flavor. The latter may be a major problem if saccharin is withdrawn from the market and no suitable intense sweetener is available to replace it.

At present, xylitol appears to be from many aspects other than cost to be the polyol of choice as a sweetener in the range of products for which it is suitable. Xylitol's cost at present makes it no competitor for sugars in the field of economics. Xylitol at present, relatively low production levels costs about $6.00/kg in bulk. Thus the cost to the manufacturer of 1.5 gm xylitol in a stick of chewing gum would be about one cent, and would result in an approximate doubling of the cost of chewing gum to the consumer. If the product is not just non-cariogenic but, indeed, anti-cariogenic, this increase in cost should not be prohibitive to well-informed consumers. However, if a two-ounce candy bar containing 50% sucrose or other sugar had to have all its sugar replaced by xylitol to be non-cariogenic, the xylitol alone would cost the manufacturer about 17 cents at the current price. However, if only 20% of the sucrose had to be replaced to produce a non-cariogenic candy bar, the cost of the xylitol would be about seven cents. At the extreme end of the spectrum, if the sucrose in many food and confectionery products had to be replaced by xylitol, as in the Turku dietary study, in order to have a low-caries-producing human dietary, the cost of the xylitol at the food processor's level would cause prohibitively large increases in food costs to be useful on a population-wide basis.

If a greatly increased demand for xylitol should develop, production costs probably could be reduced appreciably. However, there appears to be no likelihood in the foreseeable future that xylitol can become competitive on a simple cost basis with sucrose or other sugars. The enormous yields of sucrose per acre from sugar cane or sugar beets and the relative ease with which the sucrose can be extracted and purified keep its cost low. Likewise the ease of producing glucose or fructose from corn enable their prices to be competitive with sucrose. While such raw materials as wood chips or sawdust, cottonseed, oat hulls, and corn cobs, by-products of other industries, are plentiful, the extraction of xylan, hydrolysis to xylose, oxidation to xylitol, and thorough purification seem likely to keep xylitol considerably more expensive than sucrose for the foreseeable future. An offshoot, of course, would be a new source of food for man from otherwise unavailable materials nutritionally.

From the nutritional standpoint, concern must be expressed that the polyols, like the mono- and disaccharides, are ordinarily supplied for food manufacturing in relatively pure form. As such they provide calories but do not provide other nutrients -- protein, essential fatty acids, vitamins, minerals, or dietary fiber. Therefore, the incorporation of sugars and/or polyols into foods dilutes the other nutrients proportionately; the remaining components of the diet must provide a proportionately greater amount of the nutrients other than calories. "Empty" calories is not an ideal term for the calories supplied by the sugars and polyols because of the complete dependence of all bodily functions upon an adequate source of energy. From a nutritional standpoint, calories from such sources are in themselves readily usable and not undesirable per se; the problem nutritionally arises as the amount of food components providing only calories is increased to the point where the other foods cannot supply sufficient amounts of the other nutrients. This point is likely to be reached at a lower level of sugar and polyol ingestion where the individual makes no attempt to plan a wise diet of varied and nutritious foodstuffs than in individuals who strive to plan and adhere to a wise diet plan. The current average level of caloric consumption of sugars from all sources is about 24% in the United States, of which about 6% is from sugars in milk, fruits, and vegetables. Thus 18% of the caloric supply on the average is from sugars of highly purified nature with very low amounts of any other nutrients than calories. In order to provide for nutritional needs, the foods which provide the remaining 81.5% of the calories must supply all nutrients other than calories. Obviously, for individuals whose consumption of sugars averages one or two standard deviations more than the average, the foods providing the remaining calories must provide proportionately more nutrients per unit of food consumed. Current food fortification procedures cannot counteract this problem completely, since only the several best-recognized vitamins and minerals are increased, and adequate amounts of the rest are still dependent upon what is consumed in food.

REFERENCES

1. Touster, O. in Sugars in Nutrition, edited by H.L. Sipple and K.W. McNutt, Academic Press, New York, pp. 229-239, 1974.
2. Touster, O., Reynolds, V.H., and Hutcheson, R.M. J. Biol. Chem. 221, 697-709, 1956.
3. Hollmann, S. and Touster, O. J. Biol. Chem. 225, 87-102, 1957.
4. Arsenis, C. and Touster, O. J. Biol. Chem. 244, 3895-3899, 1969.
5. McCormick, D.B. and Touster, O. J. Biol. Chem. 229, 451-461, 1957.
6. Hollmann, S. and Touster, O. Non-glycolytic Pathways of Metabolism of Glucose, Academic Press, New York, p. 107, 1964.
7. Lang, K. in Metabolism, Physiology and Clinical Use of Pentoses and Pentitols, edited by B.L. Horecker, K. Lang, Y. Takagi, Springer-Verlag, Berlin, pp. 151-157, 1969.
8. Bässler, K.H., Toussaint, W., and Stein, G. Klin. Wschr. 44, 212-215, 1966.
9. Bässler, K.H., Prellwitz, W., Unbehaun, V., and Lang, K. Klin. Wschr. 40, 791-793, 1962.
10. Prellwitz, W. and Bässler, K.H. Klin. Wschr. 41, 196-199, 1963.
11. Mehnert, H., Summa, J.D., and Förster, H. Klin. Wschr. 42, 382-387, 1964.
12. Bässler, K.H., Stein, G., and Belzer, W. Biochem. Z 346, 171-185, 1966.
13. Müller, F., Strack, E., Kuhfahl, E., and Dettmer, D. Z. Ges. Exp. Med. 142, 338-350, 1967.
14. Bässler, K.H. in Metabolism, Physiology, and Clinical Use of Pentoses and Pentitols, edited by B.L. Horecker, K. Lang, and Y. Takagi, Springer-Verlag, Berlin, pp. 190-196, 1969.
15. Hosoya, N. and Iitoyo, N. in Metabolism, Physiology, and Clinical Use of Pentoses and Pentitols, edited by B.L. Horecker, K. Lang, and Y. Takagi, Springer-Verlag, pp. 197-200, 1969.
16. Dehmel, K.H., Förster, H., and Mehnert, H. in Metabolism, Physiology and Clinical Use of Pentoses and Pentitols, edited by B.L. Horecker, K. Lang, and Y. Takagi, Springer-Verlag, Berlin, pp. 177-181, 1969.
17. Pitkänen, E. and Sahlström, K. Ann. Med. Exp. Fenniae 47, 143-150, 1968.
18. Pitkänen, E. Clin. Chem. Acta 38, 221-230, 1972.
19. Donahoe, J.F. and Powers, R.J. J. Clin. Pharm. 14, 255-260, 1974.
20. Thomas, D.W., Edwards, J.B., Gilligan, J.E., Laurence, J.R., and Edwards, R.G. M. J. Aust. 1, 1238-1246, 1972.
21. Thomas, D.W., Edwards, J.B. and Edwards, R.G. in Sugars in Nutrition, edited by H.L. Sipple, and K.W. McNutt, Academic Press, New York, pp. 567-590, 1974.
22. Brin, M. and Miller, O.N. in Sugars in Nutrition, edited by H.L. Sipple and K.W. McNutt, Academic Press, New York, pp. 591-606, 1974.
23. Dubach, U.C., Feiner, E., Forgó, I. Schweizer Med. Wschr. 99, 190-194, 1969.
24. Förster, H., Boecker, S., and Walther, A. Fortschritte der Medizin 95, 99-102, 1977.
25. Mäkinen, K.K. Int. J. Vit. Nut. Res. Suppl. 15, 92-104, 1976.

POSTSCRIPT

After this conference ended and the above manuscript was typed for publication, preliminary information from toxicity tests being conducted at the Huntingdon Research Center, Huntingdon, England under contract with Hoffmann-La Roche, Inc. became available. These data were forwarded prompt-

ly by Hoffmann-La Roche, Inc. to the Bureau of Foods and the Bureau of
Drugs of the Food and Drug Administration. I am indebted to Dr. M. Brin
for a copy of the Investigational Drug Brochure pertaining to xylitol
(RO 6-7045) dated November 22, 1977 that contains the preliminary results
on these toxicity tests. The following information on the mice, rat, and
dog studies has been carefully excerpted from the Hoffmann-La Roche brochure.

Unlike most toxicity tests where only the test agent is evaluated,
groups of experimental animals were also included where either sucrose or
sorbitol were simultaneously compared to the controls. The preliminary data
at the conclusion of these two-year toxicity studies with mice, rats, and
dogs warrant an updating of my conclusions on the safety of xylitol for man
and serve as an example of the problems associated with current animal evaluations of toxicity that were referred to in the discussions after Dr. Ronk's
paper.

Mice. In the chronic study with mice, xylitol was evaluated at levels
of 2, 10 and 20% in the diet. In addition to the untreated controls, an
additional group of mice was fed 20% sucrose in the diet. An increased incidence of calculi was observed in the urinary bladders of male mice fed the
10 and 20% levels of xylitol. Associated with the calculi in some mice,
inflammatory changes, epithelial hyperplasia and malignant neoplasms of the
urinary bladder were observed. "According to an expert pathologist on
bladder tumors, the findings are therefore entirely consistent with the hyperplasia and neoplasia being a non-specific consequence of stone formation
and it is most unlikely that the test agent (xylitol) is a primary cargino-
gen for the bladder." In addition, 20% dietary sucrose was associated with
an increased incidence of renal tumors at a borderline level of significance
in comparison to both the control group and the xylitol groups.

Rats. In the two-year chronic study in rats, xylitol was fed at levels
of 2, 5, 10, and 20% in the diet. In addition, controls included a group
fed 20% sorbitol in the diet, and another fed 20% sucrose in addition to the
group fed only the control diet. Neither calculi or neoplasms of the urinary bladder were observed in the xylitol treated rats--a clear difference
from the mice in the parallel study. Adrenal hyperplasia was significantly
increased in males fed 10 and 20% xylitol, in females fed 5, 10, and 20%
xylitol and in rats of both sexes fed 20% sorbitol. An increase in adrenal
medullary neoplasia was observed in male rats fed 20% xylitol; this increase

was beyond the levels routinely seen in untreated control rats and in other treated rats of this study.

Dogs. No abnormalities of the urinary bladder or other treatment-related neoplasia were observed in the preliminary findings after two years.

Thus with the availability of these preliminary data, xylitol is at least temporarily added to the substantial list of compounds that directly or indirectly have caused cancer in mice. Under the same conditions in a parallel group of mice, sucrose at 20% in the diet was associated with a borderline increase in renal tumors. The final evaluations for these toxicity tests are awaited with interest. If the preliminary data are substantiated and today's standards for evaluating safety with respect to cancer are uniformly applied, then both xylitol and sucrose become suspect as carcinogens. The Food and Drug Administration likes to have toxicology tests conducted at levels up to 100 times the likely level of human ingestion. In these xylitol and sucrose tests the preliminary data from the mice suggest the possibility of direct or indirectly induced pathology at the 20% levels; the highest test level was thus approximately the same as the average sucrose consumption of 18% calories currently in the United States.

It is clearly too early for final judgments. However, anyone concerned about the evaluation of the safety of compounds for human use must wonder whether this is a true evaluation that sucrose is dangerous to health in ways other than in the established relationship to dental caries, or whether this is evidence that the rodent tests are producing false positives and cannot be relied upon for the evaluation of agents for human use. If the rodent tests are demonstrated to be inadequate, what tests can be developed to replace them for the determination of acceptable risk?

Xylose: a modifier of animal cell phenotype

George Ev. Demetrakopoulos, Michael S. Radeos, and Harold Amos

Department of Microbiology and Molecular Genetics, Harvard Medical School, 25 Shattuck Street, Boston, Massachusetts 02115

INTRODUCTION

D-xylose (XE) and its reduction product, xylitol (XL), are among those sweeteners which have attracted increased attention from the medical community recently[1-3]. In the past both were associated with the "dietetic fiber" and had been considered as non-utilizable[4], inert[5], useless[6], and even harmful[4,7]. Both played a significant role in the human diet long before we became so discriminating in our dietary habits. Both D-xylose and xylitol are intermediary products of normal human metabolism and are responsible for a variety of important metabolic processes[2].

Furthermore XE, and to a lesser degree XL, have been found capable of supporting excellent cellular growth (in vitro) when present as sole carbon and energy sources[8,9]. Both have been shown to exert characteristic effects on cellular morphology[8-10], physiology and metabolism not only when present as the sole carbon source but also when provided in combination with other carbon sources[8-10]. Certain of those effects have been reported earlier. In this communication XE and XL effects on mucopolysaccharide production, acidogenesis and cellular ATP levels are presented. Furthermore the mode of their transport, utilization and metabolism at the cellular level (in vitro) are reviewed in summary fashion.

MATERIALS AND METHODS

The methodology and materials used in the experimental work described in this paper have been already reported in earlier communications from this laboratory[8-11]. Specific experimental conditions are given in the legend under each figure or table.

RESULTS AND DISCUSSION

XE and XL Transport   In the small intestine (lower duodenum and uppermost jejunum) XE is transported across the mucosa by an active mechanism[12] sharing the same carrier with glucose and galactose[12]. XL is absorbed primarily by diffusion[13-14], downhill transport, which explains its slow absorption rate as well as its ability to be absorbed by any part of the gastrointestinal tract including the colon[14]. XE transport into human erythrocytes is also mediated by an active mechanism[2,15], probably the same one as glucose.

In several species XE/inulin clearance ratios ranged from 0.6 to 0.82, but when plasma glucose was raised to 300 mg %, a high enough concentration to saturate the glucose reabsorbing mechanism, this ratio increased to 1.0[16]. It appears therefore that D-XE can be actively reabsorbed in the renal tubules, but this is partially inhibited by glucose, which has a much higher affinity for the carrier.

Although no reabsorptive mechanism has been found for XL in the kidneys[13] far less than expected is excreted in the urine, probably due to the fast diffusion of XL from blood to the tissues[13].

The mode of entry of D-XE and XL into fibroblasts in culture does not demonstrate saturation kinetics even at concentrations as high as 50 mM[2,8,9]. This finding suggests that either the entry occurs by a diffusion mediated process or that a carrier has a Km so high that its saturation does not occur even at 50 mM of Pentose[2,8,9]. Similar findings have been reported by Kipnis and Cori on isolated rat diaphragm[17]. In accordance with the above hypothesis XE does not demonstrate competition for uptake with glucose and vice-versa[11]. Indeed, in the presence of low concentrations of XE the glucose uptake by fibroblasts in culture is almost doubled (Table 1); a similar result was obtained in slices of rat liver in vitro[18].

None among several pentoses, hexoses, pentitols and hexitols tested demonstrated competition for uptake with XE or XL. Furthermore, neither the presence of energy depleting agents (such as DNP or NaN$_3$) nor of thiol reagents affected XE uptake by fibroblasts in culture (Table 2). Finally insulin that has been shown to stimulate by 10 to 15-fold the glucose uptake by fibroblasts in culture has no significant effect on XE uptake[11] (Table 3). A non-specific effect of insulin on XE uptake was reported earlier on various tissue preparations[18] and in intact animals[15].

Effects on Transport of Glucose and Amino Acids   Shaw and Amos reported earlier a several fold derepression of glucose uptake in fibroblasts deprived of a carbon source[19]. Fibroblasts grown on XE or XL demonstrate a similar

Table 1: Simultaneous entry of ^{14}C-D-xylose and ^{3}H-D-glucose into chick embryo fibroblasts in culture

Picomoles/mg prot./5 min.			
^{14}C-D-Xylose		^{3}H-D-Glucose	
(5 x 10^{-5}M)	(5 x 10^{-4}M)	(5 x 10^{-5}M)	(5 x 10^{-4}M)
200 ± 15	--	--	--
--	1645 ± 40	--	--
--	--	305 ± 10	--
--	--	--	1040 ± 20
173 ± 2	--	488 ± 50	--
--	1718 ± 100	290 ± 15	--
172 ± 20	--	--	1629 ± 100

Primary cultures of chick embryo fibroblasts were grown from small inocula in multiwell tissue culture plates in Eagle's minimal medium[8,26] supplemented with glucose (1 mg/ml) and 5% fetal calf serum. Sugar uptake assays were performed for 5 min. as previously described[8,9]. The values presented are the results obtained in a single experiment (3 samples per point). They are representative of three such experiments.

Table 2: Effects of thiol reagents on glucose and xylose uptake

Inhibitor	Picomoles/mg prot./5 min.	
	^{14}C-D-glc 2 x 10^{-4}M	^{14}C-D-xyl 3 x 10^{-4}M
--	6289 ± 900	968 ± 64
P-chloro-Hg-benzene-sulphonic acid (10^{-3}M)	5602 ± 1800	916 ± 82
N-Ethylmaleimide (10^{-3}M)	1632 ± 200	755 ± 60

Primary chick embryo fibroblasts were grown as described under Table 1. At near confluency the medium was drained, the monolayers washed with phosphated buffer saline (PBS), and reincubated (at 37°C) in Hanks' balanced salt solution[27] containing P-chloro-Hg-benzene-sulphonic acid (10^{-3}M), N-ethylmaleimide (10^{-3}M) or nothing. At 5 minutes the labeled sugars, ^{14}C-D-glucose and ^{14}C-xylose, respectively were added to the media and the monolayers were left in the incubator for an additional 5 minutes.

Table 3: The effect of insulin on glucose and XE uptake by "preconditioned"[15] chick embryo fibroblasts in culture

Insulin units/ml	^{14}C-xylose		^{14}C-glucose
	10^{-4}M	3×10^{-4}M	2.5×10^{-5}M
	Picomoles/mg prot./5 min.		
0	298	1030	400
0.05	305	1461	1750
0.1	367	1431	1975

Primary chick embryo fibroblast cultures were grown to near confluency in BME supplemented with glucose (1 mg/ml) and fetal calf serum (5%). At that point the monolayers were transferred to glucose-free BME serum-free but supplemented with insulin as indicated above. 24 hours later glucose and xylose uptake were assayed as previously described[8,9].

derepression for glucose uptake[8,9]. A 2 to 3 fold increase in the uptake of glutamine and certain amino acids has been observed in cells deprived of sugar. The amino acid and glutamine transport of cells cultivated on XE or XL is not significantly different from that of cells grown on glucose (Table 4).

Effects on Cellular Growth and Phenotype   In previous communications Amos and his associates have reported the striking morphological changes occurring in cells grown in XE or in the combination of XE and glucose compared to the same cells grown on glucose[8,12]. XE has consistently been found to be capable of supporting excellent cellular growth when present as the sole carbon source. This is true even when the xylose media is pretreated with glucose oxidase in order to abolish any contaminating trace of free glucose (unpublished data). On the contrary XL has not proven consistently supportive of good cellular growth when present as the sole carbon and energy source. Certain factors such as careful monitoring of the pH of the medium, supplementation with pyruvate (1 mg/ml) as well as preconditioning of the cells in media containing both glucose and XL may influence XL ability to support cellular growth.

Acidogenesis   The amount of both lactate and pyruvate that accumulates in the medium of cells growing on XE or XL is significantly (10 to 20 fold) lower than that of cells growing on glucose alone or on the combination of glucose and xylose[2].

Cellular ATP Levels   A sharp reduction in cellular ATP content is observed

Table 4: Amino acid uptake as a function of carbon source

	Picomoles/mg prot./2 min.			
Glutamine	373 ± 20	510 ± 63	483 ± 72	840 ± 36
φ'alanine	4900 ± 1100	6300 ± 900	6800 ± 300	17050 ± 2700
Preconditioning				
glc	+	-	-	-
XE	-	+	-	-
XL	-	-	+	-

Primary chick embryo fibroblast cultures were "preconditioned" on glucose, XE or XL media and uptake measured 24 hours later. Amino acid and glutamine uptake were assayed as reported earlier[8,9].

immediately after the cell culture is shifted from glucose containing medium to medium containing XE or XL in the absence of glucose. Such reduction does not occur when normal cells are shifted from glucose to a medium free of carbon source[19]. After 24 hours growth on XE medium the steady state ATP levels of chick embryo fibroblasts rise to equivalence with those of cells grown on glucose, while the ATP levels of Nil hamster fibroblasts become comparable to those of glucose grown cells only after 48 hours. (See Figures 1 and 2.) Inhibitory effects of XE on hexokinase[20] have been cited as accounting at least in part for the loss of ATP by erythrocytes and other animal tissues[21].

Mucopolysaccharide Synthesis   XE occupies a key position in the chondroitin-4-sulfate molecule[22]. Actually the carbohydrate-protein linkage is a β-glycosidic linkage between D-xylose (pyranose form) and the hydroxyl group of L-serine of the protein-core (xylosylserine)[22]. It is firmly established that xylose plays a similar role for several other mucopolysaccharides (heparin, dermatan sulfate, etc.) and glycoproteins[22,23]. Growth on XE alone or on a combination of glucose and XE causes a several fold increase in the synthesis and net production of soluble and cellular glycosaminoglycan[24]. This result confirms and extends earlier reports, in which exposure of glucose grown cells to high XE concentrations (42 mM) caused enhancement of both glycosaminoglycan and collagen production[22,25]. Similar overproduction has been observed in cells grown in XL-containing medium[24]. The stimulatory effect of XE on mucopolysaccharide production although somewhat compromised, is still evident when glucose is provided as a co-carbon source.

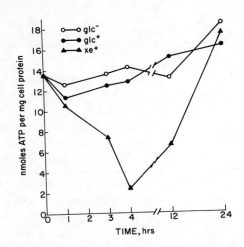

Figure 1. ATP concentrations in CEF cells starved for glucose and given D-xylose (6 mM) in lieu of glucose. Cells in primary culture were grown on glucose-containing medium until near confluence. The monolayers were washed and the cells were reincubated with BME medium containing glucose (5.5 mM) ●——●; no sugar o——o; and D-xylose ▲——▲ replacing glucose. ATP was measured by the luciferase-luciferin assay as described by Wilson et al.[28] at intervals during 24 hours.

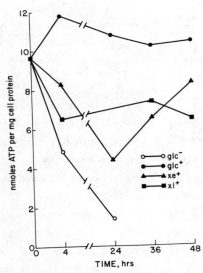

Figure 2. ATP concentrations in Nil cells starved for glucose and with D-xylose (6 mM) or xylitol (6 mM) in lieu of glucose. Nil cells were grown on MEM containing glucose as carbon source. The monolayers when 2/3 confluent were washed and reincubated with medium containing glucose (5.5 mM) ●——●; xylitol (6 mM) ■——■ ; D-xylose (6 mM) ▲——▲ ; and no sugar o——o.

ACKNOWLEDGMENTS

George Ev. Demetrakopoulos is a National Institutes of Health Post-doctoral Awardee (NIAMDD). This work was supported by National Cancer Institute contract NO1-CM53825 and N.I.H. grant RO1-CA-19015-01.

REFERENCES

1. Sipple, H.L. and McNutt, K.W., Ed. Sugars in Nutrition, Academic Press, New York, 1974.
2. Demetrakopoulos, G. Ev. and Amos, H. World Rev. Nutr. Diet. in press, 1977.
3. Scheinin, A. and Mäkinen, K.K., Ed. Suppl. 70, 1975.
4. Burns, R.L., Rosenberger, P.G. and Klebe, R.J. J. Cell. Physiol. 88, 307-316, 1976.
5. Miller, M.M. and Lewis, H.B. J. Biol. Chem. 98, 133-150, 1932.
6. Snowden, J.C. in The Carbohydrates, edited by W. Pigman, Academic Press, 1957, New York.
7. Food & Drug Administration Drug Bulletin: D-xylose approved as diagnostic agent, Washington D.C. 5:11, July-Aug. 1975.
8. Demetrakopoulos, G. Ev. and Amos, H. Biochem. Biophys. Res. Comm. 72, 1169-1178, 1976.
9. Demetrakopoulos, G. Ev., Gonzalez, F., Colofiore, J.C. and Amos, H. Exp. Cell Res. 106, 167-173, 1977.
10. Amos, H., Leventhal, M., Chu, L. and Karnovsky, M. Cell 7, 97-103, 1976.
11. Demetrakopoulos, G.Ev., Linn,B. and Amos, H. Manuscript in preparation
12. Alvardo, F. Biochem. Biophys. Acta 112, 292-306, 1966.
13. Lang, K. in Metabolism, Physiology and Clinical Use of Pentoses and Pentitols, edited by B.L. Horecker, K. Lang, Y. Takagi, Springer-Verlag, Berlin, pp. 151-157, 1969.
14. Bässler, K.H. in Metabolism, Physiology and Clinical Use of Pentoses and Pentitols, edited by B.L. Horecker, K. Lang, Y. Takagi, Springer-Verlag, Berlin, pp. 190-196, 1969.
15. Horecker, B.L., Lang, K. and Takagi, Y. Metabolism, Physiology and Clinical Use of Pentoses and Pentitols, Springer-Verlag, Berlin, 1969.
16. Wyngaarden, J.B., Segal, S., and Foley, J.B. J. Clin. Invest. 36, 1395-1407, 1957.
17. Kipnis, D.M. and Cori, C.F. J. Biol. Chem. 224: 681-693, 1957.
18. Hosoya, N. and Machiya, T. in Metabolism, Physiology and Clinical Use of Pentoses and Pentitols, edited by B.L. Horecker, K. Lang, Y. Takagi, Springer-Verlag, Berlin, pp. 248-249, 1969.
19. Shaw, S. and Amos, H. Biochem. Biophys. Res. Comm. 53, 357-365, 1973.
20. DelaFuente, G. Europ. J. Biochem. 16, 240-243, 1970.
21. Van Heyningen, R. Biochem. J. 73, 197-207, 1959.
22. Rodén, L. in Metabolism, Physiology and Clinical Use of Pentoses and Pentitols, edited by B.L. Horecker, K. Lang, Y. Takagi, Springer-Verlag, Berlin, pp. 124-134, 1969.
23. Calligani, L., Hopewood, J., Schwartz, B.N. and Dorffman, A. J. Biol. Chem. 250, 5400, 1975.
24. Demetrakopoulos, G. Ev., Lussier, M., and Amos, H. In Vitro 13:197, 1977.
25. Schwartz, B.N. and Dorffman, A. Biochem. Biophys. Res. Comm. 67:1108, 1975.
26. Eagle, H. J. Biol. Chem. 214, 839, 1955.
27. Hanks, J.H. and Wallace, R.E. Proc. Soc. Exp. Biol. Fed. 71, 196, 1949.
28. Wilson, D.M., Alderete, J.F., Maloney, P.C. and Wilson, T.H. J.Bacteriol. 126, 327, 1976.

Evidence for and speculations about xylose and xylitol metabolism as co-carbon and energy sources

Harold Amos, Joseph R. Colofiore, and George Ev. Demetrakopoulos

Department of Microbiology and Molecular Genetics, Harvard Medical School, 25 Shattuck Street, Boston, Massachusetts 02115

INTRODUCTION

The metabolism of D- and L-xylose and of xylitol by bacteria, and to a lesser extent by yeast, has focused principally upon the biochemical pathways involved in the catabolism of those compounds and related pentoses and pentitols. For the most part extensive examination of regulations of the enzymes induced by exposure to this family of 5-carbon sugars has revealed patterns of cross induction and a convergence of pathways to closely related and interconvertible pentulose-5-phosphates. Relatively few bacterial species have been studied for their ability to use pentoses and pentitols, and virtually nothing is known about the organisms implicated in dental caries in this regard despite the importance attached to recent findings of dramatic responses of caries in humans to xylitol by Scheinin and Mäkinen[1].

A recent review by Mortlock summarizes the extent of investigations dealing with the metabolism of pentoses and pentitols by this limited group of bacteria and yeasts[2]. Much more has been learned about their utilization by animal cells, tissues in vitro, and in situ, and by the whole animal[3].

POTENTIAL OF XYLOSE AND XYLITOL TO SERVE AS CO-CARBON SOURCES

From recent studies in animal cells D-xylose and xylitol appear to serve as sole carbon and energy sources for the growth of mammalian cells from small inocula[4-6]. The extent to which the term "sole" source is legitimately applied is unclear largely because of the uncertainty of the stored energy of the cells in question and of the availability as an energy source of oligosaccharides in serum as it is provided to cells in culture. Among the significant facts emerging from the studies cited is the phenotypic changes in morphology observed when D-xylose is introduced into cultures growing on glucose without removal of the glucose. The same changes that transpire when glucose is replaced by xylose take place but somewhat more slowly when xylose is added to glucose-containing medium. If the changes observed result from the metabolism of xylose, the implications are that the xylose metabolic products have a dominant role for the cell phenotype. That in fact can be interpreted to mean a dominant metabolic role even when glucose is a co-carbon source. More recent work throws this possibility into sharper focus[7]. Hamster fibroblasts transformed by SV40 or polyoma virus are unable to clone on glucose as the carbon source when carefully dialyzed serum is used in the cloning effort. Nor will D-xylose or D-fructose permit clones to form under the same conditions. The combination of D-xylose and D-glucose will, however, support a respectable cloning efficiency. This we take to be evidence for cooperativity and of co-utilization of glucose and xylose by the cells.

Concomitant metabolism of glucose and xylitol has been demonstrated to occur in the perfused rat liver of both starved and fed animals[8-9]. Jakob and coworkers with rat liver perfusates have shown very dramatic changes induced by xylitol in the balance of glycolytic intermediates resulting in more than a 100-fold increase in α-glycerophosphate and 10-fold reductions in other intermediates[10]. Lactate uptake by the perfused liver and its utilization for glucose formation were sharply inhibited by the addition of xylitol to the perfusion buffer. The first reaction of xylitol is initial oxidation to D- and L-xylulose by the NAD- and NADP-linked dehydrogenases of the liver. Various metabolic consequences are attributable to changes in the ratios of oxidized to reduced NAD and NADP that occur.

POTENTIAL IMPORTANCE OF CO-UTILIZATION OF D-XYLOSE AND XYLITOL UPON BACTERIAL FERMENTATION

Acid production by bacteria of the oral flora, especially of the plaque formation on the teeth and gums, has for some time been implicated in dental caries[1]. Thus, the Turku Sugar Studies report of reduced dental caries in patients provided with dietary xylitol is of major importance. This report includes data from two groups of subjects who showed reduction of caries:

(1) Those whose dietary sucrose was essentially replaced by xylitol, and
(2) Those on a regular diet who chewed 4 or 5 sticks of xylitol-containing gum per day.

In the first group the oral and plaque bacterial flora did not change. However, a decided reduction occurred in the viable Streptococcus mutans population of the plaque with much less acid production by the collective oral flora. Adaptation or mutant production resulting in higher acid formation by the oral flora did not occur at a significant frequency.

The group of subjects eating a regular diet supplemented with xylitol chewing gum also experienced production of much less acid by the oral flora. This discussion is directed toward the latter finding in particular, since the potential effect of xylitol as a modulator of acid production from the fermentation by bacteria of glucose or fructose may account for the lowered acid formation observed.

Although there is little direct evidence for such a metabolic effect on the plaque and oral bacteria themselves, other microorganisms have been shown to utilize D-xylose and glucose simultaneously and to utilize D-xylose more effectively under anaerobic than under aerobic conditions.

GENERAL FEATURES OF XYLOSE AND XYLITOL METABOLISM IN BACTERIA

Much of the experimental evidence for bacterial metabolism of xylose and xylitol has come from studies employing strains of <u>Aerobacter aerogenes</u>, <u>Escherichia coli</u> and several species of <u>Lactobacillus</u>[2]. Initial entry in all species investigated is by a common series of enzymatic steps to D-xylulose-5-phosphate (Figure 1). Recently certain strains of <u>Lactobacillus casei</u> have been shown to possess an inducible substrate-specific phosphoenol-pyruvate (PEP) phosphotransferase system for the transport of xylitol. Xylitol is phosphorylated in the 5 position[11]. Further metabolism from D-xylulose-5-PO$_4$ (Figure 2) through the non-oxidative pentose phosphate pathway leads to the formation of fructose-6-PO$_4$ and glyceraldehyde-3-PO$_4$. In <u>Lactobacilli</u> there is apparently no transketolase[12] but a phosphoketolase that cleaves D-xylulose-PO$_4$ into acetyl phosphate and glyceraldehyde-3-PO$_4$ (Figure 3).

EVIDENCE FOR CONCOMITANT UTILIZATION OF D-XYLOSE AND GLUCOSE OR FRUCTOSE IN BACTERIA

The entry of xylose and xylitol into bacteria is either through simple diffusion or a carrier that does not overlap in specificity the glucose or

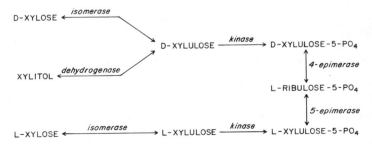

Figure 1.

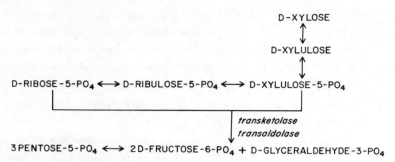

Figure 2.

a) D-XYLULOSE-5-PO$_4$ $\xrightarrow{\textit{phosphoketolase}}$ ACETYL-PO$_4$ + H$_2$O
                                                    +
                                     GLYCERALDEHYDE-3-PO$_4$

b) ACETYL-PO$_4$ + ADP $\xleftrightarrow{\textit{acetokinase}}$ ATP + ACETATE

Figure 3.

fructose carriers. Thus one can rule out competition for transport as a
factor influencing concomitant utilization of xylose or xylitol when the
pentoses are provided in mixtures of sugars.

There is considerable evidence for the simultaneous use of xylose and
other sugars in a variety of organisms. The wild type Escherichia coli K-10
grows more rapidly on the combination of D-xylose and glycerol than on
glycerol alone[13]. Certain transketolase mutants of the parent strain are
inhibited in their growth on glycerol or gluconate by D-xylose, which alone
does not support growth. It appears that little pentose is metabolized and
the nature of the inhibitor step has not been clarified. Large quantities
of phosphorylated products of xylose, prime suspects as inhibitors, do not
occur.

The homofermentative lactic acid organism Pediococcus pentosaceus can
grow on D-xylose very poorly aerobically and somewhat better anaerobically.
Growth is sharply stimulated by the addition of low concentrations (5.5 mM)
of glucose to the D-xylose (22 mM)-containing basal medium[14]. The addition
of glucose to medium containing D-xylose and D-ribose derepressed further
the xylose fermentative pathway while each of the sugars alone or any combination of two was more repressive than the combination of the three.
Thus a synergism of metabolic interaction of all three sugars is apparently
operating.

An unusual regulatory hierarchy prevails in <u>Clostridium</u> <u>thermoaceticum</u>, an obligate thermophilic anaerobe. This bacterium ferments xylose, fructose and glucose with acetate as the only product[15]. In fermentations where mixtures of the three sugars are provided, xylose is first fermented, then fructose and finally glucose. Whereas in the enterics glucose is the prime catabolite repressor[16], this unexplored example may ultimately provide evidence for catabolite repression of glucose-utilizing enzymes by metabolic products of xylose. It seems safe to conclude from the examples cited that some species of bacteria continue to metabolize D-xylose under conditions of nutrition that provide mixtures of sugars. Since the latter is the condition of mouth and tooth bacteria, the importance of xylose and xylitol addition to the mixture of sugars may reside in their ability to modify the fermentation products of resident bacteria involved in the etiology of caries.

BACTERIA OF THE ORAL CAVITY AND PARTICIPANTS IN GENESIS OF TOOTH DECAY

Evidence has accumulated in recent years implicating specific species of bacteria as etiologic agents of dental caries and periodontal disease[17]. Among those most strongly suspected are several species of <u>Streptococcus</u> of which <u>Streptococcus</u> <u>mutans</u> can be considered a prototype. Its adherence to teeth is promoted by dietary sucrose and the mechanism of its strong attachment is a polysaccharide whose synthesis is specifically enhanced by sucrose, not by glucose or other sugars. The <u>Streptococci</u> are acidogenic as are the <u>Lactobacilli</u> which also may contribute to caries by lactic acid production.

Such varied organisms comprise the oral flora that it is impossible to assign metabolic responsibility for acidogenesis or other specific product synergism in the cariogenic process.

The proposal intended to result from this analysis is that xylose and xylitol may exert an abnormally powerful influence on a nutritionally complex environment by modulating the fermentation of "acidogenic" sugars by caries-inducing bacteria of the mouth flora.

REFERENCES

1. Scheinin, A. and Mäkinen, K.K., Editors: The Turku Sugar Studies I-XXI, Acta Odontologica Scandinavica, Vol. 33, Supp. 70, 1975.
2. Mortlock, R.P. Utilization of Uncommon Sugars. Advances in Microbial Physiology, Vol. 13, 1976.
3. Horecker, B.L., Lang, K. and Takagi, Y., Editors: Pentoses and Pentitols Springer-Verlag, Berlin, 1969.
4. Amos, H., Leventhal, M., Chu, L. and Karnovsky, M. Cell. 7, 97-103, 1976.
5. Demetrakopoulos, G. Ev. and Amos, H. Biochem. Biophys. Res. Comm. 72, 1169-1178, 1976.

6. Demetrakopoulos, G. Ev., Gonzalez, F., Colofiore, J.R. and Amos, H. Exp. Cell. Res. 106, 167-173, 1977.
7. Colofiore, J.R., Wright, B. and Amos, H. A metabolic lesion in virus transformed cells revealed by cloning with dialyzed serum (manuscript in preparation).
8. Woods, H.F. and Krebs, H.A. Biochem. Jour. 134:437-443, 1973.
9. Ross, B.D., Hems, R. and Krebs, H.A. Biochem. Jour. 102: 942-951, 1967.
10. Jakob, A., Williamson, J.R. and Asakura, T. Jour. Biol. Chem. 246: 7623-7631, 1971.
11. London, J. and Chace, N.M. Abstracts of the Annual Meeting of the ASM, K2, p. 186, 1977.
12. Heath, E.C., Hurwitz, J., Horecker, B.L. and Ginsburg, A. Jour. Biol. Chem. 231: 1009-1029, 1958.
13. Josephson, B.L. and Fraenkel, D.G. Jour. Bact. 118:1082-1089, 1974.
14. Dobrogosz, W.J. and DeMoss, R.D. Jour. Bact. 85:1356-1364, 1963.
15. Andreesen, J.R., Schaupp, A., Neurater, C., Brown, A. and Ljungdahl, Lars G. Jour. Bact. 114:743-751, 1973.
16. Magasanik, B. Catabolite Repression. Cold Spring Harbor Symp. Quant. Biol. 26:249-256, 1961.
17. Gibbons, R.J. and Van Houte, J. Ann. Rev. Microbiol. 29, 19-44, 1975.

DISCUSSION OF PAPERS BY DRS. SHAW, DEMETRAKOPOULOS AND AMOS

DR. NAVIA: The topic that Dr. Amos discussed is a very important one because the problem of dental caries must be looked at in the context of the microorganisms in the oral cavity. Dr. Bowen has already discussed the importance of different kinds of diets and dietary ingredients in terms of the kinds of oral microorganisms.

The article by Mortlock in Dr. Amos' reference list is a very interesting one describing the catabolism of the so-called unnatural carbohydrates. One of the very interesting things was that, in addition to certain bacteria utilizing natural carbohydrates, some fungi, candida, utilized those sugars very well.

In view of the fact that so many different organisms are present in the oral cavity, what is your feeling with regard to the impact of such carbohydrates as the polyols in the overall microbiological complex in the oral cavity? What would be the impact of having an abundance of any one sugar or sugar alcohol that is not normally present in appreciable amounts in the diet?

DR. AMOS: That is a question I am not prepared to answer. Several others, including Dr. Bowen, who have been more intimately involved in this area may be able to answer that. That is a question in many minds and may not be answerable today.

The discussion by Dr. Bowen about the sharp decreases in pH of plaque about 20 to 30 minutes after a meal tells us that the critical time for the lowering of the pH for tooth destruction is relatively short. Any inhibitor of that pH decrease must operate during that 20 to 30 minute interval to be effective. I think that this sharp drop and return to neutrality is very interesting and hopeful for the control of pH decrease by an inhibitory material.

DR. BRIN: I would like to call the attention of this group to work that L.J. Machlin and I have been doing at the Roche Research Center. Some years ago (Brin, M., Toxicol. and Appl. Pharm. 6: 631, 1964), we reported that another polyol, sorbitol, had a slight sparing action for thiamine deficien-

cy. With the current interest in xylitol, it was desirable to first compare
the two polyols for thiamine-sparing activity, and then to extend the observation to other vitamins.

The data in Table 1 indicate that based upon body weight, xylitol
spared thiamin more effectively than sorbitol. The vitamin B_1 levels in
livers of thiamin deficient rats was six times higher for the rats fed the
xylitol diet than for the controls in comparison with only a threefold
increase for the rats fed the sorbitol diet.

More recent work extended the inquiry to riboflavin (B_2) and vitamin B_6.
Some of the data are shown on Table 2. In this case, feeding xylitol at 20%
of the diet resulted in prevention of weight loss on a thiamin-deficient
diet, and it normalized liver weight values. Erythrocyte transketolase
values were virtually normal. Body weight was protected by xylitol on a
riboflavin deficient diet, and values for liver weight and riboflavin content
were also maintained. For vitamin B_6, there was partial protection.

Table 1. Effect of Xylitol and Sorbitol on Growth and Liver Thiamin of Rats[A]

Treatment	Weight Gain gm/Rat	Liver Thiamin μ/gm	μgm/Liver
$-B_1$	11 ± 2	$0.3 \pm .1$	$1 \pm .3$
$-B_1$ + xylitol	124 ± 8	$1.9 \pm .6$*	21 ± 6**
$-B_1$ + sorbitol	81 ± 10	$1.1 \pm .2$*	9 ± 2**
$+B_1$	110 ± 8	$6.4 \pm .6$**	61 ± 4**
$+B_1$ + xylitol	107 ± 9	$5.0 \pm .6$**	56 ± 5**
$+B_1$ + sorbitol	111 ± 17	$4.9 \pm .6$**	54 ± 6**

*P < .05; **P < .01; [A]8 per group

Table 2. Effects of Xylitol on Sparing Deficiencies for Vitamins B_1, B_2,
or B_6 in the Rat[A]

Diet	Weight Gain − Xyl	+ Xyl	Liver Weight − Xyl	+ Xyl
Complete:	138 ± 7	149 ± 7	$10.2 \pm .5$	10.4 ± 4
Deficient:				
Vit. B_1	20 ± 2	140 ± 4**	$4.2 \pm .2$	$10.3 \pm .3$**
Vit. B_2	60 ± 6	138 ± 4**	$7.6 \pm .4$	$10.3 \pm .4$**
Vit. B_6	72 ± 4	114 ± 6**	$8.0 \pm .3$	$10.0 \pm .4$*

Significant differences from "minus Xylitol": *(p = < 0.01),
**(p = < 0.001)

[A]8 rats per group. Diets with 8% carbohydrate replacement fed for 2 weeks,
followed by 20% replacement for 2 weeks.

These data demonstrate that when xylitol is fed to rats at 20% of the diet, there may be protection against vitamin deficiency as a consequence of its presence. We consider this to be a health benefit resulting inadvertently from consuming this material. Although the mechanism of action remains obscure, reasonable explanations might include increased gastrointestinal synthesis and/or absorption of these essential nutrients.

DR. HERBERT: Dr. Brin, I'm trying to think of a simple explanation for your findings. I question your use of "sparing" because you have not demonstrated that the polyol-fed rats at the cellular level needed any less of these vitamins. Is it possible that the sugar alcohols simply resulted in an increased coprophagy and thereby made more B-complex vitamins available for absorption than was in the diet?

DR. BRIN: The rats were maintained on wire bottom cages to reduce the opportunity for coprophagy. However, we didn't use tail cups, so there was no absolute control. The rats in all groups had the same access to their feces.

DR. HERBERT: Yes, but the polyols certainly alter the consistency and possibly the taste of the feces to the point where the feces may even more accessible and palatable to the rat. I think that is something you would certainly want to rule out.

DR. BRIN: Yes. We are studying this matter now by determining whether a larger amount of vitamins is present in the gastrointestinal tract in the animals fed polyols; thn we will also do absorption studies to determine rates of transfer from the gut.

DR. HERBERT: I hope that you also put the tail cups on so the rats can not ingest their own feces.

The use of xylitol in nutritional and medical research with special reference to dental caries

Kauko K. Mäkinen

Institute of Dentistry, University of Turku, Lemminkäisenkatu 2, SF-20520 Turku 52, Finland

SUMMARY

The number of dentally acceptable sugars that also meet most other requirements for a sweetener in the human diet is limited. The experience obtained to date suggests that xylitol possesses many properties which make it of promising value in dentistry and nutrition. Perhaps the most important of these properties is, from the dental point of view, the absence of fermentation of xylitol in the dental plaque. The apparent mechanism of xylitol action suggests that it is effective in all age groups. As a matter of fact, the gerodontological value of xylitol has not been sufficiently emphasized. The suitability of xylitol for the above purposes is strengthened by its advantageous organoleptic and technochemical qualities.

The literature also suggests that the carbohydrate portion of the diet exerts selective effects on the functions of exocrine glands, thereby influencing the final enzyme, mucopolysaccharide, glycoprotein and electrolyte concentrations of exocrine secretions. The exocrine glands implied here are those which are directly involved in digestion. The response of other exocrine glands is not yet known with certainty. The effects of xylitol on exocrine gland function ought to be viewed simply as those which are typical of one particular type of the many natural carbohydrates, rather than as harmful side effects. This kind of inherent selectivity of the influences of carbohydrates and other dietary ingredients is necessary for the body.

The biochemistry of xylitol in the humans is sufficiently well known to justify its use as a regular sweetener for those for whom sucrose substitution is deemed necessary. Xylitol is mainly metabolized in the liver, the metabolic capacity of which for xylitol is great. Due to the slow rate of absorption of xylitol, this capacity will never be seriously approached in peroral administration.

INTRODUCTION

The present article deals with xylitol within the framework of three lines of research which have been and are being pursued at the Institute of Dentistry, University of Turku. Chronologically the first subject, the Turku sugar studies, has been previously surveyed in several contexts[1-6]. The second subject, the relationship between exocrine gland function and xylitol administration, and the third, the effect of xylitol on clinicochemical parameters and the metabolism of xylitol, developed as logical extensions from the two-year Turku feeding study. A number of supplementary plaque and saliva studies will be mentioned in conjunction with the first subject. The results obtained will be considered in light of findings made in other laboratories.

THE EFFECT OF XYLITOL ON THE CARIES INCIDENCE AND ON THE MICROBIAL AND BIO-CHEMICAL COMPOSITION AND DEVELOPMENT OF DENTAL PLAQUE IN HUMANS

Turku Sugar Studies in Perspective. The Turku sugar studies were a collaborative series carried out during the years 1972-1975 to determine the effects of chronic consumption of sucrose, fructose and xylitol on dental and general health[1]. The impetus to this trial was obtained from previous short-term plaque studies[7-10] which suggested that the consumption of xylitol in moderate amounts strongly reduced dental plaque (by 50%). The initial material of the main contributing part of the trial, a two-year feeding study, consisted of 125 subjects, mean age 27.6 years. The subjects were divided into three experimental groups: sucrose group (35 subjects), fructose group (38 subjects) and xylitol group (52 subjects). The xylitol group was planned larger due to the possibility of loss of subjects; however, no significant loss took place. A varied dietary regimen to be followed by the subjects for two years was developed in cooperation with twelve food and other factories in Finland. The subjects were investigated numerous times during the two-year period according to a continuous analysis program which comprised clinical, radiographic, biochemical and microbiological determinations. Details of the trial were described previously[1].

One of the most important results of the two-year feeding study was the massive reduction in the incidence of dental caries in the xylitol group compared to that in the sucrose group. Fructose was found to be less cariogenic than sucrose. An example of the results is shown in Figure 1.

The magnitude of the reduction in caries incidence caused by the consumption of high amounts of xylitol, which was revealed already during the first half of the two-year trial, necessitated further studies involving partial substitution of sucrose by xylitol. Therefore, a one-year chewing gum study was conducted. The initial material comprised 102 subjects, mean age 22.2 years, divided at random between the sucrose- and xylitol-sweetened gum chewing groups. The average consumption of xylitol gum was 4.5 chicles per day per subject, equalling 6.7 g xylitol per day. The number of chicles sweetened with sucrose was 4.0 per day. Subjects followed their usual dietary and dental hygiene habits, and the frequency of intake of sucrose in solid

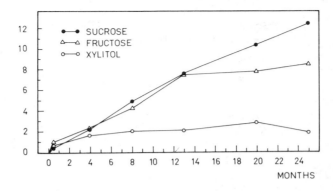

Figure 1. Turku sugar studies. Development of Caries Activity Index comprising all clinical and radiographic quantitative and qualitative changes during the two year feeding study. Vertical scale is dental caries activity caries index.

form between meals was approximately the same in both groups. The caries incidence approximated the corresponding one-year observation in the full substitution study (Figure 2).

The microbiological and periodontal findings of the Turku sugar studies have been described in detail[1]. The following results deserve attention:

(1) The two-year dietary regimen involving practically full substitution of sucrose by xylitol did not affect the major microbial categories occurring in plaque and saliva.

(2) The mean values of viable Streptococcus mutans in plaque were lower in the xylitol group than in the other two sugar groups during the course of the trial.

(3) The geometric and arithmetic means of the colony-forming unit values on selective Rogosa S.L. and Sabouraud antibiotic agar were significantly lower in the xylitol group than in the fructose and sucrose groups.

(4) The acidogenic and aciduric oral flora were reduced in the xylitol group.

(5) No evidence was obtained during the course of the study of adaptation or mutation enabling acidogenic decomposition of xylitol.

(6) The sugar groups did not differ significantly in the reactions of the periodontal tissues. When gingival exudate was tested intravitally in the hamster cheek pouch microvasculature, it was found that clearly more increasing blood cell velocity values were found in the sucrose and fructose groups. In the xylitol group the situation was opposite. It is thus possi-

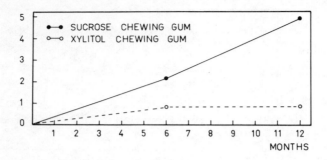

Figure 2. Turku sugar studies. Total caries activity after 6 and 12 months of sucrose and xylitol chewing gum intake. Vertical scale is dental caries activity caries index.

ble that xylitol caused less inflammatory changes in periodontal tissues than sucrose and fructose.

(7) Throughout the two-year study the consumption of the xylitol diet maintained plaque fresh weight values which were 50% lower than those maintained by the consumption of sucrose or fructose diets.

No significant changes were observed in the levels of most whole saliva and plaque enzymes and chemical compounds. Certain important differences between the sugar groups were evident, however:

(1) The ratio of proteins to carbohydrates tended to increase in plaque during consumption of xylitol.

(2) Plaque apartate transaminase activity was higher at most analysis phases in the xylitol group than in the other two groups.

(3) The activity of whole saliva proteinases appeared to increase in the xylitol group.

(4) The concentration of amino acids in whole saliva increased in the group consuming the xylitol diet. This concerned most other ninhydrin-positive compounds as well, except for urea. The amino acids which were most increased were basic (arginine, lysine), hydroxy amino acids (serine),amides (glutamine, asparagine), or imino acids (proline). The concentration of acidic amino acids was increased least.

(5) The above four points (1-4) indicated that the consumption of the xylitol diet increased the nitrogen metabolism of plaque and whole saliva. This was reflected in the increase of the concentration of various nitrogen compounds in these research subjects.

(6) Xylitol consumption tended to increase the levels of plaque glycosidases. The types of enzymes which were increased included α-glucosidase,

β-galactosidase and α-fucosidase. In saliva the levels of these enzymes
were not increased in the xylitol group, suggesting their bacterial origin.
As the salivary glycoproteins contain glycosidic linkages similar to those
present in the synthetic substrates used to determine the above enzyme activities, it is possible that the increased activity of certain glycosidases
in the xylitol group reflects the search of oral bacteria for nutrients during the shortage of fermentable hexose-based sugars.

(7) Xylitol consumption decreased the ratio of glucose to proteins in
plaque.

(8) The concentration of lactic acid in whole saliva and plaque decreased
in the xylitol group. The concentrations of lactic acid in plaque aqueous
extract and the supernatant fluid of saliva are shown in Figure 3. The
concentration of lactic acid was clearly lowest in plaque and saliva obtained from the xylitol group, although, in case of plaque, the same amount
of fresh plaque material from each sugar group was initially suspended in
0.9% NaCl. These values are consistent with other values given for plaque
and whole saliva lactic acid concentrations[11]. It is likely that the reduction of the incidence of aciduric and acidogenic microorganisms in plaque in
the xylitol group was partly reflected in the above-mentioned decrease of
lactic acid concentrations. This property of xylitol is important in relation to dental plaque, although other acids (formic acid, acetic acid,
propionic acid, etc.) may also be important in the formation of the minimum
pH values within plaque.

(9) Xylitol consumption decreased the ability of plaque and whole saliva to
produce reducing sugars from sucrose. This has been called a decrease in
the activity of invertase-like enzymes. In general, this result indicates
that there is a lessened utilization of sucrose in plaque.

(10) The levels of the salivary lactoperoxidase were increased in the xylitol group.

(11) The consumption of xylitol tended to increase the dehydrogenation potential of plaque and whole saliva samples.

It is not yet known what the minimum amount of xylitol is, when compared to total substitution, that would be able to elicit all the biochemical changes listed. However, even the use of moderate amounts of xylitol-
containing chewing gums seems capable of reducing the sucrose-splitting
capacity of dental plaque[12-14]. The biochemical results of the Turku sugar
studies have been described previously in greater detail[1].

The oral safety of xylitol in man has been fully documented in several

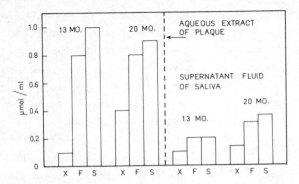

Figure 3. Turku sugar studies. Concentration of lactic acid in the aqueous extract of plaque and whole saliva at two stages of the two year feeding study. X, xylitol; F, fructose; S, sucrose[11].

clinicochemical trials[6]. The clinicochemical determinations carried out in conjunction with the Turku sugar studies showed no differences in serum tri-glycerides, glucose, insulin, uric acid, lactic acid, pyruvic acid and cholesterol, or in the urinary secretion of uric acid among the experimental groups. Other chemical determinations, including liver function tests and hematological studies, showed no differences between the sugar groups. The data obtained should thus be interpreted as an additional evidence for the organism's tolerance of xylitol. The safety of xylitol was also shown by the normality in pregnancies, deliveries and infants in the xylitol group. No abnormality in the development of these children has been detected. Some participants involved in the xylitol group have continued regular consumption of xylitol over a period of five years. No side effects have been reported.

The results of the Turku sugar studies thus showed a strong reduction of the caries increment in connection with partial and total substitution of sucrose by xylitol. The authors of the study concluded that "the accumulated clinical, radiographic, biochemical and microbiological findings have provided the evidence that xylitol is considered non- and anticariogenic."

The nature of the above claims and certain details of the experimental design of the studies have given rise to both constructive and biased criti-cism. Therefore, it is proper to consider a few aspects of the practical conditions and the methodology of the Turku sugar studies:
(1) In view of the strict requirements concerning participation in a two-year feeding study involving chronic consumption of xylitol, with numerous clinical examinations, only cooperative and voluntary adult subjects could be used. A similar study only with children would have been impossible in view

of the complex dietary regimen and requirement for full cooperation. However, a few young subjects were included in the two-year study. They belonged to families from which one or both parents participated in the trial. This situation to a certain extent enhanced the surveillance of the children's xylitol consumption.

(2) The Turku sugar studies was a blind clinical trial; the clinical and radiographic recordings were carried out by examiners who were not aware of the subjects' sugar groups. A double-blind study would have been impossible because of ethical aspects. Furthermore, the subjects quickly would have learned to identify their group as a result of certain unavoidable differences between products variously sweetened with sucrose, fructose or xylitol.

(3) The results of the clinical and radiographic examinations were based on accepted and conventional methods only. Occasional criticism that the methods used in the caries registrations were too sensitive has most likely resulted from the misconception that the final results had been based on microscopic evaluations of caries according to the method of von der Fehr et al[15]. This microscopic registration method was tested in conjunction with the Turku studies, but its usefulness for the present purpose was considered questionable. Consequently, the microscopic registrations were not included in the final data. Nor were the results based on the use of special dyes, another erroneous assumption which has been voiced. Basic fuchsin, a widely accepted and used plaque-disclosing dye, was used only in a separate plaque study[12-14], which did not belong to that series of investigations actually regarded as Turku sugar studies. No caries claims were presented in the above plaque studies.

(4) The fact that both total and partial substitution of sucrose by xylitol, involving 6.7 and 70 g xylitol, respectively, per day and subject, resulted during one year in approximately the same reduction in caries increment, contradicts the argument that mere omission of sucrose would have led to the same result. In the one year chewing trial the mean number of xylitol chicles used per day and subject was 4.5. Being a mean frequency, this value of course implies that certain subjects on selective days used more or less than 4.5 chicles per day, say, for example, three per day. Yet the overall incidence of caries was reduced very strongly (Figure 2). The pertinent point would seem to be that the between meal consumption of xylitol was relatively frequent and regular.

(5) It has been suggested in several discussions that a more logical pair for comparison would have been xylitol versus sorbitol. This might have been

so in an intramural and academic study in which it would not have been necessary to consider many practical requirements regarding the physical and chemical properties of the sweeteners used. The inclusion of sorbitol was actually considered at a very early stage of the planning of the Turku studies. A careful evaluation of the following aspects contraindicated the inclusion of sorbitol in the full substitution study:

a   Sorbitol is much less sweet than xylitol (about 0.6 versus 1.0). Achievement of products of similar sweetness and acceptability would have implied either greater intake of sweetener-derived calories or the use of artificial sweeteners in conjunction with sorbitol.

b   Sorbitol causes stomach disorders at lower concentrations than xylitol.

c   The metabolic capacity of the human body for xylitol is much better than for sorbitol.

d   The sweetened foods which were to be produced for use in the study had to be consumed voluntarily over a period of two years. While xylitol and fructose showed themselves in early trials to be very well suited as substitutes for sucrose in taste and quality as well as in most food manufacturing procedures, the same was not true of sorbitol.

e   Furthermore, the one year chewing gum study was carried out deliberately on sucrose and xylitol only, as the extraordinarily strong reduction in caries incidence during the first stages of the two year study made it absolutely pertinent.

Consequently, the decisive factors which suggested a long-term human trial on sucrose, fructose and xylitol only, were determined on the one hand by considerations of insuring subjects' satisfaction and cooperation and, hence, their continuation of the trial, and on the other hand, by the degree of the possible applicability of the results to come. Xylitol and fructose, after thorough theoretical and practical considerations, evoked the greatest hopes.

Supplementary Studies with Plaque and Whole Saliva

Binding of Labeled Xylitol to Plaque. It was considered that in case a low binding of xylitol to plaque could be shown, the unsuitability of xylitol as a readily fermentable sugar in plaque could be more easily understood. Two subsequent and simple experiments[16,17] indicated that, in contrast to the case with sucrose or glucose, human dental plaque binds labeled xylitol only to a very small extent. This fact results from the evolutionary preference of plaque bacteria for hexose-based carbon skeletons. A pentose-based sugar, particularly if it is in the polyol form, serves a poor carbon source for plaque microorganisms.

Effect of 4.5 Year Use of Sorbitol and Xylitol on Plaque. A small group of subjects who have regularly used xylitol and sorbitol for 3.2 to 4.5 years has been recently studied with regard to the ability of plaque to form acids from xylitol and sorbitol[18]. Mixed plaque was incubated in the presence of 0.0167 M xylitol, sorbitol or glucose. Plaque samples obtained from xylitol-consuming subjects did not change the pH values of the incubation mixtures from those measured for mixtures containing no added sugars. In prolonged incubations up to 22 hours in the presence of xylitol the pH values tended to exceed those determined at the zero-hour stage. Sorbitol, on the other hand, produced significantly more acidic incubation mixtures, the lowest pH values being close to 4.0 (Figure 4). The activity of plaque xylitol dehydrogenase was practically undemonstrable, whereas the activity of sorbitol dehydrogenase was higher and easily demonstrable. The results were interpreted as showing the nonacidogenic nature of xylitol even during prolonged consumption. Thus, this study does not claim that there would not be any metabolism of xylitol in dental plaque. Slow formation of certain basic equivalents during long incubations is, in fact, possible and presumably due to some metabolic process.

Effects of High Amounts of Sorbitol and Xylitol Chewing Gums on Plaque and Whole Saliva. Three different chewing gums (A, B and C), variously sweetened with sorbitol and xylitol were tested in a four week chewing experiment[19]. The numbers of subjects in the test groups were: A, 12; B, 11, C, 11. The chewing gums contained the following average concentrations of

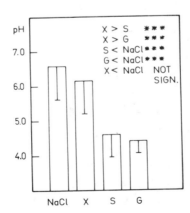

Figure 4. pH values of incubation mixtures containing plaque suspension and the sugars indicated. Plaque was obtained from human subjects who have regularly used xylitol and sorbitol for 3.2 to 4.5 years. X, xylitol; S, sorbitol; G, glucose; NaCl, 0.9% sodium chloride solution; ***P $\lesssim$ 0.001[18]

sorbitol and xylitol; A, sorbitol 56%; B, sorbitol 49% and xylitol 7%; C, sorbitol 10% and xylitol 63%. The numbers of chewing gums used daily were rather high: 7.8 to 7.9 chicles per day and subject. The weight of the chicles was approximately 1.4 g. The highest daily amounts of sorbitol and xylitol consumed by the subjects were thus approximately 6.1 and 6.9 g, respectively. In this study stimulated whole saliva samples were obtained by paraffin stimulation before the experimental period, and by stimulation with the respective chewing gums at the end of the four week chewing period. Resting saliva samples were always collected immediately before the above stimulations. All sample collections were carried out at the end of a four day period with no oral hygiene. The following findings deserve attention:

a   Most of the chemical parameters did not differ between the three groups.

b   All chewing gums significantly increased the concentration of calcium in plaque.

c   All chewing gums tended to increase the concentration of ethanol in plaque.

d   The chewing gum designated as C decreased the concentration of lactic acid in plaque more than products A or B.

e   Stimulation with chewing gums resulted in lower concentrations of inorganic phosphate in saliva than stimulation with paraffin. This is in full accordance with previous findings[20]. The four week use of chewing gums also significantly reduced the concentration of inorganic phosphate in resting saliva.

f   The concentrations of calcium in resting saliva were slightly lower after the four week experimental period than before it.

g   The levels of reducing sugars in whole saliva stimulated with chewing gums increased significantly in all three groups compared with paraffin stimulated saliva. This finding is not in disagreement with those which have shown a decrease of reducing sugars in plaque or whole saliva after consumption of nonfermentable sugars such as xylitol, as the reducing equivalents have in the present case most likely stemmed from saliva itself and not from plaque. It is possible that this increase in the concentration of reducing sugars was interrelated with another similar difference between stimulations with paraffin and chewing gums: the concentration of sialic acid in stimulated saliva was higher after use of chewing gum than after stimulation with paraffin (Figure 5). These analyses were conducted on acid-hydrolyzed saliva; the data thus represented total sialic acids in parotid saliva. It is possible that

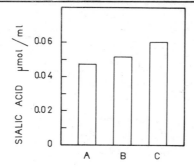

Figure 5. Concentration of sialic acid in stimulated human parotid saliva after stimulation either with paraffin (A), sucrose chewing gum (B) or xylitol chewing gum (C). The data shown were not obtained from the whole saliva experiment described in the text, but from a continuation study on parotid saliva. Parotid saliva from 17 subjects was collected during a 15 minute period following stimulation and pooled for the sialic acid determination[34].

chewing of the polyol chewing gums also increased the concentration of sialoproteins and other glycoproteins in saliva. A related observation was previously made with cannulated saliva from monkeys fed xylitol. The concentration of protein is parotid and submandibular saliva increased significantly compared with saliva obtained from sucrose-fed monkeys[21,22]. However, a more detailed study on the response of salivary sialoproteins to diet is needed. The two year feeding study did not show any xylitol dependent changes in the content of sialic acid in whole saliva[11]. All whole saliva samples in the latter study were collected by paraffin stimulation.

h   It was not possible to show any polyol associated increase in the activity of (lacto)peroxidase in resting or stimulated whole saliva. On the contrary, the levels of this enzyme seemed to be lower after the four week chewing period than before the experiment. This study did not, however, compare the effects caused by sucrose and xylitol, as was the case in the previous studies[11], whereas in the present case comparisons were made between paraffin and polyol chewing gums on one hand and between products A, B and C on the other.

i   Chewing of product C caused a lower activity of sucrose splitting enzymes (invertase-like) in stimulated saliva than products A or B.

j   The activity of sorbitol dehydrogenase tended to increase in stimulated whole saliva during the four week use of polyol chewing gums.

k   The colony forming unit values determined on Rogosa S.L. and mitis-salivarius agar decreased during the use of the experimental chewing gums, but the products did not differ significantly.

Effect of Xylitol Mouth Rinses on Whole Saliva. In order to elucidate more closely the immediate biochemical reactions in whole saliva following stimulation with various sugars, a mouth rinse experiment was carried out. Rinsing the mouth with xylitol solutions increased whole saliva pH values more than sucrose or water (Figure 6). The effect of xylitol on pH values was quick: a significant increase took place within 120 seconds following stimulation. In addition to a possible quick response of oral microorganisms to the presence of fermentable (sucrose) or nonfermentable sugars (xylitol), it is also likely that the pH changes partly resulted from translocation of HCO_3^- ions from blood or extracellular compartments to saliva. Other basic equivalents or pH-rise factors may also have been liberated in saliva. In this type of study no significant differences between sucrose, xylitol and water rinses were found for the concentrations of calcium and inorganic phosphate. It was likewise not possible to show any significant differences in the activity of peroxidase in whole saliva. This indicates that the 32-

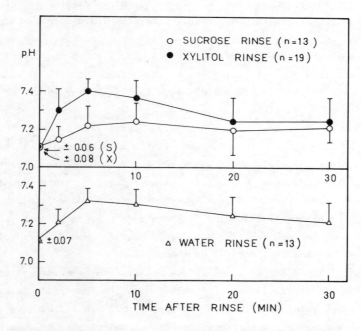

Figure 6. Effect of xylitol, sucrose and water rinses on the pH values of human centrifuged whole saliva. Several two minute saliva samples were collected by paraffin stimulation after the rinses indicated.

minute period during which saliva samples were collected after stimulation, was too short to show enzyme changes based on de novo synthesis of enzyme proteins in the salivary glands.

Effect of Single Dose of Polyol-Sweetened Chewing Gums on Whole Saliva. The studies described above showed the effect of frequent use of chewing gums on whole saliva and plaque. It was therefore of interest to know what the salivary effects of a single, acute exposure to variously sweetened gums would be, and accordingly, this type of investigation has also been carried out[23]. Single doses of chewing gums sweetened with different sugars resulted in pH changes of whole saliva shown in Figure 7. The samples studied were obtained as follows: the salivary flow was stimulated for two minutes with the experimental chewing gums, followed by the collection of several two-minute samples with paraffin stimulation during a 32 minute period. The polyol-sweetened chewing gums resulted in higher initial pH values in whole saliva than those sweetened with fermentable sugars. The differences were clearly demonstrable within the first 120 seconds following stimulation. No consistent differences between the experimental chewing gums were found for the concentrations of inorganic phosphate, calcium and (lacto)peroxidase during the 32-minute collection period. This study also suggested, as did the mouth rinse experiment, that the short collection period (32 minutes) and single dose of xylitol do not necessarily show changes in the activity of lactoperoxidase compared with effects caused by sucrose.

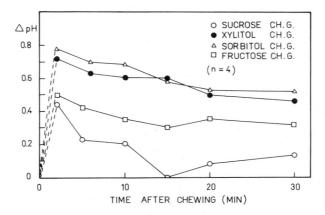

Figure 7. Effect of chewing of chewing gums on the pH values of human whole saliva. The results are expressed as differences of the pH values compared with the initial values determined immediately before a two minute stimulation with the experimental gums. Several two minute saliva samples were collected and the pH values were determined.

There has been occasional speculation that the increased pH levels of saliva after stimulating the flow with sugar alcohols like xylitol, would increase chances for calculus formation. This would be unrealistic, however, in view of the following aspects:

a   The regular diet also causes almost similar and repeated pH rises daily.

b   The innate defense mechanisms, such as the pH-rise factors described by Kleinberg[24] also tend to cause natural pH-rises.

c   It would be more realistic to assume that calculus formation largely depends on the subject's inherent quality of saliva (still emphasizing the significance of oral hygiene). A few additional pH-rises daily caused by xylitol or sorbitol should not be regarded as a particularly detrimental factor in relation to periodontal diseases.

d   Also sucrose, glucose and many other sugars and countless products containing these sweeteners cause initial pH-rises in saliva, but their ready fermentation quickly decreases the pH values with formation of extracellular organic acids.

e   The subjects who have regularly consumed xylitol over a period of five years have not shown any detectable increase in the incidence of calculus formation and periodontal diseases.

Synopsis of Other Relevant Plaque and Microbiological Studies   When considering the metabolism of xylitol by oral microorganisms one has to recall such basic phenomena as enzyme induction and repression. Because these are very common in microorganisms, one is forced to assume that oral bacteria undergo such phenomena exceedingly frequently. It is possible, however, that even during longer periods only a small number of these induction- or mutation-like phenomena are revealed due to their insignificance in oral biology. The sporadically xylitol fermenting species, with their weak acid production, seem to have no chance to succeed in the microbial plaque system.

One of the starting points of the Turku sugar studies was the assumption that human oral microorganisms would gradually become adapted to effectively utilize xylitol. It was simultaneously emphasized, however, that adaptation may be unnecessary as long as glucose and other easily available and soluble hexose-based carbohydrates are consumed[4]. Microorganisms capable of utilizing xylitol occur elsewhere in nature, but only rarely has microbial utilization of xylitol been encountered in human dental plaque[6]. The biochemical and microbiological findings mentioned in this article should be viewed against this background.

Since the previous exposure of human oral microorganisms to xylitol and their metabolism of it are both insignificant, the literature dealing with

these aspects is scant. In vitro studies on Streptococcus mutans (Ingbritt)
showed that xylitol inhibited the growth of the cells in liquid media[10,25].
This growth retardation was accompanied by a delay in the attainment of the
stationary growth phase of the cells and in the formation of aldolase in the
cells. These changes were associated with lessened lactic acid production by
the cells and higher final pH values of the growth media than in the presence
of mannitol and sorbitol which are regularly metabolized by the Streptococcus
mutans strain used. Some growth occurred in the xylitol medium, but this
resulted from the fermentable carbohydrates of the trypticase soy broth used.
It was suggested at this stage of research[25] that xylitol would induce the
formation of aminopeptidase-like enzymes in the cells and that after an
adaptive period of a few months xylitol could be used by the cells. Later
studies showed, however, that the question was not of adaptation of xylitol.
Some growth was possible because the medium contained small amounts of fermentable carbohydrates. The cells additionally formed increasing amounts of
peptidases which they used in the degradation of extracellular proteins and
peptides to nutrient amino acids. This increased the cellular metabolism
and acidity of the culture. The increased peptidase activity can be interpreted as the cells' search for nutrients in the absence of sufficient
amounts of glucose. Whether the increase in the enzyme activity should be
called induction or not is uncertain, as other possibilities have also been
discussed[6].

One of the first biochemical plaque studies with xylitol suggested that
the consumption of moderate amounts of xylitol (40 to 50 g per day and subject) for four days involving full neglect of oral hygiene decreased the
concentration of hexosamines and total sugars in plaque compared to the
effects caused by sucrose[7,8]. The effects of fructose and glucose in this
respect were nearly similar to that of xylitol. This difference may have
been due to greater synthesis of plaque polysaccharides on sucrose diet.

The redox potential (E_h) of whole saliva and plaque suspensions were
more reducing following a four day use of xylitol and fructose than after
consumption of glucose or sucrose[7]. In other words, the consumption of
xylitol and fructose decreased the electron (e^-) activity of whole saliva
and plaque. It can be assumed that the observed differences in E_h values
reflected the different response of plaque microorganisms to different
sugars. This type of correlation was not, however, shown in the Turku sugar
studies in which the E_h values were recorded after 16.5 month consumption of
the experimental diets[26]. In both studies the stimulation of saliva was

performed with paraffin and the E_h values after the xylitol diet were
+332 $\pm$ 74 mV[7] and +272 $\pm$ 43 mV[26] respectively. In view of the efficacy of
the lactoperoxidase system, it would apparently be advantageous to have high
positive E_h values in saliva, which often result from effectual formation of
hydrogen peroxide, the substrate of lactoperoxidase. Although the consumption of xylitol was associated with elevated levels of salivary
lactoperoxidase, the E_h values did not differ between the diet groups. The
nexus between the salivary redox potential and oral health, thus, is still
more or less obscure. The possible causal relationship between these factors clearly belongs on the list of future research needs.

Several reports have been published on the invertase-like enzyme
activity during xylitol consumption. The first of these reports[9] suggested
that as xylitol is not a substrate of invertase (EC 3.2.1.26; β-D-fructofu-
ranoside fructohydrolase), the significant xylitol associated reduction in
the levels of this enzyme activity can be easily understood. The basic
explanation includes the fact that xylitol does not promote the growth of
plaque, but rather inhibits it, with the concomitant lowering of the
enzyme levels. The determination of this enzyme activity has later been
utilized as a tool to demonstrate the unsuitability of a sugar to plaque
metabolism. This idea was exploited in several studies all of which have
consistently demonstrated that xylitol causes lower enzyme activity than
sucrose[10-14]. A finding which is of some importance in this sense is
depicted in Figure 8. The consumption of a xylitol diet virtually eliminated
from whole saliva an enzyme capable of splitting sucrose into two reducing
moieties. This has been regarded as one of the fundamental enzymatic
changes in plaque and whole saliva obtained from xylitol-fed subjects. It
may be mentioned that an invertase assay with crude enzyme preparations also
measures transglycosidase activity.

If four day old plaque is collected from subjects on different sugar
diets and the pooled plaque material from each sugar group is suspended in
equal volumes of buffer, the material is sonicated, and the protein concentration is then determined, it is found that xylitol apparently does not
increase the protein levels of these plaque samples. Sugars that promote
plaque growth, on the other hand, increase the concentration of protein[8].
However, if the protein values are expressed in mg protein per mg plaque used
(fresh weight), it can be observed that xylitol, in comparison to sucrose,
increases the nitrogen metabolism in plaque and makes it more protein rich
(Figure 9.)

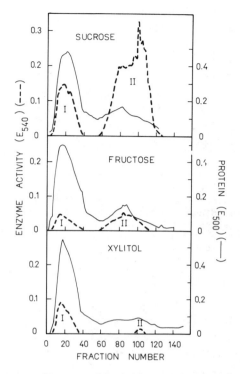

Figure 8. Turku sugar studies. Molecular permeation chromatography of centrifuged whole saliva obtained from the sucrose, fructose and xylitol groups. II (broken line) represents an enzyme yielding reducing sugars from sucrose and whose levels were strongly reduced in the xylitol group[11].

One of the previous short-term plaque studies suggested that the consumption of xylitol decreases the specific dextranase activity in plaque extracellular phase compared to consumption of sucrose[10]. A similar result was obtained in the Turku sugar studies[11]. This type of enzyme response to diet has been observed with sonicated plaque, sonicated salivary sediment or whole saliva, suggesting that the earlier observations were a question of cell wall enzymes which dissolved rather easily during preparation of the aqueous plaque extracts.

It is not advisable to apply in vitro findings directly to in vivo conditions, but it may be mentioned that in an in vitro experiment xylitol had no strong effect on the uptake of glucose by the cells of Streptococcus mutans (Ingbritt)[27]. In this study 3% xylitol inhibited by 40 to 50% the activity of dextran-hydrolyzing extracellular enzymes. Low concentrations of xylitol (up to 55 mg per liter, i.e. 0.36 mM) increased the formation of

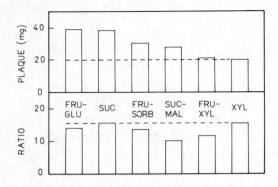

Figure 9. Relationship between plaque fresh weights and the protein concentrations of sonicated plaque material. Human subjects were given various sugars and sugar mixtures for five days during complete neglect of oral hygiene. Plaque was collected, sonicated and centrifuged. The upper graph shows the amount of plaque and the lower graph shows the ratio of plaque fresh weight (in mg) to protein concentration of centrifuged sonicates (in mg per ml). The broken line compares the values between sucrose and xylitol[8].

extracellular polysaccharides by the cells, but higher concentrations were inhibitory. These latter _in vitro_ findings may have reflections in _in vivo_ conditions: although xylitol reduced the amount of dental plaque by 50% compared to sucrose, the concentrations of plaque extracellular polysaccharides did not vary proportionately[7]. These findings, showing relatively high amounts of polysaccharides in "xylitol plaque", should not be interpreted as contradicting the noncariogenic properties of xylitol, as it is the chemical nature of the plaque polysaccharides which is decisive.

A number of recent xylitol studies deserve attention:

a   The cells of Candida albicans isolated from human mouths were not shown to use xylitol for growth. When xylitol was removed from the broth (and no other carbohydrates were added), the growth rate of the cells was the same as in the presence of this polyol[28]. The cells were thus shown to suffer more from the shortage of glucose than from the presence of xylitol. Xylitol behaved in this experiment as an inert compound.

b   Telemetric recordings of interdental plaque pH showed that 10% xylitol rinses and xylitol chewing gum were non-acidogenic[29]. This study thus confirmed the numerous previous reports about the non-acidogenic nature of xylitol in human dental plaque.

c   Xylitol chewing gum significantly reduced the plaque formation compared with sucrose chewing gum. Plaque scores were smaller in the xylitol

chewing gum group than in the placebo group, although the latter difference was not significant[30].

d   A total of approximately 200 strains of plaque bacteria divided over
ten species were tested as regards metabolism of xylitol. None of the
strains could ferment xylitol. There was no adaptation to xylitol[31]
(Havenaar et al. 1977).

Several reports dealing with xylitol in relation to plaque biochemistry
and microbiology will be published in the near future, and numerous independent laboratories are initiating their own xylitol projects. During recent
years a number of review articles covering various dental aspects of xylitol
have been published[6, 32-33].

RELATIONSHIP BETWEEN EXOCRINE GLAND FUNCTION AND XYLITOL ADMINISTRATION

Lactoperoxidase   The relationship of xylitol to the physiology of exocrine
glands comprises a particular line of research which, apart from the exocrine
action of pancreas, has been recognized and pursued only for a short period.
The information concerning these aspects is thus understandably scant. In
spite of this it may already be warrantable to assume that this research
approach will not reveal any pathologically alarming findings. The consumption of moderate amounts of xylitol, for example, has not been shown to cause
any harmful effects on any of the exocrine glands so far studied. Recent
investigations suggest that the effects of very heavy loading of xylitol or
sucrose on the parotid glands do not differ[34]. There are, however, certain
findings which suggest that the study of the above-mentioned relationship
may deserve more attention.

The impetus to this series of investigations developed during the course
of the two year Turku feeding study: the consumption of a xylitol diet was
associated with elevated levels of peroxidase in whole saliva. No similar
effect was found with plaque, salivary sediment or gingival exudate[26].
Chromatographic and biochemical studies revealed that the enzyme whose
activity was increased was the salivary lactoperoxidase, which is the enzyme
secreted from the salivary glands. This enzyme acts as a part of the body's
innate defense mechanisms and it occurs in many exocrine secretions, such as
saliva, milk and lacrimal fluid. It was therefore suggested that the xylitol-
associated strong decrease in the caries incidence could be partly explained
in terms of the elevated levels of lactoperoxidase which kills certain microorganisms in the mouth. The lactoperoxidase system is, however, rather
complex. In addition to the enzyme itself, it also comprises the oxidizable

substrate, H_2O_2, and thiocyanate ions (SCN⁻). The latter can to a certain extent be replaced by iodine ions. The enzyme acts by first oxidizing SCN⁻ ions in the presence of H_2O_2 to form hypothiocyanate ions (OSCN⁻) which, it has been suggested, act as the final inhibitory compound[35]. The mere determination of the activity of lactoperoxidase as such in a biological fluid may thus not necessarily provide sufficient information about the efficacy of the whole system. As to the oral conditions, H_2O_2-producing mechanisms are constantly available in the form of plaque bacteria. In an in vitro reaction mixture the bacteria are usually not present and H_2O_2 has to be added. From what is already known it is obvious that the xylitol-associated increase of the levels of lactoperoxidase in saliva can be demonstrated if most of the following conditions are simultaneously considered:

Magnitude of the dose of xylitol
Duration of administration
Form of xylitol consumed (liquid, solid)
Simultaneous mastication or sucking
Time between last stimulation and sample collection
Placing of meals between stimulation and sample collection
State of oral hygiene
Type of final stimulation of sample collection (paraffin, chewing gum
   sweetened with xylitol, etc.)

Admittedly, the exact conditions for the effect of lactoperoxidase to appear are not known. However, inspection of Table 1, which lists all available studies of this approach, suggests that the following requirements were met in experiments showing xylitol-dependent increase in lactoperoxidase levels (the effects of xylitol and sucrose are compared):

Prolonged use (at least 2 or 3 days) of rational amounts of xylitol
A single dose taken in conjunction with a sufficiently long mastication or
   sucking period (15 minutes) preceding the sample collection (at least 5 to
   6 g xylitol dose)
Single dose of xylitol (at least 5 to 6 g) in the form of pastils (for a
   delayed effect 2 to 3 hours following stimulation)

The Table also indicates that no lactoperoxidase effect was shown, if only one piece of chewing gum was used as the stimulator, which indicates a concentration dependent effect. A single mouth rinse with 10% xylitol solution may not produce a positive effect. On the other hand, very heavy loading of the subject with xylitol (viz. 200 g per day and subject) did not show this effect either, presumably because of changes caused by strong dehydration of the body's various compartments after such a great dose of xylitol[34].

It is also clear that certain experimental conditions may cause the salivary lactoperoxidase to decrease[21] (Table 1). In this particular case the exact reason for the decrease of the enzyme levels is not known, but it

Authors	Description of Study	Results
Mäkinen, Tenovuo and Scheinin[26,38]	Man. Total substitution of S for X over 2 years. Collection of whole saliva by paraffin stimulation after one night neglect of oral hygiene	The levels of LPO increased noticeably in the X group
Harper, Poole and Wolf[39]	Man. Single dose of X in the form of fruit pastils sweetened with a mixture of X and So. Collection of parotid saliva simultaneously during a 15-min. stimulation period	The activity of LPO was increased by the pastils
Mäkinen, Bowen, Dalgard and Fitzgerald[40]	Monkey. Total substitution for 2 to 3 days. Collection of cannulated parotid and submandibular saliva by pilocarpin stimulation on the 2nd and 3rd day	Feeding with xylitol was associated with significant increase of the levels of LPO
Mäkinen, Virtanen, Mäkinen and Neva[34]	Man. Single dose of fruit pastils sweetened with X only (5.9 g per subject). Collection of parotid saliva by stimulation with citric acid at 0, 1, 3 h	X pastils caused a higher activity (in U/ml). The differences between S and X ranged from indicative to significant, particularly in the 2- and 3-h samples
Scheinin and Mäkinen[1]	Man. Partial substitution in the form of X-chewing gum (6.75 g/day and subject corresponding to approximately 4.5 chicles, over a period of one year). Collection of whole saliva by paraffin stimulation	The X and S groups did not differ significantly
Mäkinen, Mäkinin, Söderling and Tenovuo[21]	Man. Single dose (0.6 g per kg in 1 min in the form of a 200 ml drinking solution). Collection of whole saliva by paraffin stimulation during period of 6 h	No differences between glucose and xylitol over a period of 6 h after ingestion
Mäkinen (to be published)	Man. Single dose. Rinse with 10% X solution or chewing one chicle sweetened with X (1.5 g). Collection of whole saliva by paraffin stimulation during a period	X. S, So and F did not differ. Stimulation caused a short drop in the levels of LPO
Mäkinen, Läikkö, Rekola and Scheinin	Man. Chewing of polyol-sweetened chewing gums over a period of four weeks (almost eight chicles per day and subject.) Collection of whole saliva by stimulation of the experimental gums X	The chewing gums, containing various portions of X and So, did not differ. The gums lowered the LPO levels compared with paraffin
Mäkinen, Tuori and Poutiainen (to be published)	Cow. Feeding a mixture of sugar alcohols over a period of 11 weeks (0.5 kg per kg). Whole saliva, milk and lacrimal fluid were analyzed	The polyol, molasses and control groups did not differ significantly
Mäkinen, Tenovuo and Bäckman (to be published)	Man (lactating mothers). Feeding xylitol for two days (0.6 g per kg). Milk was studied	The effects caused by X and G did not differ significantly

Table 1. List of lactoperoxidase studies involving peroral administration of xylitol. Abbreviations: LPO = lactoperoxidase; X = xylitol; Su = sucrose; So = sorbitol; G = glucose; F = fructose

is conceivable that during intensive mastication (about eight chicles per day
for four weeks) the machinery responsible for the synthesis of the enzyme in
the salivary glands was not able to cope with high secretion of saliva
caused by intensive mastication. Furthermore, in this study the baseline
values were determined on paraffin stimulated saliva and the values four
weeks later on saliva stimulated with the respective chewing gums. The
three gums did not differ, suggesting that sorbitol and xylitol may have
similar effects on the levels of lactoperoxidase in saliva.

Greatly varying results have been obtained when studying the effects of
various methods of xylitol administration on the activity of lactoperoxidase
in saliva. The initial decrease of the enzyme activity after chewing a
single chicle or rinsing the mouth once only with a xylitol solution should
not be regarded as a harmful reaction. The changes in the lactoperoxidase
activity were very similar no matter what sugar was used (Table 1). Rinsing
the mouth with water yielded the same result. The acinar units seem to
empty a part of their contents following a stimulus. This leads to transitory decrease of certain salivary ingredients, like lactoperoxidase. The
stimulus also sets into action the synthesis of protein to cope with the
increased needs of enzymes and other compounds in saliva. The situation is
different depending on whether prolonged use of xylitol is involved or a
single dose is administered in a prescribed manner. A concomitant elevation
of lactoperoxidase activity in saliva has in such cases been observed both in
humans and monkeys.

The exact physiological mechanism involved in the xylitol-induced increase of the lactoperoxidase activity is not known.
A number of possible alternatives was listed in a recent review[6]. One interesting suggestion is that thyroxine may act as a regulator of the iodinase
and/or peroxidase enzyme(s) of the submaxillary gland[36]. It has also been
suggested that the addition of an exogenous intermediate of the glucuronate
cycle would stimulate the efficacy of the pathway, thereby possibly enhancing
the synthesis of glycoproteins and/or mucopolysaccharides. It must be clafi-
fied, however, whether a relatively rapid turnover of xylitol to glucose
after absorption, as in the fasting state, would allow a stimulation
described above to occur. The lactoperoxidase molecule contains a carbohydrate moiety, and the above idea is therefore attractive, but clearly
confirming experiments should be carried out. Monkeys fed xylitol produced
viscous saliva compared with those fed sucrose. This result could not be
explained in terms of dehydration of the acinar units only on xylitol diet.

The viscosity of saliva partly stems from polysaccharides as prosthetic groups in proteins.

α-Amylase α-amylase is a typical salivary enzyme, which randomly hydrolyzes the inner α-1, 4-linkages in starch. Conceivably the amounts of amylase secreted from the pancreas depend on the concentration and chemical type of substrate carbohydrates consumed. This seems also to be the case with salivary amylase, the levels of which are higher on a predominantly carbohydrate diet than on a carbohydrate-poor diet. The consumption of a xylitol diet significantly increased the concentration of α-amylase in monkey parotid saliva[22]. The content of protein was simultaneously increased and the specific activity of α-amylase did not differ significantly between sucrose and xylitol diets. The specific activity of α-amylase thus behaved in an opposite manner compared with specific lactoperoxidase activity determined in the same monkey study. α-amylase is, however, one of the major protein components in saliva, whereas the molar concentration of lactoperoxidase is clearly smaller. Hence, changes in the specific lactoperoxidase activity by sugar diets may be more readily observable than is the case with α-amylase.

Chronic consumption of xylitol lowered the levels of α-amylase in serum compared to consumption of sucrose[11], but all values were within the physiological range. A single xylitol dose of 0.6 g per kg body weight caused the same serum α-amylase level as equivalent dose of glucose[21].

A recent study on 17 human adult subjects showed that a small dose of xylitol (5.9 g in the form of fruit pastils) slightly increased the levels of parotid saliva α-amylase compared with sucrose pastils, if the enzyme activities were expressed in enzyme units per flow rate of saliva, i.e., in U/min. No differences were found if the activities were given in U/ml or U/mg[34]. The available literature on the effect of the oral administration of xylitol on exocrine glands in humans includes studies of only the pancreas, the parotid and the submandibular glands. No pathological changes have been detected in these studies.

Carbonic Anhydrase Carbonic anhydrase became an object of attention in this laboratory after it was found that the chewing of xylitol chewing gums resulted in higher concentrations of HCO_3^- ions in human whole saliva than chewing gums sweetened with other sugars[20]. Since then one verifying study has been carried out on human parotid saliva stimulated with fruit pastils sweetened either with sucrose or xylitol[34]. The preliminary results suggested that the sugar groups differed significantly only for the first milliliters of saliva collected immediately after stimulation (the enzyme activity was expressed in U/ml). There was thus an apparent correlation between the levels of carbonic anhydrase in parotid saliva and the previously measured concentrations of HCO_3^- ions in whole saliva; both types of samples were collected immediately following stimulation. The individual deviations between subjects were, however, very high, with a few subjects showing no activity at all in some instances. Therefore, further verifying studies are being carried out to determine the exact reactions of the enzyme. The reversability of the catalysis caused by carbonic anhydrase must also be considered.

In the present assay the reaction was followed according to the formation of carbonic acid:

$$CO_2 + H_2O \rightharpoonup H_2CO_3$$

This is possible by maintaining a highly saturated CO_2 concentration in the reaction mixtures. The formed H_2CO_3 will dissociate to HCO_3^- and H^+ ions. It is thus possible to explain chemically the above mentioned correlation between the two studies. The investigation of carbonic anhydrase in saliva may deserve more attention, as the bicarbonate system forms approximately 85% of the total buffering capacity of saliva at the physiological pH values. It must be considered, however, that a part of the HCO_3^- ions in saliva following stimulation with sugars may actually be translocated through the oral mucosa from blood and the extracellular compartment.

Other Relevant Exocrinological Studies   At present the following studies which attempt to elucidate the relationship between xylitol administration and exocrine gland function, are being carried out in this laboratory:

Protein, Amino Acid and Sialic Acid Content of Human Parotid Saliva Stimulated with Xylitol and Other Sugars   It has not been possible so far to show significant differences between xylitol and sucrose in their effects on the levels of Sakaguchi-positive compounds in saliva. The Sakaguchi reaction measures arginine, but it also gives a positive, but insignificant reaction with glycine in the conditions used. Proteins also react, chiefly because the guanidine group in most proteins is free.

Enzyme Content of the Lacrimal Gland   This study has been initiated with rats. The study also investigates the enzymology of the gut wall, pancreas, the submandibular gland, and oral mucosa. The data will be published later.

Metabolism of Exocrine Glands in the Cow   In this study animals were given a mixture of sugar alcohols, chiefly consisting of arabinitol, mannitol, sorbitol, xylitol, galactitol, rhammitol and small amount of other short chain sugar alcohols. The study attempted to elucidate the biochemical composition of serum, milk, lacrimal fluid and whole saliva of animals fed the polyol mixture. The effects were compared with those found in molasses and control groups. The only biochemical parameter which showed significant differences during 11-week feeding of the animals, was the levels of milk lactoperoxidase, which were higher in the polyol group compared with the other two groups. This study needs verification, as the number of animals was eight in each group and cross-over design was not involved.

Effect of Two-Day Xylitol Feeding on Human Milk   Lactating mothers were given either xylitol or glucose (0.6g per kg body weight) for two days. The biochemical parameters studies (peroxidase, alkaline phosphatase, several metals, glucose and proteins) did not differ significantly between xylitol and glucose.

EFFECT OF XYLITOL ON CLINICOCHEMICAL PARAMETERS

There is a vast and growing literature on the clinicochemical aspects of intravenous and oral administration of xylitol in man and animals. Its metabolism in normal and diabetic humans has been thoroughly elucidated, and the safety of oral xylitol is generally recognized[6].

The following points characterize the main features of xylitol metabolism in humans:

The absorption of xylitol is slower than that of glucose.
At least 80% of the overall xylitol metabolism takes place in the liver.
The initial metabolism of xylitol is independent of insulin.
The final metabolism of xylitol depends o n insulin (as xylitol is
   converted to glucose in the liver cells; the utilization of this glucose
   naturally requires insulin).
Changes in the energy metabolism of the liver are quick (in case of intra-
   venous administration only; due to slow absorption, peroral administration
   does not normally cause quantitatively significant changes).
Xylitol causes small changes in blood sugar concentration.

Effect of Xylitol on Blood Cells   The correlation between the levels of salivary lactoperoxidase and xylitol administration necessitated a study on the effects of xylitol on the peroxidative and dehydrogenation potential of human lymphocytes and granulocytes[21]. Adult subjects were given xylitol or glucose perorally 0.6g per kg body weight as a single dose. Leukocytes and erythrocytes were collected as a function of time following administration of xylitol. The results showed that glucose and xylitol did not differ significantly with regard to any of the chemical parameters studied (granulocyte and lymphocyte peroxidase, glucose 6-phosphate dehydrogenase, lactate dehydrogenase, alcohol dehydrogenase and ionized iodine, and erythrocyte catalase and glucose 6-phosphate dehydrogenase). The serum parameters studied (lactic acid, pyruvic acid, lactate dehydrogenase, α-amylase) did not differ significantly. The blood cell findings may deserve attention, in spite of the slow rate of absorption of xylitol, since the concentration of xylitol in blood can occasionally be higher (as in the portal vein) and since xylitol can readily be transported into erythrocytes in which it is regularly metabolized. Blood cells collected during infusion therapy may yield different results.

A separate study in this series investigates the preservation of aminopeptidases in human erythrocytes stored in the presence of xylitol, sorbitol or glucose, and will be reported in the near future. This study was necessitated by the suggestions that erythrocytes stored in xylitol would be functionally ready at blood transfers. The aminopeptidases concerned play an important role in the metabolism of red blood cells.

Relationship Between the Metabolism of Glucuronic Acid, Glycoproteins and Gut Wall

This study uses at the first stage rats as the experimental object. This investigation was considered necessary because of previous suggestions that peroral administration of xylitol would affect the secretion of glycoproteins from certain exocrine glands.

APPLICATION OF XYLITOL

The use of xylitol in parenteral nutrition as a source of calories has until quite recently been the most important application of this sugar alcohol. Another important way xylitol has been used is as a sweetener for diabetics in countries where the utility value of sugar alcohols in nutrition has been better realized. The value of xylitol in diabetic nutrition should be fully exploited. This is also motivated by the impaired dental condition of many diabetic subjects. The upper rational limits of xylitol dosages in the nutrition of "average" diabetic subjects are generally known (40 to 80 g per day and subject); this also concerns juvenile diabetes[37]. Other medical and nutritional applications have also been considered; a list of such uses of xylitol has been given in a review article[6].

At present perhaps the xylitol application which seems to have the most far reaching significance is its use as a sweetener which does not promote dental caries. The strong caries reducing properties of xylitol in man are founded on its practically absolute nonfermentability by cariogenic microorganisms. As a sugar, xylitol still increases the salivary flow thereby enhancing the efficacy of the innate defensive factors of saliva. The quantities of xylitol needed to elicit caries-reducing effects are thus fairly low, as even approximately 1 g single doses taken regularly between meals in place of traditional (sucrose) sweet snacks cause sufficient pH and other chemical changes in saliva. Regular use of proper xylitol-containing products which allow mastication or sucking has therefore been recommended as a caries reducing method. It has been suggested that a repetitive use of this type of product at least 4 to 5 times per day leads to advantageous results. Due to the apparent mode of xylitol's action, its use as a panacea against dental caries is irrevelant. Unfortunately, theoretically erroneous research protocols have been planned and accomplished without appreciation of the practical consideration that xylitol acts best used immediately after fermentable sugars or as a general and frequent saliva stimulator between snacks. Thus used in a rational way, xylitol has been demonstrated to lead to results which can be characterized as enhancement or facilitation of remineralization. This in turn has been the basis for claims about therapeutic effects of xylitol[1].

REFERENCES

1. Scheinin, A. and Mäkinen, K.K. Acta Odont. Scand. 33, Suppl. 70, 1975.
2. Scheinin, A. Int. Dent. J. 26, 4, 1976.
3. Scheinin, A. Int. J. Vit. Nutr. Res., Suppl. 15, 358, 1976
4. Mäkinen, K.K. Int. J. Vit. Nutr. Res., Suppl. 15, 92 and 368, 1976.
5. Mäkinen, K.K. Int. Dent. J. 26, 14, 1976.
6. Mäkinen, K.K. Experientia, Suppl. 30, 1978, in press.
7. Scheinin, A. and Mäkinen, K.K. Int. Dent. J. 21, 302, 1971.
8. Scheinin, A. and Mäkinen, K.K. Acta Odont. Scand. 30, 235, 1972.
9. Mäkinen, K.K. and Scheinin, A. Int. Dent. J. 21, 331, 1971.
10. Mäkinen, K.K. and Scheinin, A. Acta Odont. Scand. 30, 259, 1972.
11. Mäkinen, K.K. and Scheinin, A. Acta Odont. Scand. 33, Suppl. 70, 129, 1975.
12. Mouton, C., Scheinin A. and Mäkinen, K.K. Acta Odont. Scand. 33, 27, 1975.
13. Mouton, C., Scheinin, A. and Mäkinen, K.K. Acta Odont. Scand. 33, 33, 1975.
14. Mouton, C., Scheinin, A. and Mäkinen, K.K. Acta Odont. Scand. 33, 251, 1975.
15. von der Fehr, F.R., Löe, H. and Theilade, E. Caries Res. 4, 131, 1970.
16. Mäkinen, K.K. and Rekola, M. J. Dent. Res. 55, 900, 1976.
17. Mäkinen, K.K. in Microbial Aspects of Dental Caries, Stiles, H.M., Loesche, W.J. and O'Brien, T.C., Eds., Sp. Supp. Microbiology Abstracts, Bol. II, pp. 521-538, 1976.
18. Mäkinen, K.K. and Virtanen, K. J. Dent. Res. 1978, in press.
19. Mäkinen, K.K., Bowen, W.H., Dalgard, D. and Fitzgerald, G. J. Nutrition 1978, in press.
20. Söderling, E., Rekola, M., Mäkinen, K.K. and Scheinin, A. Acta Odont. Scand. 33, Suppl. 70, 337, 1975.
21. Mäkinen, K.K., Mäkinen, P-L., Söderling, E. and Tenovuo, J., Abstracts, 1st Nordic Congress in Nutrition, Stockholm, Sweden, #25, June 1977.
22. Bird, J.L., Baum, B.J., Mäkinen, K.K., Bowen, W.H. and Longston, R.W. J. Nutr. 107, 1763, 1977.
23. Mäkinen, K.K. unpublished data.
24. Kleinberg, I., Kanapka, J.A. and Craw, D. in Microbial Aspects of Dental Caries, Stiles, H.M., Loesche, W.J. and O'Brien, T.C., Eds. Sp. Supp. Microbiology Abstracts, Vol. II, pp. 433-464, 1976.
25. Mäkinen, K.K. J. Dent. Res. 51, 403, 1972.

26. Mäkinen, K.K., Tenovuo, J. and Scheinin, A. Acta Odont. Scand. 33, Suppl. 70, 247, 1975.
27. Knuuttila, M.L.E. and Mäkinen, K.K. Caries Res. 9, 177, 1975.
28. Mäkinen, K.K., Ojanotko, A. and Vidgren, H. J. Dent. Res. 54, 1239, 1975.
29. Mühlemann, H.R., Schmid, R., Noguchi, T., Imfeld, Th. and Hirsch, R.S., Abstracts, 24th Congress of the European Organization for Caries Research, p. 87, Megeve, France, June 1977.
30. Plüss, E.M. submitted for publication in J. Clin. Periodont, 1977.
31. Havenaar, R., Huis in 't Veld, J.H.J., Backer Dirks, O., and de Stoppelaar, J.D. Abstracts, 24th Congress of the European Organization for Caries Research, p. 84, Megeve, France, June 1977.
32. Mäkinen, K.K. Int. Dent. J. 22, 363, 1972.
33. Mäkinen, K.K. in Sugars in Nutrition, Sipple, H.L. and McNutt, K.W., Eds. Academic Press, Inc., New York, N.Y., 1974.
34. Mäkinen, K.K., Virtanen, K., Makinen, P-L. and Neva unpublished data,
35. Hoogendoorn, H., Piessens, J.P., Scholtes, W. and Stoddard, L.A. Caries Res. 11, 77, 1977.
36. Chandra, T., Das, R. and Datta, A.G. Eur. J. Biochem. 72, 259, 1977.
37. Förster, J., Boecker, S. and Walther, A. Fortschr. Med. 95, 99, 1977.
38. Mäkinen, K.K., Tenovuo, J. and Scheinin, A. J. Dent. Res. 55, 652, 1976.
39. Harper, L.R., Poole, A.E. and Wolf, S.I., Abstracts, 55th Session of the Internat. Assoc. for Dental Research, Copenhagen, Denmark, #8, 1977.
40. Mäkinen, K.K., Bowen, W.H., Dalgard, D. and Fitzgerald, G. submitted for publication in J. Nutr.

DISCUSSION AFTER DR. MÄKINEN'S PAPER

DR. DEMETRAKOPOULOS: Dr. Mäkinen. What was your control in the chewing gum study? Was there a control group using the chewing gum base without any carbohydrate addition?

DR. MÄKINEN: The group of individuals who used the sucrose chewing gum were our controls. We deliberately wanted to compare a xylitol-sweetened gum with a sucrose-sweetened gum. We did not have any group chewing gum base or paraffin as control groups because we considered that it would be impossible in practice to maintain their interest and participation in a long-term study.

We did our utmost to create for this study gums which would resemble each other as closely as possible with regard to sweetness, taste, appearance, and other properties. Our purpose was to compare sucrose with xylitol in the frame of reference of a chewing gum, not sweetened versus unsweetened gums, nor people who chew versus people who do not chew gum. Furthermore, whatever control groups are used in chewing gum studies, one must consider the fact that sweetness itself stimulates saliva. An unsweetened gum does not stimulate salivary flow as much and this difference would cause difficulties in interpretation of the data.

DR. NAVIA: In the most recent issue of the Journal of Nutrition (Bird, J. L., et al., J. Nutrition 107,1743, 1977) Dr. Bowen and you reported that the feeding of xylitol to monkeys increased the amylase activity of saliva. Is there a similar effect in rodents and in humans?

DR. MÄKINEN: The consumption of xylitol for two or three days increased the amylase, lactoperoxidase, and protein levels in the saliva of these monkeys. Similar data were obtained by Dr. Poole at the University of Connecticut. If rats are given xylitol or glucose in the drinking water for three months, the lactoperoxidase levels are slightly increased in the salivary and some other exocrine glands, but we have not been able to investigate rodent saliva.

DR. NAVIA: When we do animal experiments with a replacement of 5% starch by xylitol, I have not seen any cariostatic effect. When we use 10 to 20% xylitol, there is a reduction in caries. At these high levels, the rats have to become accustomed to the xylitol by stepwise increases for the initial low levels. To what extent do these collateral effects perhaps modify the overall experience in our animal model?

DR. MÄKINEN: I do not know about rodent saliva, only about the glands. This preliminary finding has to be verified. There seems to be a correlation between the consumption of sugars and the levels of certain exocrine gland enzymes and proteins, including some glycoproteins.

DR. AMOS: I'd like to ask what the timing is in the various experiments, when do you give what, what follows what?

DR. NAVIA: In the animal experiments we provide both the starch and the xylitol together in the diet so they are present in the mouth together whenever the rats eat.

DR. MÄKINEN: The study with monkeys was shorter, from one to three days only, in monkeys. Xylitol was given mixed in food. So it is very difficult actually for me to specify the kind of conditions required in order to see these lactoperoxidase, amylase, and protein effects. We know that the effect is dose dependent and time dependent: you have to allow enough time for protein synthesis. And it seems to be pulsatory and controlled by certain cation levels. The exact mechanism is unknown, but we know that the effect is real, although its significance is obscure.

DR. DEMETRAKOPOULOS: In your dietary study with human subjects, did you calculate the caries increments per unit sucrose consumed, because the subjects used less xylitol than the subjects in the sucrose group used sucrose?

DR. MÄKINEN: We recognized several reasons why the xylitol-consuming subjects used less xylitol than the sucrose-consuming ones used sucrose. I do not have time enough to explain it now, as it is a very long story. I can tell you that this somewhat lower consumption of xylitol did not have any significant effect on the results when compared with the strong reduction in caries which was observed. The difference between the consumption of xylitol and fructose or sucrose was high enough to explain the caries reduction.

DR. BOWEN: Just one or two points about the gastric intubation study with xylitol that we did with Dr. Mäkinen in primates. The observation of increased level of amylase in the parotid saliva is in addition to his observation made in the human studies where a decrease in the amylase content of whole saliva occurred.
We found the increase in lactoperoxidase, whether the xylitol was in the diet or given by gastric intubation. Lactoperoxidase was found also in the submandibular gland as well as in parotid saliva. In a short experiment that we did with sorbitol, enhanced levels of lactoperoxidase were observed also.

DR. MÄKINEN: Yes, that's certainly true. So far we have to figure that polyols like sorbitol and xylitol may have common effects in this sense.

DR. MACKAY: Dr. Mäkinen, you stressed the importance of structure when you showed the pentitol structure. Have you any indication of what other pentitols might do? Have you any indication what would happen with pentose sugars such as xylose, ribose, arabinose, or with other structures similar to xylitol? Incidentally, the three pentoses are the only simple carbohydrates left in beer; in that sense they are truly non-fermentable.

DR. MÄKINEN: I would say that all carbohydrates with a similar structure as xylitol would behave at the molecular levels in the same way, but there are indications, for example, from studies carried out on human pancreas that sorbitol and xylitol are transported through the pancreatic cells at different rates and cause different types of effects on insulin secretion, while all sugars which resemble xylitol may behave in the same way. Sorbitol behaved differently.

DR. MACKAY: Is there a difference due to the stereo-chemical confirmation, then, of xylitol, knowing its symmetry, rather than to a size effect?

DR. MÄKINEN: We have to take into account both the size effect and the configuration. The size of the substrate becomes very important in enzyme reactions.

DR. ROUSSOS: Dr. Mäkinen, what significance do you ascribe to the so-called negative reversals in your human studies? In calculating caries activity, you have measured "positive reversals" and "negative reversals," and then calculated net activities that are the algebraic sum of the two types of reversals. What does a negative reversal mean to you?

DR. MÄKINEN: A negative reversal means an effect which we see in practice as a remineralization and rehardening.

DR. ROUSSOS: Would you expect that xylitol, particularly in the two year study during which sucrose substitution by xylitol was almost complete, that the negative reversals in the xylitol group would considerably outnumber the negative reversals in the fructose and sucrose groups? However, contrary to expectations, the observed cumulative negative reversals were practically of identical magnitude in all three groups.

DR. BOWEN: Dr. Mäkinen, in your data from the feeding experiment, you showed graphically at the end of one year no difference between the fructose and sucrose groups. However, there was a difference between the sucrose and fructose groups at the end of two years. And then when you broke down the data into the various categories of caries from C's and CR-2's and so on, it appeared that the sucrose group had less caries than the fructose group. I wonder if you care to clear up my confusion.

DR. MÄKINEN: You are absolutely right. When we did the calculations we found that, too. This just indicates, as you mentioned, that sucrose and various monosaccharides like fructose, glucose, etc. may be more or less similar with regard to the cariogenic potential, but depending on circumstances, a reduction may be observed in the overall caries incidence, even with some monosaccharides. The calculation shown below (Acta Odont. Scand. Suppl. 70, p. 76, 1975) demonstrates that the two types of changes corresponded to a 65.5% reduction in the incidence of dental caries in the xylitol group compared with the sucrose group. Consequently, the strong reduction in the incidence of caries reported in the above article was not totally based on changes from 0 to Cl + CRl only. However, when these changes were taken into account, a reduction in the incidence of dental caries was observed in the fructose group after the first study year.

DR. SHAW: Earlier in this discussion Dr. Demetrakopoulos asked whether Dr. Mäkinen had adequate controls in the Turku chewing gum study. Also, after Dr. Swango's paper, Dr. Nizel asked about the suitability of

	S	F	X
$0 \rightleftharpoons (C2 + CR2)$	0.8	1.3	0.3
$C1 \rightleftharpoons C2$ }			
}	2.1	2.0	0.7
$CR1 \rightleftharpoons CR2$ }			

$$2.9 \quad 3.3 \quad 1.0$$
$$\underbrace{\qquad}_{65.5\%}$$
reduction

the controls planned in the xylitol chewing gum study supported by the
National Institute of Dental Research.

Everyone involved in the conduct of clinical trials realizes that
this whole subject, the design, and especially the decisions about suitable controls, is complex and difficult The dilemma is especially clear
in an area such as the evaluation of whether a product formulated hopefully to be non-cariogenic is indeed less cariogenic than a standard sugar-sweetened product or possibly is non-cariogenic or even anticariogenic.
Some investigators are inclined to say that no satisfactory clinical test
of the cariogenicity of a specific product is possible in our current
milieu.

In addition to the usual determination of the necessary size of the
groups, the length of the study and the procedures for clinical assessment,
it is of the utmost importance that the control and test products be equally
acceptable to the subjects in view of the frequency and length of time that
the products must be used to obtain a satisfactory test; otherwise the different usage of the products between groups or the loss of excessive numbers
of subjects from one or more groups may invalidate the test. The investigator must also decide what frequency of product usage is likely to be effective in the current milieu of numerous daily uses of sweet, presumably
cariogenic products. If the latter products are used eight or 10 or more
times daily, how often will a completely non-cariogenic product have to be
used to exert a detectable influence on the clinical expression of caries?
In addition, for ethical and moral reasons, a control group cannot be included in which the subjects are required to use the standard sugar-sweetened product with the same frequency that the modified test product is to
be used by the experimental group. The presumption is that the standard
sugar sweetened product would cause a higher incidence of carious lesions
among the controls than they would otherwise have. The guidelines of the
Bureau of Drugs eliminate the use of such a sugar control group. In addition, under the provisions of the National Research Act (Public Law 93-348)
all behavioral and biomedical research involving human subjects conducted
under university auspices must be evaluated and approved by a properly constituted committee. If my colleagues and I at Harvard were to submit a
clinical research protocol that included a sugar control group to the Committee on Human Studies of the Faculty of Medicine, I would expect that
the proposal would be disapproved for that reason. Therefore, instead of a
sugar control group, the current practice in a clinical study of the influence of products on caries is for the subjects in the control group to use
a similar product presumed to be non-cariogenic, for example, sweetened by
sorbitol instead of sucrose. From this type of control the investigator is
likely only going to be able to conclude that the test product under the

condition of the trial is or is not more cariogenic than the presumably non-
cariogenic sorbitol-sweetened product.

In many ways the above controls leave much to be desired. Obviously
clinical investigators today have to be creative and resourceful in order to
obtain the necessary information under the many strictures under which clinical studies have to be designed and conducted. Another type of control that
is not used as frequently as I believe to be desirable is the ecological
control, namely well-matched subjects in the same environment who are clinically examined at the same intervals but whose lifestyle is not altered in
any way. In other words, these controls would have the caries activity typical of the population. Comparison of their caries data with that of both
the controls and experimentals would allow evaluation of whether either the
controls or the experimentals deviated from the community's baseline.

Obviously the long, laborious, and expensive nature of clinical trials
points up the need for short and reliable ways to approximate the cariogenicity of a product. Clearly knowing only the sugar concentration of a product is a grossly inadequate predictor. What combination of in vitro laboratory tests, animal assays and non-destructive human oral measurements
can be devised to remove or reduce the need for the traditional types of
clinical tests? Obviously information obtained from the former tests would
have to be standardized and corroborated by appropriate clinical trials before general acceptance of the former as satisfactory evaluators of cariogenicity. I look upon such a development as one of the most important to
be achieved in this area. The dental profession needs to know what old and
new products are anti-cariogenic or non-cariogenic in order to be able to
make wise recommendations to patients. The manufacturer needs an easy and
inexpensive way to evaluate where current products fit in the spectrum of
cariogenicity, and whether efforts to reduce cariogenicity by reformulation
of products have been effective. Such information would also enable the
producer to fairly represent the quality of his products with regard to
oral health to the consumer and give the consumer who is concerned about
his or her oral health the ability to select products accordingly.

The current status of research with xylitol: a review of dental caries trials in animals and human subjects

Philip A. Swango, D.D.S., M.P.H.

Community Programs Section, Caries Prevention and Research Branch, National Caries Program, National Institute of Dental Research, Bethesda, Maryland 20014

SUMMARY

Although abundant data now indicate that xylitol lacks significant cariogenicity when used as a sugar substitute, additional confirming studies are still needed, particularly in younger and more caries-prone groups of human subjects. The available information with respect to the anti-caries properties of xylitol is less clear. The data from animal experiments have not consistently confirmed such an effect, and conflicting results have been reported. This effect should be further investigated in coordinated trials in rodents, especially if some of the variables presented by the different animal models can be controlled. Additional studies in human subjects should also be conducted to supplement the findings reported in the Turku sugar studies.

The National Institute of Dental Research is now funding, under contract, a study to assess the effect upon dental caries of the daily use of chewing gum sweetened with either xylitol or sorbitol. This trial will differ from the Turku chewing gum study in that the subjects will be school children rather than young adults, and a sucrose-containing product will not be included in the comparison. The study, which is just beginning, is planned to extend for three years.

Other issues that should be addressed in future research include the possibility of microbial adaptation to xylitol after long-term use, the utilization of xylitol by oral pathogens other than Streptococcus mutans, and the amount and frequency of xylitol ingestion that can be comfortably tolerated by humans. The eventual usefulness of xylitol as a non-cariogenic sweetener cannot be determined until additional studies have answered these and other questions.

INTRODUCTION

The sugar alcohols, or polyols, are among the many sweeteners that have been suggested as dentally acceptable substitutes for sucrose. Although these nutrient substances provide the same number of calories as sucrose, they are of interest because they reportedly lack the ability to promote the development of dental caries[1]. Of the polyols, sorbitol, mannitol, and xylitol are particularly suitable for use in some types of food and confection products, and are now used, either singly or in combination, in a variety of commercially-available items. Sorbitol and mannitol are the most widely used, although probably as much for their hygroscopic and emollient properties as for their sweetness, which is only about 60% that of sucrose. Because of their relative lack of sweetness, these substances are often used in combination with saccharin. The sweetness of xylitol is roughly comparable to that of sucrose, so that it may be substituted for an equal amount of sucrose in some food items. Although the wholesale use of polyols as substitutes for sucrose is probably limited by their inherent tendency to produce osmotic diarrhea or laxation when consumed in quantity, they may be safely used as sweeteners in limited consumption products such as chewing gum, vitamin tablets, or dentifrices.

Dental interest in the use of xylitol as a sweetener has increased during the past few years, in spite of the lower cost and wider availability of sorbitol and mannitol. This interest is due to xylitol's sweetness, its reportedly low cariogenicity, and reports by some investigators that it may actively inhibit the development of dental caries, even in the presence of dietary sucrose[2,3]. In the present atmosphere of concern about the use of saccharin, and in the absence of other immediately available non-cariogenic sweeteners, the increased interest in xylitol assumes even greater significance.

LITERATURE REVIEW

Xylitol is a naturally-occurring pentose alcohol that can be derived from various types of cellulose products, such as wood, straw, cane pulp, or seed hulls. Like mannitol and sorbitol, it is a straight-chain carbohydrate, as distinguished from the ring-like molecular configurations of the saccharides. In dental plaque, microbial fermentation of most polyols proceeds at a slower rate than the fermentation of sucrose and results in the production of little or no acid. In testing the ability of several polyols to lower the pH recorded in human dental plaque, Muhlemann and deBoever found that mouthrinsing with 10% solutions of mannitol or sorbitol had significantly less effect in lowering interdental pH than did a similar solution of sucrose, and that xylitol rinses had no effect on pH[4]. This finding is consistent with a report by Noguchi and Muhlemann[5] that xylitol is neither fermented nor utilized by Streptococcus mutans, an acidogenic organism closely associated with dental caries formation. Fermentation of xylitol by other oral streptococci has not been reported, although London observed the utilization of xylitol by Lactobacillus casei[6].

In studies on dental caries in rodents, xylitol has usually been observed to support little or no caries activity in the absence of other dietary sugars. In a study of rat fissure caries conducted by Muhlemann, Regolatti, and Marthaler[7], xylitol and sorbitol were used to supplement a wheat-flour diet that was of low cariogenicity. The cariogenicity of the diet was not changed by the addition of xylitol at levels equal to 10, 20, or 30% of the total diet. Similar levels of sorbitol slightly but significantly increased the number of dentinal carious lesions detected. Karle and Buttner[8] also reported that a diet supplemented by up to 30% xylitol was non-cariogenic, whereas a similar diet supplemented with 30% sucrose was highly cariogenic. In their study, xylitol was also significantly less cariogenic than sucrose even in rats in which the salivary glands had been removed. Recent reports by Muhlemann et al.[9], Gehring and Karle[10], and Moll and Buttner[11], appear to confirm that xylitol is non-cariogenic in rats.

In contrast to these findings, Navia, Lopez, and Fischer[12] reported that the replacement of 5% xylitol, sorbitol, or mannitol for starch in a cornstarch-based diet in rats resulted in caries scores similar to those obtained by replacement with sucrose. The authors concluded that on the basis of animal tests, sugar alcohols offered no advantages as substitutes for sucrose in snack foods. These unexpected findings have not been replicated by others.

Xylitol was evaluated for its cariogenic potential in studies in which the ability to produce caries was simulated in human subjects. Ostrom and Koulourides[13] developed a caries model based upon the use of an intra-oral appliance bearing a slab of bovine dental enamel covered with Dacron mesh. This appliance is worn by the test subject and dental plaque is allowed to form on the surface of the bovine enamel. At fixed intervals during the day, the device is removed by the subject and immersed for 10 minutes in a solution of the substrate under test. After a suitable number of days, the slab of enamel is removed and subjected to a micro-hardness test to determine if softening of the enamel occurred as a result of acid production and demineralization. Using this model, Koulourides et al.[14] assessed the "cariogenicity" of nine sugars and polyols. Among the substrates tested, sorbitol was determined to be significantly less cariogenic than sucrose, while xylitol was non-cariogenic. These findings are in accord with most animal data discussed previously.

To date, the only experiments that directly assessed the cariogenicity of xylitol in human subjects are the Turku sugar studies conducted in Finland by Scheinin, Makinen et al.[15]. These investigations consisted of a related group of experiments, ranging from microbiological and biochemical studies to assessments of the effect of sucrose, fructose, and xylitol diets on the incidence of dental caries. The studies were accomplished largely by means of two clinical trials. The initial trial studied the effect of nearly total replacement of dietary sucrose by either fructose or xylitol[2]; the second trial studied the incidence of dental caries in relation to the daily use of chewing gum sweetened with either sucrose or xylitol[3]. In the replacement study, 125 subjects of both sexes, ranging in age from 12 to 53 years, were divided into three groups. One group was instructed to maintain customary dietary practices, including the consumption of sucrose in the usual manner. The other two groups were instructed to adhere to a regimen in which fructose or xylitol replaced sucrose in the diet to the greatest extent possible. Food products containing the test substances were provided to subjects free of charge, and consumption was ad libitum. Registrations of dental caries were made at the baseline and at frequent intervals through-out the two-year trial. In addition to the conventional decayed, missing and filled surfaces (DMFS) caries index, recurrent caries and pre-carious lesions were also recorded. Both clinical and radiographic examinations were conducted. By the end of the second year, caries in the xylitol group had developed only slightly beyond the level recorded at the baseline, while caries in the sucrose group had developed extensively, resulting in an average of about 7 new DMF surfaces per subject. Caries development in the fructose group was moderate, reaching a level about midway between the xylitol and sucrose groups after two years. With the combined diagnostic criteria that were used, it was reported that the development of new caries was about 90% less in the xylitol group than in the sucrose group.

The magnitude of this difference in incidence should be taken with the understanding that the techniques and indices used to record caries were not strictly comparable to those commonly used in clinical trials in the United States. Incipient or pre-carious lesions are usually excluded from caries assessments that use the DMF index because they are difficult to detect reliably and because not all such demineralized areas progress to the stage of frank cavitation. Many of these lesions, in fact, can be observed to remineralize spontaneously and are not detectable at subsequent examinations. The difference in caries activity between the xylitol and sucrose groups reported in Turku cannot, therefore, be directly compared with the levels of caries reduction typically reported from clinical trials of other caries-preventive regimens because the methods of measurement reflect different

levels of disease detection. Nevertheless, the low cariogenicity of xylitol
is consistent with the results of most other in vivo studies, and the high
cariogenicity of sucrose was again confirmed.

Beyond the issue of cariogenicity, however, Scheinin and his co-workers
concluded that the caries findings, along with other observations of the
physico-chemical effect of xylitol upon plaque and saliva, raised the
possibility of an anti-cariogenic effect for xylitol. These observations
led to the design of the second clinical trial of the Turku studies, in
which the use of xylitol in chewing gum was superimposed upon the normal
sucrose-containing diets of test subjects. In this trial[3], 102 young adult
subjects were divided into two groups and instructed to chew from three to
seven sticks of chewing gum per day at spaced intervals. The subjects were
predominantly medical and dental students of both sexes and had a mean age of
22 years. One group was provided with a xylitol-sweetened gum, and the
other with a sucrose-containing product. Dietary histories revealed the
groups to be similar with respect to their daily intake of sucrose, excluding
the use of the chewing gum. After a test period of one year, subjects in
the sucrose group had experienced an average of about three DMFS of new
decay, while the xylitol users had experienced little detectable caries
activity. The authors reported, in fact, that the mean caries prevalence
in the xylitol group after one year was lower, by an average of one DMFS
per subject, than the level measured in the same group at baseline.
Although this unexpected finding is nullified when recurrent caries and
changes in lesion size are included in the analysis, the authors discussed
the possibility that the negative increment may have been partly due to true
caries reversals as a result of xylitol use.

Because the reported caries decrement is the only finding that directly
suggests the possibility of a remineralizing effect of xylitol, this observation requires additional discussion. When using the DMF index, negative
increments are always unexpected, as such findings imply the disappearance
of previously recorded lesions. Nevertheless, findings of this nature are
occasionally reported in clinical trials. When a decrease in DMFS occurs,
the results are usually attributed to unintentional changes in diagnostic
criteria from one examination to the next, rather than to remineralizing
properties of the agent under test. Although these findings can seldom be
explained with certainty, it does not seem appropriate to attribute negative
increments to a therapeutic effect in the absence of replicate findings or
other strongly corroborating evidence. This caution is particularly appli-

cable to the Turku chewing gum trial because the use of xylitol was compared only to the use of sucrose in gum which is known to be cariogenic
product as reported by Finn and Jamison[16], as well as in other unpublished
studies. Because a control group that used a non-cariogenic gum was not
included in the Turku study, it is not possible to determine whether the
differences recorded were due to increased caries activity in the sucrose
group, caries inhibition in the xylitol group, or a combination of these
effects. Definitive evidence of an anti-cariogenic effect of xylitol is,
therefore, lacking.

Anti-caries properties of xylitol have not been confirmed by findings
from other in vivo research. In the previously mentioned study by Muhlemann, Regolatti and Marthaler[7], the addition of xylitol to the diet of rats
neither increased nor decreased the level of caries produced by the same
diet without xylitol supplementation. Similarly, in an experiment conducted
by Bowen and Amsbaugh[17] at the National Institute of Dental Research (NIDR),
xylitol did not reduce the cariogenicity of a sucrose-containing diet. In
this trial, groups of rats were fed a sucrose- and starch-containing basic
diet that was supplemented with either 5% xylitol, 5% sorbose, 5% sorbitol,
5% mannitol, or a mixture of 5% xylitol and 5% sorbose. Supplementation
with xylitol or with a xylitol-sorbose mixture did not result in reductions
in the numbers of smooth surface carious lesions on rat molars. In an earlier experiment, also performed at NIDR[17], rats fed ad libitum on a cariogenic diet containing 56% sucrose were subjected twice daily to mouthrinsing
with a 1.3 M solution of either sucrose, xylitol or sorbitol or with deion-
ized water. Because a previous experiment with the same animal model had
shown that mouthrinsing with a 0.2% solution of sodium fluoride strongly
inhibited caries formation in rats, it was reasoned that an anti-caries effect of xylitol might be revealed in a similar manner. A reduction in the
incidence of caries was not observed, however, in either the xylitol or sorbitol groups.

In contrast to these reports, Grunberg, Beskid, and Brin[18] observed
that the percentage of rats developing caries in lower molars was markedly
reduced when xylitol was used to replace 10% of a cariogenic diet that did
not contain sucrose. In the same trial, however, the percentage of animals
showing caries in second and third molars was also reduced by supplementing
the diet with 10% sucrose, although the reduction was not as pronounced as
in the xylitol group. An interpretation of the results of this trial is
further hindered by the fact that in the xylitol group, only 7 out of 20

animals survived the 21-week experiment. Survival rates were generally
poor in the other sugar groups as well. This observation is of interest because poor animal health is often associated with changes in the frequency
and amount of feeding, which affect the pattern of caries development.

Other more recent in vivo studies of the partial substitution of sucrose
by xylitol have been conducted, although some of these reports are available
only in abstract form at the time of this writing. In a study by Mundorff
and Bibby[19,20], in which various sucrose and xylitol combinations were fed to
groups of rats by means of an automatic feeding machine that delivered 10
separate feedings per day, a control group was fed a cariogenic diet
containing 67% sucrose. In 3 of the test groups, xylitol was provided in
place of sucrose in some of the daily meals, and comprised either 5, 10, or
25% of the daily sucrose ration for these groups. In a fourth group,
xylitol replacing 10% of the sucrose was mixed with the basic diet and fed at
all ten daily feedings. Significant differences in the number of carious
lesions that developed during the six-week period were not observed between
any of the groups. When caries severity was measured, however, severity
scores were significantly lower in two of the groups receiving xylitol at
separate feedings than in the group receiving the full sucrose regimen. The
groups with lower scores were those in which xylitol replaced either 5 or 10%
of the sucrose, and the severity scores in these groups were 22% and 23%
lower, respectively, than in the control group. The group receiving xylitol
mixed with the sucrose diet in all feedings did not experience significant
reductions in either the number or the severity of carious lesions.

Moll and Buttner[11] also reported the results of a trial involving
alternate feedings with xylitol and sucrose. In their study, six groups of
rats were fed 14 meals per day by a programmed feeder. One group received a
non-cariogenic basal diet, a second group received the basal diet supplemented with 30% sucrose, and a third group received the basal diet supplemented with 30% xylitol. In three additional groups, animals received the
xylitol-supplemented diet for either 7, 10, or 13 meals, and the sucrose
supplement in the remaining feedings. At the end of the six-week test
period, the results showed that the 30% xylitol mixture was non-cariogenic.
Between-meal feedings with 30% sucrose did not significantly increase the
cariogenicity of the xylitol diet. These results appear to indicate that in
a diet based primarily upon xylitol as the major carbohydrate, supplementary
feedings with sucrose are without cariogenic effect. Other findings of this
nature have not been reported. The authors noted, however, that animals

tended to avoid eating the xylitol rations, which may have affected the caries attack rate in these groups.

In a recently-reported trial by Muhlemann et al.[9], xylitol was non-cariogenic in rats in a 40-day test, and it apparently reduced the cariogenic potential of sucrose administered simultaneously. The animals consuming xylitol diets, however, suffered severe diarrhea and gained less weight than control animals. The possible effect of poor animal health upon dental caries development has been discussed previously.

DISCUSSION

The extent of xylitol's usefulness as a dentally acceptable substitute for sucrose will depend upon both its cariogenicity and its anti-cariogenic properties in the presence of dietary sucrose. With regard to cariogenicity, the existing data are substantially in agreement that xylitol lacks significant caries-promoting properties. This conclusion is supported by findings from in vitro experiments, animal studies, and clinical trials with human subjects. The conflicting report by Navia, Lopez and Fischer[12] of investigations conducted in rodents has not been corroborated by other investigators.

The existing data are much more equivocal, however, regarding the possibility of anti-caries properties of xylitol which may be due in part to the great variation existing between the animal models that have been used in caries trials. Differences in the animal strains used, the composition of the basal diet, the amount of sucrose or xylitol added, the frequency of feeding, the amount of food consumed, the age of the animals, the length of the experiment, the health and weight gains of the animals during the trial, and the methods used to score dental caries can probably account for much of the lack of uniformity in the reported results. However, some tentative conclusions can be made. In general, when xylitol has been mixed with sucrose-containing diets fed to laboratory animals, little or no caries inhibition has been reported. In experiments in which xylitol replaced sucrose at one or more meals during the day, some evidence has been reported of reduced caries activity in the xylitol-fed animals, when compared with groups fed a full regimen of sucrose. In short, the ability of xylitol to inhibit dental caries formation in the presence of dietary sucrose has not been consistently observed in rodent trials, although some equivocal evidence of this effect has been reported.

In human subjects, the only trial reported in which xylitol has been superimposed upon a sucrose-containing diet is the chewing gum study conducted in Turku[3]. While this study showed striking differences in dental caries development between users of xylitol and sucrose in chewing gum, it cannot be clearly shown that these results were due to an anticaries effect of xylitol. In fact, it seems plausible that much of the difference in caries attack between groups could be attributed to the known cariogenic effect of sucrose, since subjects in that group used an average of about four sticks of sucrose-containing gum per day during the one-year study. The inclusion of a control group that did not use a sucrose gum might have permitted a more definitive interpretation of the findings.

In the rodent studies, the frequent observation of poor weight gains and, in some cases, poor survival rates in xylitol-fed animals requires at least a brief discussion of the safety of xylitol for human use. Osmotic diarrhea is a common problem with some polyols, including xylitol, sorbitol, and mannitol. The occurrence of this condition is highly dependent upon the quantity of material consumed, and seems to be reported most often among animals fed diets containing more than 10 to 20% xylitol (or other polyols). It should also be pointed out that the condition can be minimized by having an adaptation period during which the polyols are gradually introduced into the diet. In the Turku sugar studies, diarrhea and flatulence were reported among some xylitol users in the sucrose replacement experiment, although these problems were usually transitory, and subsided after a period of adaptation[21]. Similar occurrences were not reported during the chewing gum trial. It seems appropriate to conclude that small amounts of xylitol are well accepted by human subjects, but that larger amounts may present problems of tolerance and adaptation.

REFERENCES

1. Shaw, J.H. J. Dent. Res. 55, 376-382, 1976.
2. Scheinin, A., Makinen, K.K., and Ylitalo, K. Acta Odont. Scand. 33, Suppl. 70, 67-104, 1975.
3. Scheinin, A., Makinen, K.K., Tammisalo, E., and Rekola, M. Acta Odont. Scand. 33, Suppl. 70, 307-316, 1975.
4. Muhlemann, H.R. and deBoever, J. in Dental Plaque, edited by W.D. McHugh, Livingstone Press, Edinburgh, pp. 179-186, 1970.
5. Noguchi, T. and Muhlemann, H.R. Schweiz, Mschr. Zahnheilk. 86, 1361-1370, 1976.
6. London, J.P., Laboratory of Microbiology, National Institute of Dental Research, personal communication, August 1977.
7. Muhlemann, H.R., Regolatti, B. and Marthaler, T. Helv. Odont. Acta 14, 48-49, 1970.
8. Karle, E. and Buttner, W. Deutsche Zahnartzl. Z. 26(11), 1097-1108, 1971.
9. Muhlemann, H.R., Schmid, R., Noguchi, T., Imfeld, T., and Hirsch, R.S.

Caries Res. 11, 263–276, 1977.
10. Gehring, F. and Karle, J.E. European Organization for Caries Research (ORCA) Abstr. 71, 1977.
11. Moll, R. and Buttner, W. European Organization for Caries Research (ORCA) Abstr. 72, 1977.
12. Navia, J., Lopez, H. and Fischer, J. J. Dent. Res. 53 (Special Issue) Abstr. 611, 1974.
13. Ostrom, C.A. and Koulourides, T. Caries Res. 10, 442–452, 1976.
14. Koulourides, T., Bodden, R., Keller, S., Manson-Hing, L., Lastra, J., and Hausch, T. Caries Res. 10, 427–441, 1976.
15. Scheinin, A. and Makinen, K.K., editors, Turku Sugar Studies I-XXI, Acta Odont. Scand. Vol. 33, Suppl. 70, 1975.
16. Finn, S.B. and Jamison, H. J. Am. Dent. A. 74, 987–995, 1967.
17. Bowen, W.H. Caries Prevention and Research Branch, National Caries Program, National Institute of Dental Research, personal communication, July 1977.
18. Grunberg, E., Beskid, G. and Brin, M. Internat. J. Vit. Nutr. Res. 43, 227–232, 1973.
19. Mundorff, S. and Bibby, B. J. Dent. Res. 56 (Special Issue B) Abstr. 339, 1977.
20. Mundorff, S. Eastman Dental Center, personal communication, July 1977.
21. Makinen, K.K. and Scheinin, A. Acta Odont. Scand. 33, Suppl. 70, 105–127, 1975.

DISCUSSION OF DR. SWANGO'S PAPER

DR. BEIDLER: When you say sucrose substitution by xylitol, was this done on the basis of gram by gram, mole by mole, or sweetness by sweetness?

DR. SWANGO: The procedure varied from one study to another. In most cases xylitol was replaced for sucrose or starch on a gram for gram basis. In others, xylitol was added as a supplement mixed into the cariogenic or stock colony diet that was provided to the controls. In a third type of study with a periodic feeder, xylitol was given in place of sucrose for a certain number of feedings per day while the animals received sucrose at the other feedings.

DR. NAVIA: It is very important to emphasize that the effect observed in a caries assay is related to the kind of microbial flora in the system and the kind of diet. In our first studies, we used a 5% level of xylitol because that was the level that our rats tolerated well. The rats were not inoculated with any one organism, but had a native flora which produced caries. Although S. mutans may be the most important, or at least a very important, organism in terms of caries, a series of organisms is capable of producing carious lesions, and the rat mouth usually harbors numerous of them. When I fed 5% xylitol in a starch diet in comparison with 5% sucrose in the same diet, I did not see any caries difference between the two groups. Later in other experiments we compared a fructose jam versus a xylitol jam versus a sucrose jam, providing these materials in a Kønig periodic feeder. At that time we saw no difference between the ability of these three products to cause caries in rats. However, when the animals were accomodated to xylitol by stepwise increases to 10 or 20% xylitol by further replacement for starch, then I started to see a reduction in caries in comparison to rats fed the same levels of sucrose.

I know that xylitol is not fermented by S. mutans 6715 and this organism does not grow when xylitol is the only carbon source. I would like to make a plea for a careful definition of the flora in rat or monkey experiments. Unless that is done, the response cannot be predicted or evaluated.

DR. AMOS: In support of Dr. Navia, I think work in which the organisms are not defined does not allow for adequate comparisons. The plaque and the microorganisms in it need to be studied intently at the beginning, during various phases of any experiment, and at the end.

DR. GEY: I would like to discuss the logistics of experimental rat caries. The most cariogenic components of the human diet are the high sugar between-meal snacks. In contrast, the diets in most assays on experimental caries in rats have a constant ratio between xylitol or other sucrose substitute, and cariogenic sucrose throughout the assay. Whenever the rat eats, both compounds (the cariogenic and the non-cariogenic one) are present simultaneously. This procedure does not reflect the normal conditions of human food consumption. In the human diet, xylitol or any other sucrose substitute may be considered of greatest value for use in confections consumed between meals. In order to mimic the intermittent consumption of food and xylitol or other sucrose substitutes in a human diet, the rat should be given xylitol alternately between feedings of a cariogenic diet.

This type of intermittent administration of xylitol and a sucrose containing diet was tested. Male Wistar rats received a restricted amount of cariogenic diet 2000 containing 56% sucrose, fed in three portions daily at fixed times. One group served as controls with no other treatment. In five other groups, twice after each feeding period, a small volume of 60% solutions of sucrose substitutes (sorbitol, sorbose or xylitol) was topically applied by dropping on to the molars and painting them, using a soft artist's brush; water or a 60% solution of sucrose was used for comparative purposes. These "snacks" representing 10% of the calories provided by the diet were given 30 and 90 minutes after each feeding period. To sum up, each regular but quantitatively defined cariogenic feeding was followed by two paintings of the sucrose substitutes, sucrose or water. After five weeks the number of molar lesions observed are shown in this table. The 60% sucrose solution increased the number of lesions significantly in comparison to both controls: by about 35% compared to controls treated with water and by about 50% in comparison to controls consuming only diet 2000. Topical applications of sorbitol or sorbose solutions had neither a positive nor a negative effect since the caries values of sorbitol- and sorbose-treated rats were in the range of both control groups. Xylitol was the only compound that diminished the number of carious lesions slightly but significantly in comparison to both control groups, i.e. by about 20 and 27% in comparison to water-treated and untreated controls, respectively. This result is in agreement with the alternate feeding experiment of Moll and Büttner (1977) which was described in Dr. Swango's review.

In the present experiment the restricted feeding pattern necessarily reduced the growth in all animal groups and diminished also the laxative effect of sorbitol and xylitol substantially. Still the final body weight in the xylitol-treated rats was 12% lower than that of water-treated controls, while the weight of sorbitol-treated animals was 15% lower. Although the feces of the polyol-treated rats showed some initial softness, intestinal disturbance by xylitol was not likely a major factor in the reduction of caries; the sorbitol-treated rats had at least the same reduction in weight gain but sorbitol lacked any significant beneficial effect on caries. In further studies, even modest laxation should be avoided by stepwise adaptation of the weanling rat to xylitol by a weekly increase of 5% in the xylitol content of the diet. The present results suggest that

Table :   Effect of frequent topical application of
            sucrose substitutes § on rat caries

Topical application	Molar lesions	Final body weight (gm)[+]
60% sucrose (45)	8.0 ± 0.2 *	66 ± 1
none (45)	5.9 ± 0.2 o	71 ± 1 *
60% sorbitol (25)	5.9 ± 0.3 o	58 ± 2
water (25)	5.4 ± 0.2	68 ± 2
60% sorbose (20)	5.0 ± 0.4 o	65 ± 2 *
60% xylitol (44)	4.3 ± 0.3 *	60 ± 1

o $p > .05$ } in comparison to water-
* $p < .01$ } treated controls
§ twice after each of 3 meals
+ initial body weight = 32 ± 1 gm

this type of protocol with intermittent topical application of sucrose substitutes or corresponding protocols for alternate feedings with xylitol may be useful to differentiate between non- and anti-cariogenic potentials of xylitol in the rat.

DR. AMOS: How did it come about historically that xylitol is used in all these experiments and not xylose?

DR. DEMETRAKOPOULOS: Xylose is not as sweet, with a value of .67 compared to sucrose and xylitol which are 1.0.

DR. MACKAY: A toxicity question about xylose and cataract formation has never been satisfactorily resolved, so any industrial interest disappeared. Xylitol attracted attention because it appears to have a better health picture.

DR. DEMETRAKOPOULOS: I have to defend xylose. Those experiments were poorly designed (Darby, W.J. and Day, P.L., J. Biol. Chem. 133, 503, 1940). In a specific strain of weanling rats, 50% xylose was added to the diet and cataracts were observed but were transient and disappeared when the diet was continued. Galactose or lactose at that level will cause cataracts also. I do not think that an adequate study of toxicity of xylose has been conducted. Actually, xylose is used now in the clinic for tests. About 40 grams can be given to a patient as a single oral dose for diagnostic purposes.

DR. NIZEL: Dr. Swango, you mentioned the fact that in a current study the effects of sorbitol in chewing gum and xylitol in chewing gum are being compared. Do you have any control group?

DR. SWANGO: There are no groups other than the two comparison groups.

DR. NIZEL: How can you come to any conclusion about anti-cariogenic effects?

DR. SWANGO: We do not intend to. The intention is to compare xylitol-sweetened gum with the present standard sugarless gum, sweetened with sorbitol.

DR. NIZEL: Your conclusions at best can only be that the xylitol gum is non-cariogenic rather than anti-cariogenic. It seems to me that you really want to know whether the xylitol gum is anti-cariogenic.

DR. SWANGO: We believe that it is important to establish whether this product is non-cariogenic in young subjects with a higher caries susceptibility than in the Finnish studies, or, in other words, the kind of clinical trial that Dr. Shaw proposed in his review. Further it is unethical to have a control group chewing sucrose sweetened gum.

DR. NIZEL: Yes, I understand that. But it would seem to me at least a non-chew group would be a basic control. Such a group could be made up from children of the same age in the community who are not in the experiment but who would represent the typical caries pattern for that community.

SESSION III.

SYNTHETIC SWEETENERS

Moderator:

Guy. A. Crosby

Chemical Synthesis Dynapol
1454 Page Mill Road
Palo Alto
California 94304

Essential properties and potential uses of synthetic sweeteners

Karl M. Beck

Chemical and Agricultural Products Division, Abbott Laboratories, North Chicago, Illinois 60064

Sweetness is a highly desirable taste for many animals, and the history of man is replete with evidence of the quest for things that taste sweet. Rock paintings in caves near Valencia, Spain show that honey was harvested and used in Neolithic times, and cane sugar was reported by Theophrastus in 300 B.C. Cane sugar was introduced in Western Europe in 1187 and historians cite the craving for sugar as a major factor in the development of slave trade as a cheap source of labor to provide the "honey from reed"[1]. Manna from heaven probably tasted so good because it contained mannitol.

It has long been known that things other than sugars are sweet. Lead acetate was called "sugar of lead" long before it was given a chemical name, and it was used by the Borgias to sweeten cocktails for their political adversaries. Early in the development of the field of organic chemistry it was observed that various chemical compounds exhibit sweetness; in 1831 it was reported that chloroform has a sweet taste[2].

Synthetic sweeteners began with the observation by Fahlberg in 1879 that o-sulfobenzimide, which we now know as saccharin, has an intensely sweet taste[3]. Dulcin (p-ethoxyphenylurea) was reported in 1883. Since then hundreds of synthetic organic chemicals have been described as having a sweet taste[2,4,5] and the search continues for new synthetic sweeteners.

How do synthetic sweeteners differ from sugars, and what uses arise from these differences? With sugars like sucrose, glucose, lactose, high fructose corn syrups all readily available, why is there a need also for synthetic sweeteners? What are the practical criteria for a commercial synthetic sweetener? What sweetening agents have met these criteria in the past, and what is the outlook for new and better synthetic sweeteners? This overview of synthetic sweeteners will address these questions.

Synthetic sweeteners and the commercially used sugars have one property in common -- they are all sweet to the human taste. Beyond that, there is

an impressive number of differences. Some of these differences are
illustrated by a comparison of the properties of cyclamate and sucrose
(Table 1).

Table 1. How sucrose and cyclamate differ

Properties	Sucrose	Sodium cyclamate
formula	$C_{12}H_{22}O_{11}$	$C_6H_{11}NHSO_3Na$
chemical class	carbohydrate	salt
sweetness	1	30 to 60
melting point	160-186° (decomp.)	480-500° (decomp.)
solubility in water	68 g/100 cc	21 g/100 cc
specific gravity	1.176 (40% soln)[a]	1.002 (1% soln)[a]
viscosity	5.187 (40% soln)[b]	1.04 (1% soln)[b]
food calories	4 kcal/g	0

a - at 20°C   b - at 25°C

Synthetic sweeteners usually are nonnutritive. Sugars are carbohydrates and they are metabolized and used as food by the body. Generally,
sugars furnish four kilocalories per gram. Synthetic sweeteners generally
cannot be used as food. Since they are non-caloric, synthetic sweeteners
can be used as replacements for sugar in the formulation of low sugar food
for diabetics and low calorie foods for people on calorie restricted diets.

Synthetic sweeteners characteristically are intensely sweet in contrast
to sugars for which sweetness ranges from 1/3 to 1-1/3 times that of sucrose
(Table 2). Since synthetic sweeteners exhibit sweetness factors from 20 to

Table 2. Relative sweetness of carbohydrate sweeteners

sucrose (standard)	1.0
glucose	0.7
fructose	1.3
lactose	0.4
maltose	0.5
sorbitol	0.5
mannitol	0.7
glycerin	0.7

several thousand times that of sucrose (Table 3), very small concentrations
of synthetic sweeteners can be used to achieve acceptable levels of sweetness.
Synthetic sweeteners have advantages over sugars where low sweetener concentration, physical weight, viscosity, density -- any of the results of a low

Table 3. Relative sweetness of non-carbohydrate sweet chemicals
(sucrose = 1)

3-methylcyclopentylsulfamate	20
p-anisylurea	18
sodium cyclohexylsulfamate (cyclamate)	30
chloroform	40
methoxy-2-amino-4-nitrobenzene	167
p-ethoxyphenylurea (dulcin)	200
6-chlorosaccharin	200
sodium benzosulfimide (saccharin)	300
n-hexylchloromalonamide	300
stevioside	300
naringin dihydrochalcone	300
2-amino-4-nitrotoluene	300
p-nitrosuccinanilide	350
p-methoxymethylnitrobenzene	500
1-bromo-5-nitroaniline	700
6-chloro-D-tryptophane	1000
perillaldehyde oxime (perillartine)	2000
neohesperidine dihydrochalcone	2000
5-nitro-2-n-propoxyaniline (P-4000)	4000

solids content -- are desirable. This property is very helpful in
sweetening the resin coating of tablets or in making smaller chewable
tablets[6]. The amount of sugar needed to sweeten the packing syrup for fruit
sections gives a syrup with specific gravity higher than that in the fruit
and causes shrinkage of the fruit. The low concentration of synthetic
sweeteners can reverse this osmotic pressure imbalance and provide a plumping
effect, preventing loss of texture and flavor during storage[7].

Sugars may be hygroscopic or difficult to dry. A synthetic sweetener
lacking this affinity for water has advantages in the production of freeze-
dried fruits. In liquid pharmaceuticals the cap lock problems encountered
with sugar syrups would be obviated[8].

Sugar does not have a high decomposition temperature and tends to
caramelize and char during high temperature cooking processes, such as
frying. A synthetic sweetener with greater heat stability could be used to
prepare bacon that can be fried crisp without charring[9].

Sugar serves as food for many microorganisms. A synthetic sweetener
which is microbiologically inert can be used to sweeten dentifrices, to make
non-cariogenic chewing gum, or to flavor a fermented product like yogurt.

Three needs currently exist for synthetic sweetening agents, and a
fourth may become important in the future.

First, there are over ten million diabetics in the United States. A
generation ago it was common to tell a diabetic that he must learn to forego

sweetness. A good synthetic sweetener allows diabetics to have palatable
diets with little or no insulin requirement. Synthetic sweeteners are
useful agents in the total management of diabetes.

Second, overweight repeatedly has been cited as the nation's leading
health problem. There is adequate statistical evidence that longevity is
inversely proportional to overweight. The Food and Nutrition Board of the
National Academy of Sciences recommended in 1968 that all adults in the
United States reduce their caloric intake. Recommended calorie levels are
based on age and height, but the average middle-aged male, for example,
should reduce intake from 3,000 to 2,600 calories and the comparable female
from 2,200 to 1,850 calories per day[10]. Deleting 100 grams of sugar a day
from the diet would be a 400-calorie reduction -- the amount recommended by
the Food and Nutrition Board. A synthetic sweetener can permit some of this
reduction without sacrificing palatability. For many persons with a
tendency toward obesity, watching caloric intake is a way of life. A
synthetic, nonnutritive sweetener can enable them to enjoy soft drinks,
desserts, and several other low-calorie foods in a calorie-restricted diet.

Third, as mentioned before, the physical properties of sugar do impose
some limitations or problems in its use. Synthetic sweeteners have technical advantages in certain products, especially in oral pharmaceutical
formulations, in toiletries like dentifrices and mouthwash, and also in
certain foods like fruits, fruit juices and bacon. Technical superiority
of products should be recognized and allowed.

The fourth almost seems like a contradiction of the second. While food
is readily available in the United States and overeating is a problem, half
the people in the world do not have enough to eat and feeding the world
population adequately is a serious problem. The earth is a planet of fixed
dimensions with a finite amount of arable land. As the total world
population expands, the challenge to find new and better ways to provide
enough food increases. One approach to this problem is the development of
new sources of protein -- single-cell protein from methyl alcohol or
petroleum, fish flour and edible algae. Flavoring these materials is going
to be important, and sweetness is a desirable flavor. Synthetic sweeteners,
which could be manufactured in abundant amounts, could be used to flavor
these proteins to make them more palatable.

In fact, remembering the sugar shortage a few years ago, and considering
the increasing need to make the available resources feed more people, thought
should be given to blending sugar and synthetic sweeteners to make the sugar

supply go farther. Corn and sorghum are cheaper and more abundant sources of sugar than cane and beets, but they produce glucose rather than sucrose. Glucose is as nutritive pound for pound as sucrose but it is only two-thirds as sweet. Adding a synthetic sweetener to upgrade the sweetness of glucose would have an effect of increasing the sugar supply. Perhaps someday we may have to consider "double-sweet" sugars as a means of stretching the available carbohydrates. Someday we may have to choose between stretching the food supply and triage.

Considering these uses, what are the required characteristics for a synthetic sweetening agent to be commercially useful in foods, drugs, cosmetics and toiletries?

CHARACTERISTICS FOR SYNTHETIC SWEETENERS

Sensory Properties

When sugar is used in foods it provides a sweet taste. People are accustomed to certain taste characteristics in quality of sweetness and a time-intensity curve for this taste. To have good acceptability in foods and beverages a sugar substitute should have a sweetness profile reasonably similar to that of sugar and should elicit only a sweetness perception, that is, be fairly free of bitterness or other tastes. Ideally, it should have a sweet taste that appears in a second or two and remains for about 30 seconds (Figure 1).

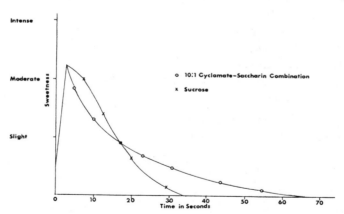

Figure 1. Both sucrose and cyclamate-saccharin in combination have above moderate sweet taste at 2 to 3 seconds. Sucrose has a rounded sweetness profile disappearing in about 35 seconds. Sweetness of the cyclamate-saccharin combination drops off somewhat faster, but total taste lasts for about 60 seconds.

A delay in appearance of sweetness in the taste profile or a lingering sweetness may be acceptable in lozenges or chewing gum, but if the sweetness response is delayed or persists, it will make most food and beverage products taste "funny". Chemicals with off-tastes or after-tastes present flavor problems. The more a sweetener tastes like sugar, the easier it will be to use it in the formulation of food and beverage products with broad consumer acceptance. Many of the synthetic chemicals that taste sweet also exhibit other tastes or they have poor sweetness profiles, so they would be difficult to use in foods and beverages.

In pharmaceuticals, ability to mask the bitterness or unpleasant taste of drugs is important. Sugar is not a particularly good masking agent, while synthetic sweeteners have demonstrated significant flavor advantages in pharmaceutical formulations.

Chemical Properties

Sweetness normally is but one component of a complex flavor system, so a sweetener usually is used along with a variety of natural and artificial flavors and colors. Therefore the sweetener must be compatible with these various chemical compounds. A synthetic sweetening agent should be chemically unreactive or inert toward all the natural constituents and other chemical additives present in products in which it is used. In pharmaceuticals the sweetener must be compatible with the drugs with which it is used.

Physical Properties

Temperature stability is an important requirement. Food processing may involve boiling, baking, pressure cooking, fast freezing or freeze drying. Stability through a temperature range of -30° to $+260^\circ$ C is a desirable property of a synthetic sweetener.

Foods, drugs and cosmetics have to be stored, shipped, and stocked on grocery or pharmacy shelves. A sweetener must be stable under various actual storage conditions. Length of stability will vary with the product, but a two-year shelf life is necessary for some, and six months is minimal for most food and beverage products.

Many products in which a sweetener is used are liquid, have a high moisture content, or go through a liquid phase in processing. Stability in solution is an important property for a sweetener. The pH of foods and beverages can range from 2.5 to 8, so stability throughout this pH range is essential for broad application.

While many foods and beverages contain sugar concentrations of 15% or less, products like jellies and syrups can be more than 50% sugar. A synthetic sweetener should be soluble enough to duplicate the sweetness of simple syrup, which is about 70% sugar.

Sugars are carbohydrates and have certain chemical and physical properties characteristic of such structures. As discussed earlier, in some technological uses a synthetic sweetener which is not a carbohydrate may have superior physical properties.

Safety

Certainly the most difficult hurdle for a synthetic sweetening agent is the proof of safety that is required for any new food additive. The old criteria of feeding 100 times maximum expected intake for two years to two species of animals, one of which is a non-rodent, seems to be passé. Longer studies and more species are being required. In the evaluation of any substance it must be remembered that nothing is absolutely safe; anything will be toxic if enough of it is ingested. The Food and Drug Administration (FDA) now seems to want studies to determine the maximum no-effect level, and to use this to establish an acceptable daily intake level to provide a 100-fold safety factor. Metabolism and excretion studies must be conducted, and separate feeding studies must be conducted with any metabolites.

Scientists are quite familiar with the great number of natural toxicants that occur in food[11,12]. Although a level of 15 parts per billion (ppb) of aflatoxin in the diet of rats will cause cancer in 100% of the animals, FDA permits 20 ppb aflatoxin in peanuts. Yet the Delaney Clause has zero tolerance for a food additive that can be shown to cause any measurable incidence of cancer at any level. It becomes exasperating to have consumer protectionists agitate for safety requirements for food additives that in effect create a system of double standards for natural substances and for synthetic chemicals. One of the most widely eaten foods is the potato; the most widely used chemical in food is salt; and the most widely used drug is aspirin. Yet with the current requirements for proof of safety, one might experience serious difficulty in obtaining FDA clearance for the potato as a new food, for sodium chloride as a new food additive, or for acetylsalicylic acid as a new drug. The risk/benefit concept seems to need some development.

A synthetic sweetening agent seems to elicit especially stringent safety requirements that certainly sucrose and lactose could never meet. Presently, for even a fairly innocuous chemical, I think one would have to plan on 10

years and more than \$20 million to collect the necessary data for a food
additive petition for a new sweetener -- and then be prepared for a
continuing battle with the self-appointed "consumer protectionists" who seem
to resist anything new.

From the long list of synthetic chemicals reported during the past 100
years to have a sweet taste, which have met these practical requirements of
chemical, physical, sensory properties and of safety, and what is the outlook
for new synthetic sweeteners?

SYNTHETIC SWEETENERS

Saccharin

Saccharin (Figure 2, I) was discovered by Remsen and Fahlberg at Johns
Hopkins University in 1879 and has been in use continuously since 1900. It
is 300 to 400 times as sweet as sucrose. Chemical and physical properties

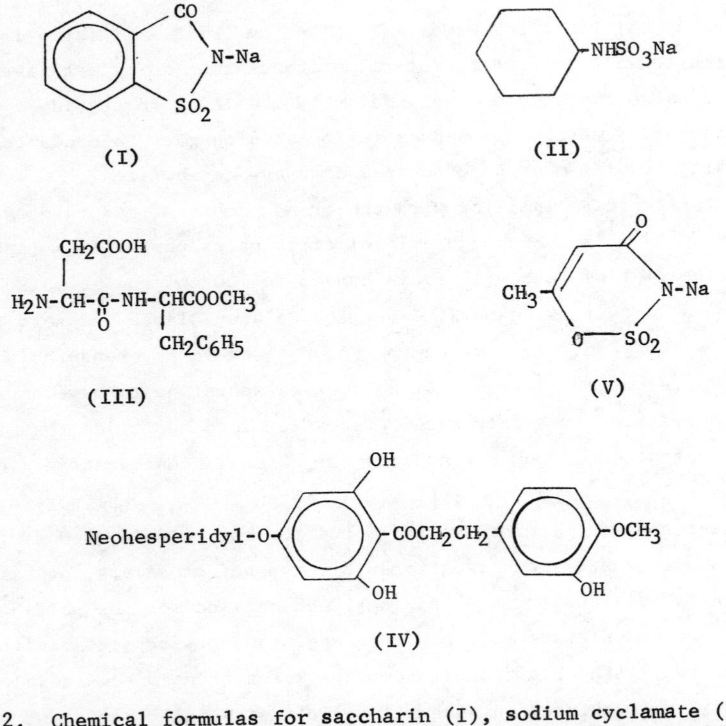

(I)

(II)

(III)

(V)

(IV)

Figure 2. Chemical formulas for saccharin (I), sodium cyclamate (II),
aspartame (III), neohesperidine dihydrochalcone (IV) and acetosulfam (V).

are quite satisfactory: it has good compatability, solubility and stability. Saccharin has a bitter after-taste for many people, which has always been a limiting factor for its use in foods and beverages[13]. Saccharin is a good component for a co-sweetener combination, where another synthetic sweetener can mask the saccharin after-taste.

Saccharin was generally considered to be safe for use in special dietary foods for over 70 years, but questions of safety caused the FDA to remove it from the GRAS (generally regarded as safe) list in 1972. A study by the Health Protection Branch of the Canadian Department of Health and Public Welfare, completed in 1977, indicated that saccharin may be carcinogenic when fed to rats. This prompted proposals in the United States and in Canada to ban all food, drug, cosmetic and toiletry uses of saccharin except single-ingredient table sweetener products.

Cyclamate

Sodium cyclohexylsulfamate (Figure 2, II) was discovered at the University of Illinois in 1937. Cyclamate was used commercially in the U.S. from 1950 until further use in this country was banned in 1970 on the basis of one study in which a few animals fed large quantities of a mixture of cyclamate and saccharin developed bladder tumors. The use of cyclamate still is allowed in more than 30 countries around the world. It has excellent physical and chemical properties and good taste characteristics. Cyclamate is 30 to 60 times as sweet as sucrose. Beginning about 1955 cyclamate-saccharin combinations, particularly a mixture in which about half of the sweetness comes from each, were used increasingly and permitted food technologists to formulate highly acceptable low-calorie and sugar-free products.

A food additive petition was filed in November, 1973 to permit cyclamate to be used in the U.S. again. This petition was denied on October 4, 1976. Abbott Laboratories requested that an administrative law judge be appointed to serve as a hearing examiner in this matter, and FDA agreed to this procedure.

Aspartame

Aspartyl phenylalanine methyl ester (Figure 2, III), developed by scientists at G.D. Searle, was approved in 1974 for use in dry products. Approval was stayed in November, 1975 and the safety data still is being reviewed. Although not completely non-caloric, it has a sweetness of 140 to

180 times sugar, so its calorie contribution is negligible. Aspartame hydrolyzes in water, so physical stability in liquid products is a serious practical limitation. The question of brain damage in neonatal mice from the aspartic acid from aspartame, acting in conjunction with glutamic acid, illustrates the complex nature of questions that can be raised in regard to the safety of sweetening agents.

NATURAL SWEETENERS

Glycyrrhizin

This is not a synthetic chemical but is a nonnutritive sweetener. Glycyrrhizin, an extract of licorice root, is 50 to 100 times as sweet as sugar. It is permitted for food use, but the pronounced licorice taste limits its use to specific products where this taste can be tolerated.

Dihydrochalcones

About 15 years ago Horowitz and Gentili at the United States Department of Agriculture in Pasadena, California developed a series of sweet-tasting dihydrochalcones which can be made from citrus bioflavonoids[4]. Two of these were interesting enough to stimulate further work -- naringin dihydrochalcone and neohesperidine dihydrochalcone (Figure 2, IV). Solubility would be a limiting factor for the naringin derivative. Taste is a limiting factor for both. The dihydrochalcones have delayed onset of taste and the sweetness persists for several minutes. Also, they have a characteristic after-taste.

A food additive petition has been filed for the use of neohesperidine dihydrochalcone in chewing gum, toothpaste, and mouthwash. The taste characteristics of this sweetener would be satisfactory in these applications.

Several sweetening agents have been studied extensively[5,15,16], and a few even used briefly in the past, but they have been found to be toxic. These include dulcin, P-4000, perillaldehyde oxime and stevioside. None of these sweeteners is likely to be approved for use in the U.S. because they do not meet the safety requirements.

Several proteins are known which have an intensely sweet taste. Although these are not synthetic sweeteners, they are sweet enough to permit formulation of low-calorie food. However, monellin and thaumatin have serious stability limitations. Miraculin is not actually a sweet-tasting substance but is a taste-modifying chemical that elicits a sweetness response from things that are tart. This characteristic would require educating consumers to accept entirely new flavor characteristics. Thus stability and taste limit the use of these proteins as sweeteners.

Some new synthetic sweeteners are currently being studied. One that has attracted particular attention is acetosulfam (Figure 2, V)[17] that was discovered at Hoechst. It is reported to have very low acute toxicity. Preliminary reports are that its taste is similar to that of saccharin and is about half as sweet as saccharin. Chronic toxicity work is underway but acetosulfam still is in an early stage of development.

The perfect synthetic sweetening agent is yet to be found. From a technical viewpoint, a cyclamate-saccharin combination is a very good, practical sweetener. Taste characteristics are close enough to sugar to allow formulation of products with excellent consumer acceptance. Chemical and physical stability are very good. The safety of cyclamate and of saccharin have been studied extensively; many scientists feel that both sweeteners are harmless under actual conditions of use. Furthermore, they are economically feasible sweeteners. Technically, this is the synthetic sweetener system to beat.

There are advantages to having several synthetic sweeteners available. From a safety consideration, daily intake of each will be lower if several sweetening agents are distributed for various uses. From an application viewpoint, it would be helpful to be able to choose the best one of several sweeteners for each specific use. If saccharin, cyclamate, aspartame, neohesperidine dihydrochalcone and some combinations of these all were allowed, a very useful spectrum of synthetic sweeteners would be available.

The alternative is for someone to develop a new and better synthetic sweetener and prove that it is absolutely safe. That is quite a challenge.

REFERENCES

1. Cantor, M.B. and Eichler, R.J. Chem. Tech. 7, 214, 1977.
2. Beck, K.M. in Kirk-Othmer: Encyclopedia of Chemical Technology, 2nd Ed., 19, John Wiley, New York,pp. 593-607, 1969.
3. Fahlberg, C. and Remsen, I. Chem. Ber. 12, 469, 1879.
4. Beck, K.M. in Symposium: Sweeteners, edited by G.D. Ingelet, Avi Publishing Co., Westport, Conn., pp. 131-140, 1974.
5. Beck, K.M. Food Prod. Devel. 9, 47, 1975.
6. Endicott, C.J. and Gross, H.M. Drug Cosmet. Ind. 85, 1976, 1959.
7. Salunkhe, D.K. et al., Food Technol. 17, 85, 1963.
8. Lynch, M.J. and Gross, H.M. Drug. Cosmet. Ind. 87, 324, 1960.
9. Beck, K.M., Jones, R.L. and Murphy, L.W. Food Eng. 30, 114, 1958.
10. Miller, D.F. and Voris, L. J. Amer. Diet. Assoc. 54, 109, 1969.
11. Coon, J.M. Food Technol. 23, 55, 1969.
12. Committee on Food Protection, Toxicants Occurring Naturally in Foods, National Academy of Sciences, Washington, D.C., 1973.
13. Helgren, F.J., Lynch, M.F. and Kirchmeyer, F.J. J. Amer. Pharm. Assoc., Sci. Ed. 44, 353, 1955.

14. Horowitz, R.M. and Gentili, B.   J. Agr. Food Chem. 17, 696, 1969.
15. Daniels, R., Sugar Substitutes and Enhancers, Noyes Data Corp., Park Ridge, N.J., 1973.
16. Farnsworth, N.R. Cosmetics and Perfumery 88, 27, 1973.
17. Clauss, K. and Jensen, H. Angew. Chem., Internat. Ed. 12, 869, 1973.

Present status and uncertainties concerning saccharin

Ferdinand B. Zienty

Food, Feed and Fine Chemicals, Specialty Chemicals Division, Monsanto Industrial Chemicals Company, St. Louis, Missouri 63166

Saccharin is 1,2-benzisothiazol-3(2H)-one 1,1-dioxide:

It is 300 to 500 times as sweet as sucrose depending upon the medium in which it is formulated.   In dilute solutions saccharin is intensely sweet, with a hint of an aftertaste that is perceived more strongly by about 15% of consumers than by the other 85%.   The aftertaste often is characterized as bitter or metallic by those sensitive to this flavor note[1].   Skillful formulation of products sweetened by saccharin results in flavors acceptable to a large body of consumers.

Saccharin occurs in white crystals or as a white crystalline powder and is odorless.   One gram of saccharin is soluble in 290 ml of water at 25° and in 25 ml of water at 100°, resulting in solutions acid to litmus.   One gram of the salt, sodium saccharin, is soluble in 1.5 ml of water, and also is white, crystalline, and odorless.   Because of its ready solubility, sodium saccharin is used much more extensively than saccharin.   Saccharin is an anhydrous product stable and non-hygroscopic under normal storage and shipping conditions.   Sodium saccharin is made in the form of its dihydrate; the powdered product tends to lose some of its water of hydration on storage.

Saccharin is compatible with most food and drug ingredients, is stable in aqueous solution and in most formulated and processed foods over a considerable pH range[2].   Thus it is unchanged in the heat processing of products like jams, jellies and canned fruits.   At pH levels below 2.5 combined with elevated temperatures, saccharin may hydrolyze slowly to o-sulfamoyl-benzoic acid and o-sulfobenzoic acid.

Priced at about $2.40 per pound in 1977, sodium saccharin is only 4 to 5% as costly as sucrose at equivalent sweetness.

Under Interim Regulation 180.37[3] saccharin and its ammonium, calcium and sodium salts are approved for use in beverages, fruit juice drinks and bases or mixes, as a sugar substitute for cooking and table use, and in processed foods. Limits per unit of food are specified. These sweeteners also may be used for technological purposes to reduce bulk and enhance flavors in chewable vitamin tablets and chewable mineral tablets, to retain flavor and physical properties of chewing gum and to enhance flavor of flavor chips used in nonstandardized bakery products. These sweeteners also are used in a variety of dose form drug products to improve flavor acceptability and in a number of cosmetics, dentifrices, mouthwashes and similar products.

The National Academy of Sciences, at the request of the Food and Drug Administration, reviewed available safety data on saccharin on several occasions and has discussed the results of pertinent studies in reports issued in 1955[4], 1970[5] and 1974[6]. Results of similar scientific studies in animals have been variable, some showing no problems attributable to saccharin, others suggesting problems and still others being inconclusive. Variations in protocols, methodology and interpretation appeared to be reflected to some extent in the results obtained. The National Academy of Sciences in 1970 recommended further long-term studies using modern protocols, more definitive metabolism studies and epidemiological studies.

A 1976 summary[7] of post-1970 scientific studies related to the safety evaluation of saccharin listed nine on metabolism (Table 1), eight on mutagenicity (Table 2), and 16 long-term feeding studies (Table 3). Reports on these studies have been submitted to the U.S. Food and Drug Administration (FDA) but few have been published. Several additional studies are in progress. Saccharin has been found to be excreted unchanged in animals and man, with little evidence of metabolism to other products.

After sustained administration to rats in the FDA study (Table 1) saccharin was reported to be widely distributed in various parts of the body with elevated levels in the kidneys, bladder and liver. In the Iowa-Illinois study saccharin was reported to cross the placenta of gravid monkeys and to be distributed widely in tissues of the fetus.

Seven of the eight mutagenicity studies (Table 2) are reported to be negative while the eighth (India) suggested the possibility of some effect.

In the long-term feeding studies (Table 3) saccharin or sodium saccharin was fed to rats and mice at levels up to 7.5% of the diet. Results from 11 of the tests were reported to show no conclusive evidence of specific potential hazard. In two additional tests, one conducted by the Wisconsin

Table 1.   Saccharin: metabolism studies

Title	Date Reported	Authors or Investigators	Type of Saccharin Used
(1) saccharin metabolism in Macaca mulatta (rhesus monkeys)	July 1971	Pitkin, Andersen, Reynolds and Filer, Jr.; University of Iowa Hospital, Iowa City, Ia., and University of Illinois at the Medical Center, Chicago, Ill.	free
(2) placental transmission and fetal distribution of saccharin	September 1971	Pitkin, Reynolds, Filer, Jr. and Kling; University of Iowa Hospital, Iowa City, Ia., and University of Illinois at the Medical Center, Chicago, Ill.	free
(3) metabolic fate of saccharin in the albino rat	April 1972	Kennedy, Fancher, Calandra, and Keller; Industrial Bio-Test Laboratories, Inc., Northbrook, Ill., and Monsanto Company, St. Louis, Mo.	free and sodium salt
(4) the metabolism of saccharin and the related compounds in rats and guinea pigs	July 1972	Minegishi, Asahina, and Yamaha; National Institute of Hygienic Sciences, Tokyo, Japan	free
(5) the metabolism of saccharin in laboratory animals	June 1973	Byard and Golberg; Albany Medical College of Union University, Albany, N.Y.	sodium salt
(6) the excretion and metabolism of saccharin in man. I. methods of investigation and preliminary results	June 1973	McChesney and Golberg; Albany Medical College of Union University, Albany, N.Y.	free
(7) saccharin: distribution and excretion of a limited dose in the rat	October 1973	Matthews, Fields, and Fishbein; National Institute of Environmental Health Sciences, National Institutes of Health, Research Triangle Park, N.C.	free
(8) excretion and metabolism of saccharin in man. II. studies with 14 C-labeled and unlabeled saccharin	April 1974	Byard, McChesney, Golberg, and Coulston; Albany Medical College of Union University, Albany, N.Y.	sodium salt
(9) the metabolism of saccharin in animals	January 1975	Lethco and Wallace; Food and Drug Administration	sodium salt

Table 2.   Saccharin: mutagenicity studies

Title	Date Reported	Author(s) or Investigator(s)	Type of Saccharin Used
(1) induction of dominant lethals in mice by sodium saccharin	April 1972	Rao and Qureshi; Osmania University, India	sodium salt
(2) study of mutagenic effects of saccharin (insoluble)	April 1972	Newell and Maxwell; Stanford Research Institute, Menlo Park, California (Food and Drug Administration contract)	free
(3) summary of mutagenicity screening studies	November 1972	Litton Bionetics, Inc., Bethesda, Md. (FDA contract)	ammonium salt
(4) summary of mutagenicity screening studies	November 1972	Litton Bionetics, Inc., Bethesda, Md. (FDA contract)	calcium salt
(5) summary of mutagenicity screening studies	November 1972	Litton Bionetics, Inc., Bethesda, Md. (FDA contract)	sodium salt
(6) dominant lethal test in the mouse for mutagenic effects of saccharin	April 1973	Machemer and Lorke; Institute for Toxicology, Bayer, Germany	sodium salt
(7) study of mutagenic effects of sodium saccharin	November 1974	Newell, Jorgenson, and Simmon; Stanford Research Institute, Menlo Park, California (FDA contract)	sodium salt
(8) progress report: study of mutagenic effects of sodium saccharin	January 1975	Oster; Bowling Green State University, Bowling Green, Oh. (FDA contract)	sodium salt

Alumni Research Foundation[8] (WARF) and the other by the FDA[9] in its own laboratories in Washington, D.C., sodium saccharin was fed to rats prior to mating and the litters from these animals were then used for the long-term studies.   Consequently, the test animals were exposed to sodium saccharin in utero and fed sodium saccharin for the remainder of the long-term test.   The WARF and FDA studies were reported in 1973 to have shown an increased incidence of tumors in the bladders of male rats as compared to the controls. The Health Protection Branch, Health and Welfare, Canada, performed a third

Table 3. Saccharin: long-term feeding studies [7]

Title	Date Reported	Author(s) or Investigator(s)	Type of Saccharin Used
(1) a comparison of the chronic toxicities of synthetic sweetening agents	November 1951 (reevaluated 1970)	Fitzhugh, Nelson, and Frawley; Food and Drug Administration	free
(2) feeding studies on sodium cyclamate, saccharin, and sucrose for carcinogenic and tumour-promoting activity	April 1970	Roe, Levy, and Carter; Institute of Cancer Research, Royal Cancer Hospital, London, England	free
(3) carcinogenic and teratogenic aspects of saccharin	August 1970	Lessel; Boots Pure Drug Co., Ltd., Nottingham, England	free
(4) long term saccharin feeding in rats	1973	Wisconsin Alumni Research Foundation Institute, Inc., Madison, Wis.	sodium salt
(5) chronic toxic research towards the carcinogenicity and teratogenicity of cyclamate, saccharin, and their mixture	March 1973	Verschuuren, Kroes, Peters and van Each; National Institute of Public Health, Netherlands	sodium salt
(6) carcinogenicity of chemicals present in man's environment	January 1973	Ulland; Litton Bionetics, Inc., Bethesda, Md. (National Institutes of Health contract)	free
(7) studies on saccharin and cyclamate (in rats)	May 1973	Bio-Research Consultants, Inc., Cambridge, Mass. (National Institutes of Health and National Cancer Institute contract)	sodium salt
(8) studies on saccharin and cyclamate (in mice)	May 1973	Bio-Research Consultants, Inc., Cambridge, Mass. (National Institutes of Health and National Cancer Institute contract)	sodium salt
(9) sodium saccharin: combined chronic feeding and three-generation reproduction study in rats	May 1973	Friedman and Taylor; Food and Drug Administration	sodium salt

(10)	co-carcinogenic action of saccharin in the chemical induction of bladder cancer	June 1973	Hicks, Wakefield, and Chowaniec; Middlesex Hospital Medical School, London, England	free
(11)	long-term toxicity study of sodium cyclamate and saccharin sodium in rats	undated	Ikeda, Horiuchi, Furuya, Kawamata, Kaneko, and Uchida; National Institute of Hygienic Sciences, Tokyo, Japan	sodium salt
(12)	chronic toxicity of sodium saccharin: 21 months feeding on mice	undated	National Institute of Hygienic Sciences, Tokyo, Japan	sodium salt
(13)	lack of carcinogenic effects of cyclamate, cyclohexylamine, and saccharine in rats	1973	Schmähl; Heidelberg, Germany	free
(14)	long-term administration of artificial sweeteners to the rhesus monkey	1975	Coulston, McChesney, and Golberg; Albany Medical College of Union University, Albany, N.Y.	sodium salt
(15)	a chronic study of artificial sweeteners in Syrian golden hamsters	1975	Althoff, Cardesa, Pour, and Shubik; The Eppley Institute for Research in Cancer, University of Nebraska Medical Center, Omaha, Neb.	free
(16)	carcinogenicity study of commercial saccharin in the rat	June 1975	Munro, Moodie, Krewshi, and Grice; Health Protection Branch, Health and Welfare, Canada	sodium salt

carefully-controlled long-term feeding test[10] using the in utero exposure protocol first used by WARF and FDA. The results from the Canadian test were reported to confirm the findings in the WARF and FDA studies.

The occurrence of bladder tumors in rats exposed to saccharin from the time of conception in the WARF, FDA and Canadian tests appears to be acknowledged by experts in the field. However, interpretation of the results still is under discussion and debate. The findings are unique in the light of current information on other long-term feeding tests on materials previously reported to cause tumors in animals. In the WARF study, the only one of the three published in some detail, the survival rate of rats on the 5% saccharin diet was greater than for the no-saccharin controls. Speculation on mechanism of tumor incidence has included causative impurities in saccharin, interaction with naturally-occurring carcinogens in the feed, hormonal imbalance, a "foreign body" effect resulting from crystallization of saccharin in tissues at the highest dosing levels, and alteration of the animals' metabolism by the heavy excretion load imposed at the high dose levels used. The major impurity in saccharin manufactured by the original Remsen-Fahlberg process is ortho-toluenesulfona- mide (OTS). Several long-term feeding studies are in progress on OTS (Table 4) with results to be available in 1978. Some remarks are recorded[11] on the finding of bladder stones in feeding studies on some sulfonamide carbonic anhydrase inhibitors. OTS also inhibits this enzyme.

The Food Chemicals Codex monograph on saccharin was revised in 1974[12] to limit OTS content to 100 parts per million (0.01%). The FDA in 1977 proposed[13] that the OTS content of saccharin and its salts be limited to 25 parts per million (0.0025%).

Five epidemiological studies on saccharin are summarized in another report[14] (Table 5). Four show no relationship between saccharin ingestion and bladder tumors in man. The fifth study (Canadian) alleges a possible low-level association of bladder tumor occurrence in human males with long- term consumption of saccharin; however, an editorial critique accompanying the publication points out serious deficiencies in this study.

Saccharin and its salts are approved for specific uses under FDA's Interim Food Additive Regulation 180.37[3]. On April 15, 1977, the FDA proposed[15] revocation of this regulation and classification of saccharin as a drug. This proposal was greeted by a storm of protest from the American public and the press. The Congress of the United States started working in

Table 4. Ortho-toluenesulfonamide: long-term feeding studies

Investigator	Date Started
(1) I.C. Munro et al., Health Protection Branch, Health and Welfare, Canada, Ottawa, Canada	1974
(2) The British Industrial Biological Research Association, Carshalton, Surrey, England	1974
(3) Eppley Institute for Research on Cancer, University of Nebraska, Omaha, Nebraska	1974

mid-1977 on legislation to postpone the proposed ban on saccharin while risk-benefit, freedom of choice and other issues are examined. The Senate passed a bill, S. 1750, introduced by Senator Edward M. Kennedy, Dem., Massachusetts, by a vote of 87 to 7 on September 15, 1977. In the House of Representatives a comparable bill, H.R. 8518, sponsored by Representative Paul G. Rogers, Dem., Florida, was approved on October 17, 1977 by a vote of 375 to 23. A joint Senate-House committee will compromise the differences between the two bills with final passage expected within a few months. This legislation places an 18-month moratorium on the FDA saccharin ban with a charge to the Secretary of Health, Education and Welfare to arrange with the National Academy of Sciences for an assessment of current issues such as toxicity of impurities in saccharin and health benefits in humans of non-nutritive sweeteners. Furthermore, HEW is expected to make a broader assessment of food additive safety evaluation as well as recommendations for needed legislation.

Table 5. Saccharin: epidemiological studies

(1)	Armstrong, B., and Doll, R., Brit. J. Prev. Soc. Med. 28, 233, 1974.
(2)	Armstrong, B. et al., Brit. J. Prev. Soc. Med. 30, 151, 1976.
(3)	Kessler, I.I., J. Urology 115, 143, 1976.
(4)	Wynder, E.L. and Goldsmith, R., Cancer 40, September 1977.
(5)	Howe, G.R. et al., The Lancet, p. 578 and editorial p. 592, September 17, 1977.

REFERENCES

1. Rader, C.P., Tihanyi, S.G., and Zienty, F.B. J. Food Sci. 32, 357, 1967.
2. DeGarmo, O., Ashworth, G.W., Eaker, C.M., and Munch, R.H. J. Am. Pharm. Assn., Sci. Ed. 41, 17, 1952.
3. Code of Federal Regulations, Title 21, Food and Drug Administration, Section 180.37, revised as of April 1, 1977.
4. Food Protection Committee, The Safety of Artificial Sweeteners for Use in Foods, National Academy of Sciences, Publ. 386, Washington, D.C., 1955.
5. Food Protection Committee, Safety of Saccharin for Use in Foods, National Academy of Sciences, Washington, D.C., 1970.
6. Food Protection Committee, Safety of Saccharin and Sodium Saccharin in the Human Diet, National Academy of Sciences, Washington, D.C., 1974.
7. Need to Resolve Safety Questions on Saccharin, Report of the Comptroller General of the United States, HRD-76-156, Washington, D.C., 1976.
8. Tisdel, M.O., Nees, P.O., Harris, D.L., and Derse, P.H. in Symposium: Sweeteners, edited by G.E. Inglett, The Avi Publishing Company, Inc., Westport, CT, p. 145, 1974.

9. Fed. Regist. 42, 19999. 1977.
10. Fed. Regist. 42, 20000, 1977.
11. Fed. Regist. 42, 1461 and 1486, 1977.
12. Food Chemicals Codex, 2nd Ed., First Supplement, National Academy of Sciences, Washington, D.C., p. 49, 1974.
13. Fed. Regist. 42, 1486, 1977.
14. Fed. Regist. 42, 33768, 1977.
15. Fed. Regist. 42, 19996, 1977.

Practical considerations for synthetic sweeteners: past, present and future — cyclamates

Ronald G. Wiegand, Ph.D.

Chemical and Agricultural Products Division, Abbott Laboratories, North Chicago, Illinois 60064

First, a review of the regulatory status of cyclamate seems in order, and then some discussion of the scientific status.

Cyclamate was discovered in 1937. The results of studies on safety and effectiveness as a sweetener were submitted to the Food and Drug Administration in the form of a New Drug Application (NDA) which was approved in early 1950. Later that same year, cyclamate was marketed in the United States. In 1953, the 10 to 1 mixture of cyclamate with saccharin was introduced. This mixture had a more acceptable taste than either sweetener alone and became the primary marketed form of cyclamate.

In 1958, Congress passed an amendment to the Food Drug and Cosmetic Act dealing with food additives and providing for the GRAS (generally regarded as safe) list of substances. Cyclamates, and also saccharin, went onto this list. Thus, cyclamate went from an NDA-approved substance to a GRAS list substance and was never a food additive in a regulatory sense.

In the 1960's the FDA required teratology, fertility and reproduction (TRF) studies on new drugs. GRAS substances were exempt from this requirement, but nonetheless, Abbott sponsored a two-year general toxicity study of the 10 to 1 cyclamate: saccharin mixture at an outside lab. Half of the animals also received cyclohexylamine, the metabolite of cyclamate. This study was being completed in October, 1969, when the first indication of urinary bladder tumors in the rats was made known to Abbott. Within a matter of days Abbott pathologists, those of the testing laboratory, experts from the National Cancer Institute, and outside consultant pathologists reviewed the slides and agreed that tumors were present in the group of animals receiving the highest dose. On October 18, 1969, the Department of Health, Education and Welfare held a press conference announcing restrictions on the use of cyclamate.

On December 31, 1969, the FDA announced that cyclamate-containing products would be handled by "Abbreviated New Drug Applications and specific medical labeling for cyclamate-containing products." This action made cyclamate available on a prescription basis for diabetic patients. However, the following August, the FDA announced that cyclamate would no longer be allowed even on a drug basis after September 1, 1970, which became the last date of availability of cyclamate as a sweetener in the United States.

Cyclamate continues to be available in other countries, including Canada where use in table-top sweeteners is allowed; West Germany, where all uses are allowed but beverages are limited to 0.8 gm/l; Denmark, Belgium, and the Netherlands, where approval is for table-top use only; France -- table-top use by prescription; and Italy, where all uses are allowed but require authorization by the Ministry of Health. Thus, while cyclamate is not allowed in the United States, its availability has been continued in other countries by several different means.

Suspected carcinogenicity was the reason for withdrawal of cyclamate in the United States, and the data showing tumors in rats caused several carcinogenicity studies to be initiated around the world. More than a dozen such studies were done at the United States National Cancer Institute, the United States Food and Drug Administration, the German National Cancer Institute, the Osaka University School of Medicine, the British Industrial Biological Research Association, the Netherlands Rijks Institute and other research institutions. These data supported the conclusion that cyclamate and its metabolite cyclohexylamine did not cause tumors. Based on these data, a Food Additive Petition was submitted to the FDA in November, 1973 by Abbott Laboratories.

Carcinogenicity has remained the central question about safety of cyclamates. When cyclamate was finally removed from the United States market in 1970, the FDA also had in hand data from an FDA study in Osborne-Mendel rats showing two tumors in the low dose group and one tumor in the high dose group, and no tumors in the mid-dose and control groups; all treated animals were fed calcium cyclamate. No tumors were found in sodium cyclamate-fed animals. When the food additive petition was reviewed, an additional carcinogenicity study in Osborne-Mendel rats was suggested by the FDA on the basis of the above data. This study was not done, and the FDA asked the National Cancer Institute to appoint a task force of eminent authorities in the field to review all the data on carcinogenicity. This group visited many of the laboratories in which the studies have been conducted and reviewed the data in detail. They concluded in their report of February, 1976[1] that:

"...the present evidence does not establish the carcinogenicity of cyclamate or its principal metabolite, cyclohexylamine, in experimental animals."

The Joint FAO/WHO Expert Committee on Food Additives reviewed the same question and most of the same data in June, 1974, and came to a similar conclusion[2], namely:

"It is now possible to conclude that cyclamate has been demonstrated to be non-carcinogenic in a variety of species."

The current situation with regard to the food additive petition on cyclamate is that the FDA has rejected the petition on the basis that the carcinogenicity question and other issues have not been satisfactorily resolved. The food additive petition is the subject of hearings within the executive branch of government, and carcinogenicity is one of four issues being considered. The other three issues relate to genetic damage and mutagenicity, acceptable daily intake, and consumption patterns and safe conditions of use.

Many mutagenicity studies have been done with cyclamate and cyclohexylamine employing microbial systems, _in vivo_ systems, and _in vitro_ cytogenetic systems. In the Ames test, the results show a lack of mutagenicity. _In vivo_ studies show a mixture of positive and negative results with the positive results showing mostly increased breaks and gaps, a non-specific lesion which is usually considered to be capable of spontaneous repair. Exchange and translocation figures, which are the cytogenetic changes associated with mutagenetic damage, were not routinely seen.

Dominant lethal studies and host-mediated tests do not show cyclamate or cyclohexylamine to be mutagenic. Similarly, _in vitro_ tests do not support mutagenicity, with breaks and gaps being the lesions most often seen and usually at concentrations exceeding the levels found in man after ingestion. Cattanach[3] reviewed the entire evidence and concluded that:

"The main conclusions from this review of the mutagenicity test data so far available on the cyclamates and their metabolites is that none of the compounds can be considered mutagenic."

Similarly, Machemer and Lorke[4] concluded:

"Only few compounds have been so comprehensively investigated for mutagenic effects as cyclamate... An overall evaluation of the published results ... have not provided evidence of cyclamates having a mutagenic effect."

If we accept that cyclamate and cyclohexylamine are neither carcinogenic nor mutagenic, which I believe is true and is certainly supported by a wealth of experimental data collected worldwide and reviewed extensively by noted authorities, then we come to the point of determining what toxic effect is first produced by cyclamate ingestion. This information then would be used to calculate the allowable daily intake (ADI). The toxicity which occurs at lowest dose is a testicular effect, specifically a decreased testicular weight in rats fed cyclohexylamine at 0.6% in their diet. The effect is not present at 0.2% CHA in the diet. A study is presently underway at constant mg/kg doses to more accurately determine the no-effect dose.

Translating this no-effect dose of cyclohexylamine to an ADI for cyclamate involves the conversion of cyclamate to cyclohexylamine. This conversion is effected by the bacterial flora in the gastrointestinal tract. About 20% of the population has the ability to convert cyclamate to cyclohexylamine, and only about 2 to 3% of the population converts 5% or more of the cyclamate. Cyclamate is poorly absorbed, with most of the ingested cyclamate remaining in the gastrointestinal tract as a substrate for conversion. At high doses of cyclamate, this substrate is excess and it appears that the amount of cyclohexylamine produced does not increase. Also, the ability to convert cyclamate to cyclohexylamine requires several days of cyclamate ingestion to reach the full rate in an individual, and is lost when ingestion is not maintained. Therefore, it is a complicated problem to calculate an intake of cyclamate which produces a given amount of cyclohexylamine. The situation would be straightforward if this were a situation in which a metabolite is formed in a reasonably uniform percentage. The current situation is that the FDA suggests use of a 30% conversion in calculating an ADI for cyclamate for the cyclohexylamine data.

The last item to discuss is patterns of use and consumption of cyclamate. In 1969, cyclamate was allowed in beverages (accounting for about 70% of usage), in dietetic foods (accounting for about 15% of usage), and table-top products (accounting for the remaining 15%). Total usage in the United States was 17 million pounds, which amounts to an average daily per capita consumption of 0.1 gram. The FDA will have to satisfy itself that the allowed uses and their levels in food and beverage will reasonably be expected to result in actual usage within the ADI for normally prudent people. This assumption seems possible at the present time.

REFERENCES

1. Report of the temporary committee for the review of data on carcinogenicity of cyclamate, National Cancer Institute, February, 1976.
2. Evaluation of certain food additives, 18th report of the Joint FAO/WHO
   Expert Committee on Food Additives, p. 27, WHO Tech. Report Series
   #557, Rome, Italy, June 3-4, 1974.
3. Cattanach, B.M. Mutation Res. 39, 1-28, 1976.
4. Machemer, L. and Lorke, D. Mutation Res. 40, 243-250, 1976.

Aspartame: a commercially feasible aspartic acid based sweetener

Robert G. Bost and Annette Ripper

Regulatory Affairs, G.D. Searle and Co., P.O. Box 1045, Skokie, Illinois 60076

INTRODUCTION

Aspartame is the generic name for the sweet compound 1-methyl N-L-α-aspartyl-L-phenylalanine. The fact that it is a dipeptide composed of two naturally occurring amino acids places it in a category all its own--it is neither an artificial sweetener nor a caloric sweetener in the traditional sense. Aspartame is a nutritive food additive with intense sweetness and flavor-enhancing properties. Though the components of aspartame are distributed widely in our food supply, the particular combination of amino acids and a methyl ester do not occur naturally.

The chemical formula for aspartame is $C_{14}H_{18}N_2O_5$ as shown below in Figure 1.

$C_{14}H_{18}N_2O_5$ molecular weight — 294.3

TASTE CHARACTERISTICS

For a sugar substitute to be acceptable for commercial use, it must have sufficient sweetening power, be nontoxic, be reasonably inexpensive, and be thermostable. Unfortunately, the terms "sugar substitute" and/or "artificial sweetener" have undesirable connotations for many people. Both terms describe compounds which are built up of several elements, or artificially produced, as opposed to extraction from a plant. While it has not yet been possible to duplicate the exact flavor, texture, density, and structural functions of sugar in a sugar substitute, it is possible to have an acceptable product.

Aspartame has a sugar-like taste and has a potency 120 to 280 times that of sugar depending on the food system in which it is used. Like sugar or protein, aspartame provides approximately 4 calories per gram; however, because of its intense sweetness, if aspartame is used as a sweetener in place of sugar, the amount needed to yield equivalent sweetness will provide only about 1/200th the calories of sugar.

In addition to imparting sweetness, aspartame also functions as a flavor enhancer and flavor extender. It is particularly effective in enhancing acid fruit flavors and in extending sweet taste. This property has been demonstrated in chewing gum where sweetness is perceived to last from five to seven times longer than sugar-sweetened gums[1].

The many functional and property differences of aspartame do not make it suitable as a blanket substitute for sugar. Conversely, the properties of aspartame make possible new products and new processes for known products and for others not heretofore available. In summary, aspartame can: (1) sweeten foods; (2) enhance flavors (particularly fruit flavors); (3) reduce calories; (4) avoid nutrient dilution because of lower bulk occupied by aspartame as compared to the same amount of sweetness from sucrose; (5) reduce volume and weight of presweetened products such as dry beverage powder; (6) super-sweeten products by adding concentrated sweetness as a small fraction of the total weight; (7) reduce viscosity, stickiness, or other properties associated with sugar; (8) lower cost of sweetening; and (9) reduce sucrose consumption where indicated such as in diabetes or predisposition to dental caries.

PHYSIOCHEMICAL CHARACTERISTICS

In its pure form aspartame is a white, odorless, crystalline powder. Since the compound is a peptide, it is amphoteric and the dissociation constants at 25°C are pk^{1} = 3.1 and pk^{2} = 7.9. Aspartame contains an ester linkage that, under certain moisture, temperature and pH conditions may hydrolyze to the dipeptide, aspartylphenylalanine. The dipeptide can then cyclize to the corresponding diketopiperazine (DKP) as shown in Figure 2. Food grade aspartame specifications are as follows:

Assay: Not less than 98% or more than 102% aspartame calculated on a dry basis as determined by lithium methoxide titration.

Purity: Not less than 95% transmission at 430mu of a 1% solution in 2N HCl. Loss on drying not to exceed 4.5% after four hours at 105°C. Not more than 2% 5-benzyl-3, 6-dioxo-2-piperazine-acetic acid (DKP) as determined by thin layer chromatography and not more than 1% other impurities.

Detailed analytical procedures are outlined in CFR 21 172.804.

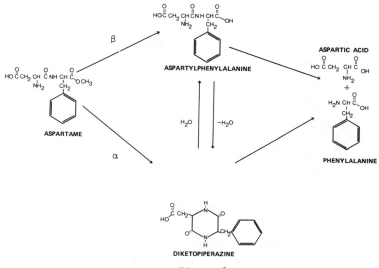

Figure 2

Aspartame contains a small amount (3 to 5%) of incidental water arising from its method of preparation, though food grade specifications permit no more than 4.5% water. Because aspartame has a tendency to take up moisture, its stability has been studied under a variety of storage conditions.

In a study to determine the stability of aspartame of various water contents it was found that moisture equilibrium level of aspartame at 88% relative humidity and 40°C is about 8% and at 56% relative humidity and 50°C it is about 7.5%. After quantification of the moisture content of aspartame, calculated amounts of water were mixed with the chemical using a mortar and pestle. The final moisture content was determined by Karl Fischer titration. The aspartame was then sealed in clear glass ampules and subjected to various conditions of chromatography with densitometry. Differences in degradation patterns were detected as a function of moisture content:

Formation of DKP from Aspartame Containing 8% Moisture

Initial	0.54%
2 weeks at 70°C	2.19%
12 weeks at 55°C	2.72%

This study demonstrated that aspartame is relatively stable even at an 8% moisture content; only negligible decomposition occurs after two weeks at 70°C. Under severe conditions--an open container in a chamber maintained at 55°C and 80% relative humidity for a period of two months--aspartame was converted to DKP at a rate of about 1.5% per month.

Solid aspartame has been stored in closed containers at 40°C for periods up to one year in order to determine the mode and extent of decomposition under storage conditions somewhat more severe than would be encountered in normal distribution. Analysis of stored samples was accomplished by examining thin layer chromatograms of the storage samples. The system used was able to distinguish between unchanged aspartame, aspartylphenylalanine, DKP and the constituent amino acids. The sensitivity of the method is approximately 0.1% for each of these substances.

Only aspartylphenylalanine and DKP were detected in the chromatograms of stored samples. After six months to a year of storage at 40°C, some increases of as much as 1% in the level of DKP and of 0.5% in the level of aspartylphenylalanine were noted. These results are based on estimations of spot intensity versus standards run adjacent to the sample chromatograms. The values reported are considered to be accurate to about 0.5% in view of this. It is noteworthy that the constituent amino acids were not seen in these chromatograms.

It is concluded that the decomposition under dry conditions of storage occurs principally by routes α and β (Figure 2) although the rate of this decomposition is extremely slow.

SIGNIFICANCE OF ASPARTAME STABILITY DATA

The conversion products of aspartame, DKP and aspartylphenylalanine, can be detected in the pure bulk chemical. In some food systems the constituent amino acids can also be detected. In all instances, the presence of DKP is dependent upon temperature, pH of product, length of storage and moisture content of the mixture. Though DKP is allowed to be up to 2% of the bulk chemical, standards for DKP levels in food products have not been determined. Studies have shown that no significant conversion of aspartame to DKP occurs even when food products have been stored under adverse conditions. The small amount of aspartame which is converted to DKP does not alter the acceptability of the foods since sweetness is not appreciably reduced, and DKP is tasteless.

DKP is not a metabolite of aspartame, as shown in studies in animals and man. Therefore, the level of DKP in the bulk chemical or in food products should not increase through metabolic processes. However, since it is known that very small amounts of DKP will be in the bulk chemical and in foods sweetened with aspartame, DKP was tested for its toxicologic, metabolic, pharmacologic and clinical effects as part of the safety evaluation of aspartame as a food additive.

SOLUBILITY

Aspartame is not readily soluble in water, has limited solubility in alcohols and is relatively less soluble in organic solvents of lower polarity. The isoelectric point (minimum charge and minimum solubility) is at pH 5.2. Aspartame's solubility is a function of both pH and temperature, and maximum solubility is at pH 2.2; solubility increases with temperature. The tendency to readily form highly soluble salts below its isoelectric point makes it simple to improve both the rate and degree of solubility either by first dissolving the food acid in the system, then adding the aspartame or dissolving both at the same time. Its use as a food ingredient has been improved by several methods: (1) encapsulating aspartame in a dextrinous water soluble agent by spray drying[2]; (2) spray drying solutions of an edible bulking agent and aspartame[3]; (3) vacuum drum drying solutions of an edible bulking agent and aspartame[4]; (4) co-drying solutions of aspartame and an edible bulking agent[5]; (5) co-drying solutions of aspartame and an edible organic acid[6]; (6) co-drying a solution of aspartame with an edible, bland, low calorie polysaccharide in defined ratios[7];(7) admixing aspartame in aqueous suspension with hydrolyzed amylaceous derivatives comprising predominantly oligosaccharide solids[8]; (8) discreetly dispersing the sweetener throughout a matrix created by melting a fusable mass and subdividing it to encapsulate the aspartame[9]; (9) co-grinding aspartame with an acid in the presence of an organic solvent in which the dipeptide is insoluble[10].

Though minimum solubility of the pure chemical can go as low as 0.5%, that concentration would nearly be equal to 100% sugar when the sweetness intensity of aspartame is considered. Solubility in various solvents is shown in Table 1. Isoelectric aspartame exhibits limited solubility in organic solvents.

Table 1. Solubility of Isoelectric Aspartame

Solvent	Solubility ($25^{\circ}C$)
Water	1.0 %
Methanol	0.87 %
Ethanol	0.37 %
Chloroform	0.026%
Heptane	0.004%

Aspartame is essentially insoluble in vegetable oil; this insolubility can prevent or retard the conversion of aspartame to its diketopiperazine. The relative solubilities of the α and β L-aspartyl-L-phenylalanine methyl ester are known in a wide range of solutions[11].

APPLICATIONS AND STABILITY IN TARGET APPLICATIONS

Aspartame is a suitable sweetener for a wide variety of foods because of its sugar-like taste and its ability to blend well with other food flavors. Generally, aspartame is said to be 200 times as sweet as sugar. However, aspartame is not used to replace sucrose or other carbohydrate sweeteners in the same 1:200 ratio. The flavor enhancing property of the sweetener as well as other ingredients in a particular food system can affect its potency[12]. Therefore, the use of aspartame as a replacement for sugars requires reformulation and not simple substitution of ingredients, as shown in Table 2.

Table 2. Aspartame Content of Typical Food and Beverage Formulations

Product Category	Amount of Sucrose Per 100 Grams	Amount of Aspartame to Replace Sucrose	Potency of Aspartame	Serving Size (U.S.Household Measure)	Amount of Aspartame Per Serving
	grams	grams	x sucrose	grams	grams
Dry					
Table top sweetener	100.0	.050	200	4 (1 tsp.)	.002
Powdered soft drinks	10.3	.069	150	226 (8 oz.)	.156
Instant pudding	17.0	.085	200	150 (1/2 cup)	.128
Gelatin dessert	18.0	.113	160	137 (1/2 cup)	.157
Presweetened cereal	37.0	.205	180	28 (1 cup)	.058
Aqueous					
Carbonated soft drinks	10.0	.055	180	340 (12 oz.)	.187
Canned fruit Cocktail	20.0	.114	175	113 (4 oz.)	.129
Flavored yogurt	17.0	.085	200	226 (8 oz.)	.181
Frozen dessert	15.0	.060	250	170 (6 oz.)	.102

As a pure chemical, aspartame does not flow easily and does not dissolve readily in cold beverages. Processes for improving the solubility and flowability of aspartame for both dry and aqueous food systems have been cited in the previous section.

Though aspartame is intended primarily as a sweetener, it can also potentiate or enhance flavors or their perception. This characteristic is most evident when aspartame is used to sweeten chewing gum. Aspartame-sweetened chewing gum has been shown to be sweet up to seven times as long as sugar-sweetened gum, or about 35 minutes[1].

The initial uses of aspartame in food applications will be in dry foods such as: table top sweeteners; dry mixes for instant coffee and tea, gelatins, puddings, fillings, and dairy product analog toppings; cold breakfast cereals; and chewing gum. The use of aspartame in aqueous systems has been successful despite the sweetener's instability under certain processing and storage conditions. Typical foods in which aspartame can be used as a sweetener are shown in Table 2.

In most dry foods, stability of aspartame is not a concern unless it is used in combination with hygroscopic ingredients. Dry stability of aspartame has been shown to be quite good in the sachet form of table top sweetener which is a mechanical blend with lactose and is packaged in standard food service paper packets. In products such as the granulated spoon-for-spoon table top sweetener (equal to sucrose in measure and in sweetness) which is spray dried with Maltodextrin, protective packaging must be used to maintain stability of both the aspartame and the bulking agent.

The stability of aspartame in aqueous systems has been studied in several food preparations under normal conditions of intended use. The sweetener's stability was determined by measuring the formation of the diketopiperazine of aspartame, aspartylphenylalanine, the constituent amino acids, and remaining aspartame. This work is particularly relevant to the use of aspartame in carbonated soft drinks where the stability of the sweetener has been shown to be a factor of pH, storage conditions and length of storage. Separate studies which measured aspartame stability in buffered solutions at various pH's showed that maximum stability was observed between pH 3.9 and pH 4.3. Since most carbonated soft drinks have pH values below this range, degradation of aspartame and loss of sweetness are expected.

Taste acceptability studies were performed with a commercially prepared aspartame-sweetened diet cola. Since the diet cola formulation had a 3.1 pH, a loss of aspartame was anticipated; after 6 months storage at room temperature about two-thirds of the aspartame remained. Even with a gradual loss of aspartame, the aspartame-sweetened cola was preferred to the commercial saccharin product for up to five months. Since most bottled or canned soft drinks are consumed within two to three months of manufacture, this five-month period of acceptability indicates that aspartame is a suitable sweetener for soft drinks.

While carbonated beverages are one application of aspartame in which stability problems had to be overcome, yogurt is an ideal food system for aspartame because it is cultured to a pH which provides optimum stability. Stability studies with aspartame-sweetened yogurt have shown that after 30 days less than one-fourth of the aspartame is converted, sweetness perception changes little and the products are acceptable overall.

Extensive experimental work by potential users of aspartame in the food industry has shown the dipeptide to be a superior sweetener in a greater variety of foods than was thought possible just a few years ago. The creativity and ingenuity of food technologists have overcome the problems which were once perceived as limitations for the use of aspartame as a food sweetener. Data show that the rate of analytical disappearance of aspartame is rarely paralleled by loss of acceptance. When aspartame is converted to DKP, the only taste effect is the loss of sweetness due to the lack of taste of DKP.

SAFETY

A primary consideration in the evaluation of a food additive is the examination of toxicity data in laboratory animals and the metabolic fate of the test substance. In the study of aspartame, the metabolic fate of the cyclized non-biologically derived conversion product of aspartame, a diketopiperazine (DKP), has also been evaluated.

Aspartame metabolism has been measured in the rat, dog, and monkey. Using radiolabeled compounds, these studies have shown that the methyl ester of aspartame is rapidly lost in the small intestine and is largely lost via expired air as CO_2 from the one-carbon pool. A small amount can be found in urea, glycogen, and other metabolic niches.

After demethylation, about 75% of the dipeptide residue is further cleaved in the small intestine at the brush border, or within the intestinal cells, into aspartate and phenylalanine. The free aspartate then reaches the portal circulation and, in the liver, largely enters the citric acid via the oxaloacetate route. The small amount that is not metabolized by that route most probably enters the urea cycle directly and possibly a much smaller amount enters the lesser metabolic pathways. The carbon of the aspartate moiety appears in the CO_2 of the expired air within a few hours. While about 70% of the moiety is lost by this end event, another 5% of the carbon is found as urinary and fecal carbon.

The phenylalanine moiety of the cleaved dipeptide also enters the portal circulation. In the liver about 75 to 80% is directly incorporated into plasma proteins. A portion finds its way into tyrosine, and a small amount of the carbon from phenylalanine is lost as CO_2 in expired air.

Metabolic studies with DKP have shown that about 15 to 20% of the ingested DKP crosses the gut and enters the portal circulation. Most of this is lost as phenylacetylglutamine, a normal urinary metabolite of phenylalanine, or returned to the gut lumen unchanged.

Metabolism studies in man using radiolabeled material have shown that man metabolizes the compounds, aspartame and DKP, in a manner similar to that of the experimental animals. Thus, the metabolism studies have shown that aspartame is broken down into its constituent moieties and that these moieties are handled in the same manner as naturally occurring constituents in the diet. In addition, DKP is not very readily absorbed. The material which is absorbed is excreted unchanged or as a naturally occurring urinary metabolite.

With this background of metabolic data, and despite the evidence that aspartame is metabolized like naturally occurring dietary constituents, an extensive assessment of the toxicologic potential of aspartame and DKP has been conducted. The toxicology studies have ranged from acute, single-dose administration studies to lifetime and multigeneration studies in laboratory animals.

In the acute toxicity studies, mature animals of both sexes received aspartame or DKP at dose levels ranging from 1.0 to 5.0 gm/kg. No deaths were observed in any of the animals nor was any aberrant behavior noted. Thus, the minimum lethal dose for both aspartame and DKP exceeds 5 gm/kg in the mouse, rat and rabbit.

Chronic toxicologic testing of aspartame and DKP have been performed in the dog, rat and mouse. In the dog, two studies, one of eight week duration, the other, 106 weeks, have been performed. Administration of up to 4 mg/kg/day every day for 106 weeks produced no biologically meaningful alterations in the normal physiologic state nor was there any evidence of histopathologic changes following termination of the administration and sacrifice of the animals.

In the rat, chronic studies have been performed ranging in length from 8 weeks to lifetime. Both neonates and adults have been tested. In the lifetime studies, the treatment of rats at levels ranging up to 8 gm/kg/day did not produce any conclusive evidence of treatment-related changes in physical appearance, behavior, clinical laboratory values, nor did it produce any unique histopathologic changes in any organ or tissue examined.

Lifetime studies have also been performed in the mouse at dose levels up to 13 gm/kg/day. No biologically meaningful alterations in body weight, food consumption, physical exam or postmortem findings were observed. DKP administration for the lifetime of rodents has also failed to produce signs of toxicity.

It has been reported that the monkey is susceptible to high dose levels of phenylalanine with grand mal-type seizures resulting[13]. Two chronic toxicity studies were performed on newborn monkeys to determine whether aspartame or equimolar amounts of L-phenylalanine would have the same effect. The first study was performed at the Primate Research Center, Madison, Wisconsin. This study was designed as a pilot study and contained only seven animals. All of these animals were given aspartame; therefore, there were no concurrent controls. Animals receiving 3.0 and 3.6 gm/kg/day exhibited grand mal-type seizures after 31 weeks of treatment. The results of this study are suspect because of an intercurrent Shigella infection in the monkey colony during the time this study was conducted. Unfortunately, the investigator died during the study; the study was terminated following his death, and the etiology of the convulsions was not determined.

Another primate study is currently being conducted at the University of Illinois. In this study there are three aspartame-treated groups at dose levels of 1, 2 and 3 gm/kg/day, a phenylalanine-treated group, and an untreated control group. Several animals in each group have completed 270 days of compound administration, and there has been no evidence of convulsions. In addition, electro-encephalograms have also been conducted with no evidence of brain dysfunction. Thus, the results of this study suggest that the convulsions seen in the previous study may have, in fact, been the manifestations of extraneous factors.

Teratology studies with aspartame were conducted in three species-- mouse, rat and rabbit--without showing evidence of aspartame per se treatment related effects. Previous studies have shown that diets containing high levels of L-phenylalanine produce a marked decrease in the ability of the maternal rat to rear offspring. This has generally been attributed to an altered maternal economy during gestation and decreased secretion of maternal milk during lactation. This condition generally produces a reduction in maternal body weight gain during gestation, a reduction in pup body weight at birth and/or weaning, and a reduction in pup survival rate at weaning.

In experiments with rats in which aspartame was administered at dose levels of 4.0 gm/kg/day or greater during gestation and 6.8 gm/kg/day or greater during lactation, decreased maternal body weight and decreased pup survival at weaning were observed. These effects were not observed at lower dose levels. None of the studies produced any evidence of fetal malformations.

Teratology studies in mice and rabbits confirm the findings of the rat studies. At high levels of aspartame and phenylalanine, reduced maternal food intake and decreased conception rates were seen in some studies. However, there was no evidence of a treatment-related embryotoxic effect in any of the studies.

Clinical studies have also been conducted with aspartame. These studies range from single, high dose administration to relatively long-time studies in which aspartame was administered daily for 27 consecutive weeks. Dose levels ranged from 30 mg/kg/day in the chronic studies to 200 mg/kg/day in the acute administration. The 200 mg/kg dose level represents an acute abuse level of aspartame. Plasma aspartate and phenylalanine levels were measured. Plasma levels of both amino acids were increased, but neither approached a toxic level and the levels returned to normal limits within 6 to 8 hours. The long-term administration at 30 mg.kg.day did not produce any perturbations in the measured hematology, clinical chemistry or urinalysis parameters.

Thus, the metabolism studies on aspartame show that the moieties of the molecule are metabolized like normal dietary constituents. Both acute and long-term toxicology studies and teratology studies in laboratory animals have shown that high doses of aspartame can be tolerated without producing adverse effects. These studies and the human studies have produced strong scientific evidence that aspartame is safe for human consumption, and that aspartame can be a useful sugar replacement.

REGULATORY STATUS

In July, 1974, the FDA approved the use of aspartame in table top sweeteners; tablets for coffee and tea; cold breakfast cereals; chewing gum; and dry mixes for beverages, instant coffee and tea, gelatin, puddings, fillings, toppings, and dairy product analogs.

During the review period following the publication of the regulation objections to this approval were filed. One objection was based on evidence that large doses of glutamic acid could produce brain lesions in newborn mice[14]. Both glutamic acid and aspartic acid are dicarboxylic amino acids, and the objector postulated that the aspartic acid moiety of aspartame might also produce brain damage. At the time the objection was raised, Searle did not have evidence to substantiate or disprove whether this observation was equally true for newborn or developing children. Searle voluntarily withheld aspartame from the marketplace until these hypotheses could be examined thoroughly. The objectors were promised a public hearing by the FDA when such further evidence was in hand.

Since the objections were raised, studies have been conducted to test the brain lesion hypothesis in neonatal mice and monkeys[15]. Studies which measured plasma levels of aspartic acid and phenylalanine in humans following administration of high aspartame levels have also been carried out[16] The evidence of a lesion in the neonatal mouse has been confirmed when plasma aspartate levels reached 60 to 70 μmol/dl. However, the presence of a lesion in the newborn monkey was not found even though plasma levels reached 500 μmol/dl. In the human, the highest mean plasma level for aspartate reached was 1.5 μmol/dl even at dose levels which can best be termed "abuse" levels. These results suggest that a large margin of safety exists for humans even at very high aspartame levels.

In August, 1975, the FDA initiated an inspection of Searle research. The FDA report revealed a variety of clerical errors, and criticized laboratory methodologies and animal handling. On December 5, 1975, the FDA stayed its aspartame approval pending a more detailed inspection of Searle research and the holding of the public hearing.

A process is now underway to review fifteen aspartame studies performed at Searle and Hazleton Laboratories. The process began with a review of one long-term toxicology study and two teratology studies by the FDA. The additional studies are currently being reviewed by Universities Associated for Research and Education in Pathology (UAREP), an independent association of university professors. This review began in August, 1977, and is expected to take 5 to 8 months.

We are hopeful that the review can be completed within this time frame and that the Public Board of Inquiry can follow shortly upon the completion of the review.

REFERENCES

1. Bahoshy, B.J., Klose, R.E. and Nordstrom, H.A., U.S. Pat. 3,943,258, 1976.
2. Pischke, L.D. and Shoaf, M.D., U.S. Pat. 3,962,468, 1976.
3. Glicksman, M. and Wankier, B.N., U.S. Pat. 4,001,456, 1977.
4. Glicksman, M. and Wankier, B.N., U.S. Pat. 4,007,288, 1977.

5. Glicksman, M. and Wankier, B.N., U.S. Pat. 3,761,288, 1973.
6. Glicksman, M. and Wankier, B.N., U.S. Pat. 3,922,369, 1975.
7. Furda, I. and Trumbetas, J.F., U.S. Pat. 3,934,048, 1976.
8. Baggerly, P.A., U.S. Pat. 3,955,000, 1976.
9. Shoaf, M.D. and Pischke, L.D., U.S. Pat. 4,004,039, 1977.
10. Berg, J.H. and Trumbetas, J., U.S. Pat. 3,868,472, 1975.
11. Ariyoshi, Y. and Sato, N., Bulletin of the Chemical Society of Japan, 45, 942, 1972.
12. Beck, C.I. in Symposium Sweeteners, Inglett, G.C., AVI, Westport, CT. 164, 1974.
13. Waisman, H.A. and Harlow, H.F. Science, 147, 685, 1965.
14. Olney, J.W. and Ho, O. Nature, 227, 609, 1970.
15. Reynolds, W.A., Butler, V. and Lemkey-Johnston, N. Journal of Toxicology and Environmental Health 2, 471, 1976.
16. Steginck, L.D., Filer, L.J. and Baker, G.L. Journal of Nutrition 107, 1837, 1977.

DISCUSSION AFTER PAPERS BY DRS. BECK, ZIENTY, WIEGAND AND BOST

DR. ROUSSOS: I would like to ask Drs. Zienty, Bost and Wiegand the following question. If your product were to be reinstated in the future, do you think the approval would be for extensive use as in the past, or limited use?

DR. WIEGAND: Certainly the most likely use of cyclamates will be in table top sweeteners and in dietetic foods. Beverage use would probably be at a restricted concentration.

DR. BECK: Gerry, part of the food additive petition was to list the proposed uses. We asked for use in 12 of the 43 FDA recognized categories of foods. We had to suggest a tolerance level for each use. We put these in order of importance to us, in order of ease of defending or justifying the use. First we have to have an estimate of the average daily intake. That's partly going to be based on what number is agreed upon for the extent of metabolism of cyclamates to cyclohexylamine and the maximum no-effect level of cyclohexylamine. Once we have that number, then we start trying to get agreements with FDA on how many uses can be allowed based on the ADI for each use and still have people not exceed the ADI. At this point nobody can guess how many uses might be approved, except that it won't be more than the 12 we asked for.

DR. ZIENTY: When saccharin was removed from the GRAS list, the new interim regulation specified a limited number of permissible uses with tolerance levels. Future revisions of this regulation undoubtedly will aim to restrict the total consumption of saccharin in foods, currently at 3.5 million pounds annually. Furthermore, the legislation before Congress is likely to require a warning label.

DR. SWANGO: Dr. Beck, you made a suggestion that perhaps some non-nutritive sweeteners could be used to help feed the starving populations in underdeveloped countries. I do not understand how that could be accomplished.

DR. BECK: One of the great needs in trying to feed people that are undernourished is to find sources of protein. There is a lot of interest in various kinds of manufactured sources of protein, such as single cell protein, that can be made by fermentation, or edible algae that could be grown on ponds or in tanks. If you have ever tasted any of these things, you noted that they did not have the delightful flavors that we associate with our favorite sources of protein. Flavoring these sources of protein is going to be an important consideration. My comment was that sweetness is a desirable flavor and that uses might be developed for flavoring these synthetic proteins with a synthetic sweetener.

DR. MOSKOWITZ: I'd like to address a question simultaneously to Dr. Bost and Dr. Beck. It concerns the measure of relative sweetness. I think it's more than an academic question. Rather, it has very practical implications.

We talk about saccharin being 200 times sweeter than sugar, or a hundred times sweeter, depending on the level. This level is really not a measure of perceived sweetness. It is that concentration of saccharin which produces a taste as sweet as a certain percentage of sucrose and should be referred to as relative potency. In the case of many of the artificial sweeteners, the dose response relation between concentration and rated sweetness is much flatter than a similar amount of sucrose. You can change the concentration of saccharin by a large ratio without dramatically changing perceived sweetness. Whereas at the same time you can change the sweetness of sugar quite dramatically by the same proportional shift in concentration.

When we talk about how much a person would take in, we say one two-hundredth in the case of saccharin; then if we are interested in how much a person would consume, there's a large range on the concentration function from high to low that will produce equivalently acceptable sweetness. We might be able to get away with, instead of a ratio of one two-hundredth that of sugar in the case of saccharin, one five-hundredth of the amount. This is dose response not usually taken into consideration, but instead is often glossed over.

DR. BECK: Many people will argue that you should measure sweetness intensity by duplicating the sweetness of 10% sucrose or 8% sucrose or some sort of average concentration. Using this technique, you might get quite a different number, that takes into effect the concentration factor, but it still doesn't take into effect the synergistic effect of flavor. For example, cyclamate is much sweeter with fruit flavors than with non-fruit flavors; I don't know any particular explanation for that. The same is true of saccharin, isn't it, Ferd?

DR. ZIENTY: Yes.

DR. BECK: It's been my general observation that saccharin is ten times as sweet as cyclamates, although the sweetness of both saccharin and cyclamates is quite variable depending on conditions. They seem to vary under the same conditions, and the factor of tenfold seems to hold up. I don't know if that's true with aspartame or not. I haven't worked with aspartame enough to know. A lot of factors, one of which you discussed, Howard, influence the relationship. As concentration increases, this sweetness curve starts flattening and you can get to a point where you can increase concentration substantially without a major increase in sweetness. With more dilution, the higher the perceived sweetness of some things becomes.

DR. MOSKOWITZ: I maintain the ideal sweetener would be one that has a nearly flat physical function. Thus, as you diminish concentration, you might lose sweetness, just as long as you don't diminish it that rapidly. The dihydrochalcones are very close to that. You can go down and down and down in concentration and scarcely change the perceived sweetness. That's the ideal sweetener if it proves to be non-toxic. You could go down as low as you want within reasonable limits.

DR. MACKAY: On the assumption that agreement is reached to market these sweeteners, has any of the speakers any reason to believe that the use of the product would not be restricted to or result in the creation of a special dietary food? Second, have you had any assurance on the question of being allowed to mix an artificial sweetener with a nutritive sweetener? Third, have you gone into the question of how any attempt to market a product to children, assuming it were a special dietary food, would be viewed in Washington? The reason for these questions is to try to show some of the problems that a food manufacturer has in trying to use these materials in today's climate, and in trying to bring about one of the objectives of this conference.

DR. BECK: Yes. I think that any synthetic sweetener that is ever approved in the United States will be only for special dietary use. FDA has taken the attitude that sugar is a food, synthetic sweeteners like cyclamates and saccharin are non-foods, and that replacement of a food with a non-food is adulteration unless it's for some special dietary purpose.

As far as I know on your third question, FDA considers that any food that is put out would be available to both children and adults, and I think that they would use that fact in trying to determine what uses would be permitted. I don't know if there is going to be a special approval for use in a product only if it's sold to adults. You can't control who consumes the product. As far as combining nutritive and non-nutritive sweeteners, FDA issued on July 17th new regulations on low calorie special dietary foods. I haven't studied these regulations, but I believe they (FDA) were taking the attitude at one point that to make low calorie claims you had to have a 50% reduction in calories compared to the ordinary product. I don't know if that's still the philosophy or not.

DR. BOST: I have read the regulations on low calorie dietary foods. I'm not sure I understand them. Aspartame, though, is not regulated as a non-nutritive sweetener. It's a nutritive sweetener and, therefore, falls outside the class of saccharin and cyclamates.

DR. BECK: Certain technological uses can be cleared for a synthetic sweetener. If you clear it as a sweetener for use in low calorie products, theoretically it is possible also to negotiate with FDA for certain technological uses; flavoring agents would be one. If you use cyclamate in a chewable vitamin tablet, for instance, as a flavoring agent, you do not make any claims for calorie reduction.

Saccharin in toothpaste and mouthwash is not there for any caloric purpose, only for flavoring. The idea is that kids will brush their teeth better if the toothpaste tastes good. If the toothpaste contains 0.2% saccharin, it tastes a lot better than without any sweetening agent.

I don't think that clearing a synthetic sweetener for special dietary uses automatically clears it for use as a flavoring agent, but it does create a background on which to discuss these things on an individual product basis with FDA.

DR. BEIDLER: Do you have any evidence that diet drinks help control obesity? I don't know of any evidence.

DR. CANTOR: I don't think they have any value in reducing weight.

DR. BECK: The difference between a food additive and a drug has been one concern when people talk about clearing saccharin or cyclamate. A food additive has to be shown to be safe. However, a drug must be shown to be both safe and efficacious for a specific purpose. Can you really prove that cyclamate or saccharin or any non-caloric sweeteners are effective in a weight reduction program or as a substitute for sugars in the control of diabetes?

Clearing cyclamate or saccharin or aspartame does not mean that everybody has to use them. What you are doing is making them available to people who want to use them. I think it would be nice if I could continue to buy a bottle of soft drink with no calories in it that I enjoy.

DR. CAGAN: Why is it necessary that efficacy be defined in terms of losing weight? Why not simply in terms of the pleasure that people get from having a sweetener? That, after all, is why people seek sweet-tasting foods.

DR. BECK: For a food additive all you have to do is prove to the FDA that the material is safe. In addition, of course, you have to have a product which the consumer likes enough to buy. However, as soon as you make a claim that your product is efficacious for some purposes, by definition, the FDA regulations require that you must demonstrate proof that the product is effective as claimed.

DR. GILKES: I think that I can clear up more of the points being discussed about food additives and drugs. I have been with the Food and Drug Administration for eleven years and have served as supervisory dental officer for the past four years in the Bureau of Drugs, Division of Surgical Dental Drug Products. The requirements for testing and approval of drugs differ considerably from those for foods. Under the law, a request for identification and approval as an investigational new drug (IND) must be filed when clinical studies are to be begun in man for both the efficacy and safety of a particular product. Then some products would progress to a new drug application (NDA) which must be reviewed and approved prior to the company's marketing the product. In contrast to the Bureau of Foods, where they are given one year, we have 180 days to review a new drug application. New drug applications are voluminous because under law the raw data must be submitted including copies of the individual patient record forms. This is not true for a food additive petition.

A new drug is defined, according to the Bureau, as any new entity or any old active ingredient or agent which is to be used for a new indication. The Bureau of Drugs does not stipulate how animal studies must be conducted, nor does it conduct animal studies or monitor same, except in rare instances.

However, guidelines for animal studies were developed with input from industry. To review the documents that we receive, we have people from four disciplines. One, a group of pharmacologists, look at the animal work. Two, dentists review the clinical portion or protocol when it is submitted as an IND, and of course they also review the results of clinical studies when they are submitted to an NDA. Three, chemists review and evaluate the manufacturing controls, stability, and so forth that are required under good manufacturing practices. And lastly, when necessary, a statistician from our Division of Biometrics takes part.

At the Bureau we have developed guidelines for caries clinical studies with the help of a Dental Drug Products Advisory Committee, three members of which are present here today, Drs. Bibby, Newbrun and Schrotenboer. They too addressed themselves to clinical studies for chewing gums. These were in the vein of drug claims. Inasmuch as claims of even non-cariogenicity must be supported by clinical studies and eventually sent to the Bureau, it would seem to me logical that interested parties would contact personnel at the Bureau for some guidance along the lines of what we would be looking for in such studies. The guidelines haven't been established as regulations. We feel that the Agency is there not only to insure that safe and efficacious products reach the consumer, but also to aid industry as they proceed in clinical studies of such products.

Lastly, I want to reemphasize that the FDA scientists didn't write that statute which bans products. Congress wrote it, and we must lay our problem at the proper doorstep if it is to be resolved.

I know I speak for Dr. Joseph Inscoe, who is the supervisory pharmacologist in our division, when I say, feel free to call on us at any time if we can be of help to you, either along the lines of animal studies or prior to submitting a protocol for clinical studies.

In summary, with regard to the FDA attitude to efficacy, efficacy must be demonstrated when any therapeutic or preventive claim is made for a product. If the consumer likes the way a product tastes, it is not considered by the FDA to be a justifiable therapeutic claim.

DR. MOSKOWITZ: Then why make an efficacy claim for any new sweetener? It seems to be opening up the door to a lot of aggravation.

DR. BECK: Efficacy automatically comes into the picture if we file a new drug application for cyclamate or saccharin. Likewise, if FDA decides that the only way saccharin can stay on the market is as a drug, then somebody has to file a new drug application to make even that route of supply possible. Part of that new drug application will be proof of efficacy for whatever claims you are making for cyclamate or saccharin as drugs that will be helpful to the consumer.

DR. MOSKOWITZ: If sweeteners are not permitted as food additives, then it really looks like a harder road down the line.

DR. BECK: If we cannot get cyclamate cleared as a food additive, I don't think we are likely to file a drug application for it. However, that's not an official decision.

DR. WIEGAND: If saccharin was reclassified as a drug in order to allow its availability for diabetic patients, the industry would have to come up with evidence within something like an 18-month period showing some therapeutic value. I don't think the FDA believes that that would be demonstrable and I think there may actually be some problem in doing that.

DR. ROUSSOS: Dr. Beck, if you are really concerned with calories, at least in certain items like a cake or other items where sucrose is also used as a bulking agent, could you visualize a situation where you are using a non-caloric sweetener in conjunction with a non-metabolizable bulking agent? The reason I ask that specific question regarding non-metabolizable bulking agents is that if you want a product to be almost 100% non-cariogenic, you would try to eliminate some of the less cariogenic substances such as cooked starch, as well as the sugars.

DR. BECK: In a soft drink, of course, you can take all the sugar out and put in a synthetic sweetener or a combination of sweeteners, and the product still looks and tastes like a soft drink. From a practical standpoint there is a limit to how much you can reduce calories from sugars in various products and still have any customary type of texture. For example, ice cream is normally about 40% solids; if you go down to less than 30% solids, the product freezes into a flavored ice cube. It's not ice cream anymore. Likewise, Chuck Stone worked on baked goods and couldn't get appreciably lower calories without major changes in texture that made the products undesirable. You have major practical problems when you begin to reduce the caloric value of many products because of the gross changes in texture.

DR. HERBERT: We are speculating. First, research hasn't been done to prove sweeteners are of any use in weight loss or are helpful to diabetics. Good workers, like Dr. Lillian Recant in Washington (personal communication), have diabetic patients who do very well on an average American diet without reducing glucose intake, although obese diabetics require caloric modification.

Such workers have further established that reducing carbohydrate intake further impairs carbohydrate tolerance on isocaloric diets with caloric restrictions for obese diabetics. Therefore, we don't know that sweeteners have a role in diabetics.

Now let's look at Scribner's letter in "Chemical and Engineering News", October 10, 1977, page 62, where he notes the point that the evidence is that cyclamates and saccharin are perhaps not carcinogens, but more likely are promoters. He refers to evidence that bladder cancer only occurs following these agents if they are given with a non-carcinogenic dose of the carcinogenic alkylating agent, methylnitrosourea. He further notes good evidence of a threshold dose for them, which would mean that they are safe below that threshold dose.

DR. ROUSSOS: I tend to disagree with Dr. Herbert with regard to his comment about diabetes, although I agree with him that this issue is highly controversial.

Even though obesity and genetic factors are considered as important ones in producing diabetes, or at least in increasing the risk of the disease, arguments have also been advanced by numerous clinical investigators that dietary sugar and fat are especially diabetogenic. Studies with humans and experimental animals have established that diets containing sucrose or fructose produce larger increases in blood lipids, especially the triglyceride fraction, and in hepatic lipogenic enzymes, than diets containing an equivalent amount of glucose polymers. The association of elevated blood triglyceride levels with risk of coronary heart disease has recently been confirmed. In general, although sucrose and fructose alone, at current consumption levels, have not as yet been shown to be directly implicated in the etiology of cardiovascular disease and diabetes in the majority of the population, together with other environmental factors, they may produce a combination that is more lipogenic or diabetogenic than any of the other factors alone. Sucrose alone may be a very important causative factor in cardiovascular disease and diabetes in that segment of the population described as "carbohydrate-sensitive." It has already been mentioned that dental caries does not develop in the absence of dietary carbohydrate and that, without a reduction in the sucrose and fat content, it would be practically impossible for a national diet of conventional foods to ensure the essential nutrient requirements of low calorie consumers. Moreover, low fat, low carbohydrate diets in which sugars are eliminated may be useful in controlling obesity, which is known to reduce life expectancy and increase morbidity and mortality from various causes, including diabetes mellitus, and perhaps cardiovascular disease.

Even though, with the exception of dental caries, considerable controversy surrounds the role played by dietary carbohydrate and sugars in particular in the pathogenesis of the diseases described above, it is apparent from this brief review that a reversal of the present trend of increased consumption of refined sugar to an increased consumption of the more complex and natural forms of carbohydrate should be encouraged. In light of the foregoing, it is also obvious that the availability of safe, nonnutritive, commercially utilizable sugar substitutes of both natural and synthetic origin is desirable, particularly for those people who simply cannot control their craving for a sweet taste.

DR. HERBERT: That's largely anecdotal. The paper in the New England Journal of Medicine 297:644-650, 1977 by George Mann on "The Diet-Heart

Controversy" presents evidence that what you said is speculation rather than fact.

DR. ROUSSOS: But there are also USDA and other studies which support the conclusions that I have just presented, although admittedly not everyone agrees with them.

DR. HERBERT: That's not quite correct. The studies indicate that increased calories are associated with increased cholesterol and increased triglycerides, but the rest of what you just stated is largely from the proposals of Yudkin which have been discounted by subsequent studies. Yudkin himself has withdrawn from much of that position because the facts do not support it.

DR. ROUSSOS: I would agree that excess calories and the resulting obesity are the common denominators in every case other than caries, but we also agree, I think, that about one-third of the adult population in Western countries is obese, and that obesity is associated with reduced life expectancy and increased morbidity and mortality from various diseases, including diabetes mellitus and perhaps cardiovascular disease.

DR. HERBERT: Agreed.

DR. ROUSSOS: Okay, but I think that you would agree that diabetologists who do not reduce the sugar intake by diabetics only do that when they are treating them with insulin. They would prefer to have their diabetic patients use the complex polysaccharides, not the mono- and disaccharides, which tend to increase insulin requirements and secretions.

DR. HERBERT: I'd certainly agree with you that it would be nice to have patients use the complex polysaccharides, but what I am holding out against is going from fact to anecdote. When we talk about sugar as toxic, we are talking anecdote. If we are talking about it as part of a problem of too many calories, I agree with you 100%.

DR. ROUSSOS: As I approach this workshop knowing that there is a limited supply of funds for investigation in this area, my preference would be to develop a sucrose substitute which is both non-cariogenic and non-caloric. In that way, we could approach both the caries and obesity problems.

DR. HERBERT: From my perspective I would decrease my fat intake since per unit weight of fat I receive more than twice as many calories as I do per unit weight of any carbohydrate or protein source. This concept is in line with the recommendations of the McGovern Select Committee on Nutrition in their proposed Dietary Goals for the United States. They proposed an increase in ingestion of the complex carbohydrates and decreases in sugar, fat and salt usage. These are good recommendations except that they are presented as dietary goals for the United States, whereas in fact they are appropriate dietary goals for the overweight and for most persons with hypertension.

DR. SOELDNER: At this point I would like to make a short comment. As Dr. Roussos and Dr. Herbert recall, we have had available since 1915 the basic and fundamental information relating body weight and calorie restriction in the form of the classic studies by Benedict and Joslin. Essentially, these very old studies clearly indicated that weight reduction does not take place until the caloric intake is less than the caloric expenditure. Since then it has become quite clear that calorie control (quantity and quality) is a key component in diabetes, and may be an important factor in dental caries and a great many other diseases.

And yet, since 1915 and the publication of that very lovely study, and in addition the accumulation of what I would guess to be at least 5,000 reports and studies, we find no easy answer to the problem of weight reduction, and the basic system which regulates body weight is essentially still unknown.

However, the basic information is available, and a very important aspect of this is nutritional education. In my opinion, we have not used the information that we have available in an appropriate public education or diabetes patient education program. Yet, I think that although the information we have is incomplete, we have a broad understanding of the basic problem; but we have not yet been able to come to grips with using the available information, as limited as it is, to any meaningful extent.

SESSION IV.

SWEETENERS OF NATURAL ORIGIN

Moderator:
H. van der Wel

Unilever Research
Vlaardingen/Duivan
Olivier van Noortland 120
Vlaardingen, The Netherlands

Citrus-based dihydrochalcone sweeteners

R.M. Horowitz and Bruno Gentili

Fruit and Vegetable Chemistry Laboratory, US Department of Agriculture, Agricultural Research Service, Pasadena, California 91101

SUMMARY

It is unlikely that the presently known dihydrochalcones could be used in all applications as a direct replacement for sucrose or existing synthetic sweeteners without some reformulation of the product. However, in a sizeable number of applications, dihydrochalcones could be used to advantage, and with minimal reformulation of the products could give acceptable and hitherto unattainable results.

INTRODUCTION

The flavonoid constituents of Citrus are abundant and easily isolated, have remarkable taste properties, yield valuable byproducts and are available in large quantity. The citrus fruit production in the United States exceeds that of all other fruits combined. Among the various flavonoids are eight flavanone glycosides, four of which are bitter, four of which are tasteless and all of which can be reduced to dihydrochalcones. It is surprising, considering their parentage, that some of these dihydrochalcones should be intensely sweet. We will review here the preparation, properties and possible applications of the more promising compounds in this group[1-3].

Three sweeteners that claim attention are neohesperidin dihydrochalcone (I), naringin dihydrochalcone (II) and hesperetin dihydrochalcone 4'-β-D-glucoside (III). These were among the earliest compounds in the series; a large body of information about their taste and toxicology has now been accumulated. Many variants of these substances have been prepared, including both glycosidic and non-glycosidic derivatives, but none of these appears to offer any major improvement in properties over those of the earlier compounds. Since of all the sweeteners in the series neohesperidin dihydrochalcone is the one most likely to gain acceptance, we will deal mainly with it and the closely related substances II and III.

SOURCE AND PREPARATION

The starting compounds, plant sources and products are shown in Table 1.

Table 1. Sources of Dihydrochalcone Glycosides

Starting compound	Plant Source	Product
Neohesperidin (IV) (bitter)	Seville orange (C. aurantium)	Neohesperidin dihydrochalcone (I) (sweet)
Naringin (V) (bitter)	Grapefruit (C. paradisi)	Naringin dihydrochalcone (II) (sweet)
Hesperidin (VI) (tasteless)	Sweet orange (C. sinensis) Lemon (C. limon)	Hesperetin dihydrochalcone 4'-β-D-glucoside (III) (sweet)

Two of the starting compounds, neohesperidin and naringin, contain as their sugar component the disaccharide β-neohesperidose (2-$\underline{O}$-α-L-rhamnopyranosyl-β-D-glucopyranose)[4]; the third compound, hesperidin, contains the disaccharide β-rutinose (6-$\underline{O}$-α-L-rhamnopyranosyl-β-D-glucopyranose). Experience has shown that many (but not all) phenolic or flavonoid glycosides are taste-eliciting (bitter, sweet or bitter-sweet) if their sugar component is β-neohesperidose or β-glucose. On the other hand, analogous compounds in which the sugar component is β-rutinose have thus far been found to be tasteless. When hesperidin is used as the starting material, it is essential to alter in some way the taste-abolishing β-rutinosyl group if a taste-eliciting substance is to be obtained. The easiest way to do this is by hydrolysis to remove rhamnose.

The conversion of neohesperidin and naringin to their dihydrochalcones[5] involves nothing more complicated than dissolving them in dilute alkali, which yields the chalcone, and hydrogenating the chalcone solution with the aid of a standard hydrogenation catalyst (Figure 1). The products are formed rapidly in almost quantitative yield and can be crystallized easily from water, methanol or acetone. The reduction of hesperidin in dilute alkali yields hesperidin dihydrochalcone (VII), a tasteless compound. Partial hydrolysis of this dihydrochalcone, either by acid or by dissolved or immobilized enzymes, gives rise to the sweet hesperetin dihydrochalcone 4'-β-D-glucoside (HDG)(III) in an overall yield of 40 to 50% (Figure 1)[6-8].

Because hesperidin could be produced in abundance as a byproduct of the orange-processing industry, hesperidin-based sweeteners, such as III, are

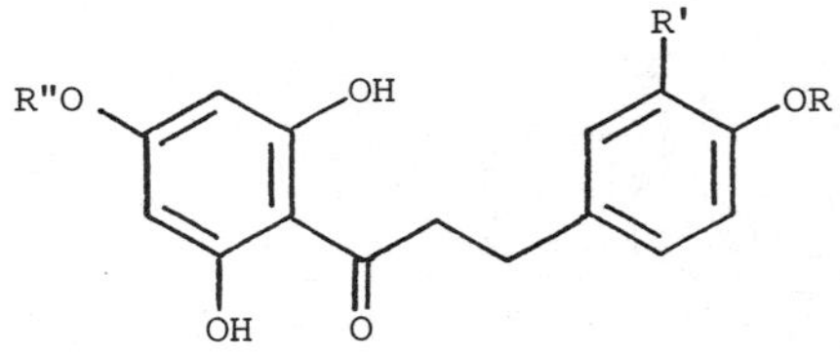

I, II, III, VII, VIII 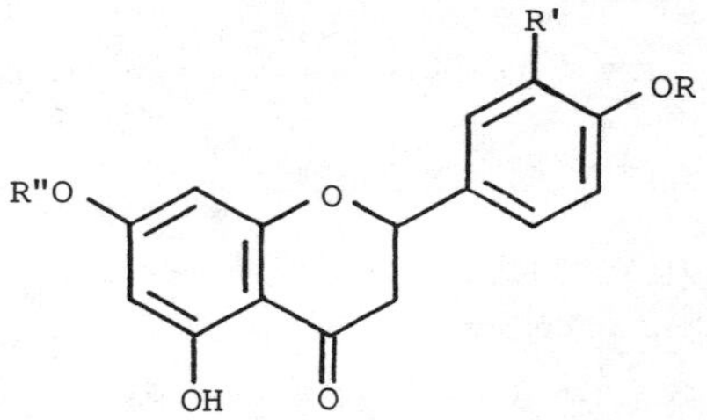   IV, V, VI

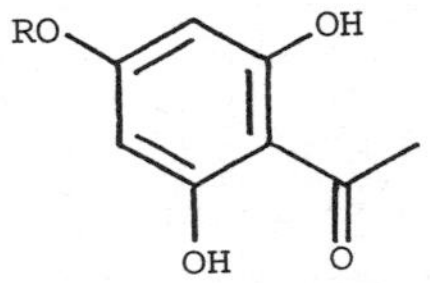

IX

IV and I: R = CH$_3$, R' = OH, R" = β-Neohesperidosyl
V and II: R = R' = H, R" = β-Neohesperidosyl
VI and VII: R = CH$_3$, R' = OH, R" = β-Rutinosyl
III: R = CH$_3$, R' = OH, R" = β-D-Glucosyl
VIII: R = CH$_3$, R' = OH, R" = H
IX: R = β-Neohesperidosyl

IV + 10% KOH + H$_2$ + Pd/C → I

V + 10% KOH + H$_2$ + Pd/C → II

VI + 10% KOH + H$_2$ + Pd/C → VII;   VII + H$^+$ or Enzyme → III

V + 25% KOH → IX;   IX + Isovanillin + 50% KOH → IV → I

Figure 1. Structures and interconversions of flavanones and dihydrochalcones

attractive candidates for further study. On the other hand, the large-scale
preparation of neohesperidin dihydrochalcone (I) is hampered by the fact that
neohesperidin (IV) is not available on a commercial scale from its best
natural source, the Seville orange. These oranges are grown in the United
States chiefly for ornamental purposes; their cultivation elsewhere for use
in the manufacture of perfume and marmalade is also limited. To get around
this impasse it is necessary to convert the readily available naringin to
neohesperidin and thence to the dihydrochalcone by the series of reactions
shown in Figure 1. The overall yield in this process averages about 20 to

30% but could doubtlessly be improved through further study[9-11]. In fact, a yield of 77% is claimed in a recent patent[12-13].

CHEMICAL PROPERTIES

Dihydrochalcones I-III are crystalline solids which are colorless and odorless when pure. Compound I, the most soluble, yields at equilibrium a solution containing about 1.2 grams of the dihydrochalcone per liter of distilled water. None of these crystalline solids dissolves very rapidly, so warming or stirring may be necessary to effect complete solution. The sodium and calcium salts of I have been prepared and are said to have a solubility in excess of 1 gm/ml[14].

Stability would probably not be a major problem in most applications of dihydrochalcones. Dilute aqueous solutions have been kept in the laboratory for periods of years with little change in appearance or sweetness. It has been reported that the compounds are resistant to hydrolysis by acids at pH's > 2 at normal room temperature[15]. At high temperature (75 to $100°C$) free sugars arising from hydrolysis can be detected in a matter of minutes or hours in the pH range 1.5 to 3.6. However, the hydrolytic products of I, i.e., III and its aglycone, hesperetin dihydrochalcone (VIII), are intensely sweet, though the latter compound is not very soluble.

Qualitative and quantitative analytical methods for the dihydrochalcones have been reported. The methods are based on paper, thin-layer and high pressure liquid chromatography as well as proton magnetic resonance[11,16-19].

TASTE

In general, dihydrochalcones such as I-III have a pleasant sweetness, somewhat slow in its onset and of varying (usually long) duration. Except for naringin dihydrochalcone (II), the sweet aftertaste is not marred by bitterness, but is often described as having a slight cooling or licorice-like quality. It has been claimed that the delay in onset of sweetness can be reduced by combining the dihydrochalcone with gluconic acid or δ-gluconolactone[20]. No way has yet been found to reduce the lingering of the sweetness. However, most tasters agree that monosaccharyl derivatives, such as III, have less of the sweet aftertaste than disaccharyl derivatives such as L

Many quantitative studies have been made of dihydrochalcone sweetness. The data of Guadagni, et al.[21] are shown in Tables 2 and 3. At threshold neohesperidin dihydrochalcone (I) is almost 1900 times sweeter than sucrose, while naringin dihydrochalcone (II) and HDG (III) are almost 300 times

Table 2. Thresholds of Sweeteners in Water Solution[21]

Compound	Mean threshold (mg/kg)	95% confidence interval
Sucrose	1300	920 - 1840
Naringin dihydrochalcone (II)	4.5	3.5 - 5.7
Hesperetin dihydrochalcone glucoside (III)	4.5	3.5 - 5.7
Neohesperidin dihydrochalcone (I)	0.7	0.5 - 0.9

Table 3. Relative Sweetness of Dihydrochalcones and Sucrose Compared at Varying Concentrations of Sucrose[21]

Sucrose (mg/kg)	Concentration of sucrose/concentration of sweetener		
	Neo DHC (I)	HDG (III)	Nar DHC (II)
1300[a]	1857	289	289
5000	927	193	156
10000	667	135	102
20000	400	111	69
25000	368	100	63
50000	250	78	50

[a]Calculated on basis of threshold values for all sweeteners. All other figures obtained on basis of sweetness equivalent to that of sucrose.

sweeter. As concentration increases, the relative sweetness of I and II decreases markedly, but that of III decreases more slowly. Thus, the relative sweetness of I, II and III at the 2.5% sucrose level is 20, 21 and 35%, respectively, of their values at threshold.

An unexpected property of these sweeteners is their ability to mask or interfere with the perception of bitterness[21]. As shown in Table 4, sucrose

Table 4. Effect of Dihydrochalcones and Sucrose on Bitterness[21]

Sweetener	Concentration	Threshold concentration in water of	
		Naringin	Limonin
None	---	20 ppm	1.0 ppm
Sucrose	10,000 ppm (1%)	25 ppm	1.0 ppm
Sucrose	30,000 ppm (3%)	30 ppm	1.0 ppm
Neo DHC (I)	16 ppm[a]	34 ppm	1.4 ppm
HDG (III)	80 ppm[a]	34 ppm	3.2 ppm
HDG (III)	300 ppm[b]	45 ppm	3.5 ppm

[a]Equivalent to 1% sucrose in sweetness
[b]Equivalent to 3% sucrose in sweetness

even at relatively high concentration is not very effective in raising the
threshold of the bitter flavanone, naringin, and it has almost no effect at
all on the bitter triterpenoid, limonin. On the other hand, neohesperidin
dihydrochalcone (I) and particularly HDG (III) give, at relatively low concentrations, striking increases in the threshold values of these bitter
principles.

COST AND CALORIC VALUE

At present neohesperidin dihydrochalcone (I) is listed by one manufacturer at $150/lb., a price that would certainly decrease if there were a
substantial demand for the compound. Even at this price the amount of I
needed to contribute sweetness equivalent to that of a pound of sucrose would
cost only about 50 cents (assuming that I is 300 times sweeter). The cost in
terms of ingested calories should be negligible. We estimate that the amount
of I needed to contribute sweetness equivalent to that of a pound of sucrose
would yield 2 to 3 calories (we assume a sweetness ratio of 300 and that only
the rhamnose and glucose moieties are metabolized).

OTHER COMPOUNDS

It was mentioned above that many dihydrochalcone variants have been
synthesized. Hesperetin dihydrochalcone 4'-β-D-galactoside and hesperetin
dihydrochalcone 4'-β-D-xyloside are both sweeter (1.5 to 2x) than the
4'-β-D-glucoside (III) and in addition have very good taste characteristics[22-23]. Both are difficult to synthesize. Other new hesperetin dihydrochalcone 4'-monosaccharides and 4'-disaccharides have been reported; all are
difficult to make and there is no evidence that any of them are superior to
the older compounds[24]. It was recognized early that the sugar moiety is
neither necessary nor sufficient to confer sweetness on a dihydrochalcone
aglycone[25]. For example, the aglycone hesperetin dihydrochalcone (VIII) was
found to be sweet, if essentially insoluble. A large number of 4'-O-carboxy-
alkyl and 4'-O-sulfoalkyl derivatives of hesperetin dihydrochalcone have been
synthesized recently[26-30]. Many of these compounds have excellent solubility
and several are good sweeteners, but again, they seem to provide no significant improvement in taste over that of compounds I-III. Moreover, because
alkyl substituents of this type are not found in naturally occurring flavonoids, their introduction does away with one of the chief attractions of
dihydrochalcones such as I-III, i.e., that they are only a short step removed
from natural products usually regarded as innocuous.

APPLICATIONS

Dihydrochalcone sweeteners obviously work best where long lasting sweetness is advantageous or, at least, not detrimental. These applications would include products such as chewing gums, confections of various sorts, toothpastes, mouthwashes and similar preparations[31]. It is likely that in these applications dihydrochalcones could be substituted for existing synthetic sweeteners without drastic reformulation of the product. The situation with regard to soft drinks is not entirely clear, but it appears likely that dihydrochalcones could be used with some success if these products were reformulated. An example of a soft drink formulation utilizing I has been reported[15].

Other applications of the dihydrochalcones would be in products where it is desirable or, indeed, necessary to reduce bitterness. These products include a very large number of pharmaceuticals and a few fruit juices, such as grapefruit juice. Grapefruit juice has a rather restricted popularity, presumably due to its bitterness. Recent work in several laboratories has shown that the addition of as little as 12 ppm of neohesperidin dihydrochalcone to moderately bitter grapefruit juice leads to a dramatic increase in its acceptability to taste panels[32].

SOME GENERAL CONSIDERATIONS

Dihydrochalcones are typical examples of the large group of plant products designated as flavonoids and defined by the presence of two C_6 aromatic groups joined by a C_3 bridge (C_6-C_3-C_6). The particular type of flavonoid (anthocyanin, aurone, flavone, flavanone, chalcone, dihydrochalcone, etc.) is determined by the oxidation level of the C_3 bridge and by whether one of the carbon atoms of the bridge is linked through an oxygen atom to one of the aromatic units to form a third ring.

The sweet dihydrochalcones I and II are not known to occur naturally. In chemical structure, however, they are simply the open-chain, reduced analogs of the substances from which they are derived, the bitter flavanones neohesperidin (IV) and naringin (V). The latter compounds are widely distributed in citrus fruit and occur, for example, in all products made from grapefruit or Seville oranges. Dihydrochalcones have apparently not been found in any plant foods, possibly because until recently they were relatively difficult to detect. Studies now underway in the authors' laboratory point to the probable occurrence of certain types of dihydrochalcones in kumquats.

As a group, the flavonoids are usually regarded as innocuous. They contain neither nitrogen nor sulfur--elements often associated with toxicity--and are believed to be metabolized to carbon dioxide and various aromatic acids. Perhaps of most importance is the fact that since flavonoids occur in all higher plants, they are, and always have been, a common constituent of the diet.

REFERENCES

1. Horowitz, R.M. and Gentili, B. J. Agr. Food Chem., 17, 696, 1969.
2. Horowitz, R.M. and Gentili, B. in Sweetness and Sweeteners, edited by G.G. Birch, L.F. Green and C.B. Coulson, Applied Science Publishers, London, pp. 69-80, 1971.
3. Horowitz, R.M. and Gentili, B. in Symposium: Sweeteners, edited by G.E. Inglett, Avi Publishing Co., Westport, Conn., pp. 182-193, 1974.
4. Horowitz, R.M. and Gentili, B. Tetrahedron, 19, 773, 1963.
5. Horowitz, R.M. and Gentili, B. U.S. Patent 3,087,821, 1963.
6. Horowitz, R.M. and Gentili, B. U.S. Patent 3,429,873, 1969.
7. Horowitz, R.M. and Gentili, B. U.S. Patent 3,583,894, 1971.
8. Krasnobajew, V. Ger. Offen. 2,402,221, 1974. Chem. Abstr. 82, 15272, 1975.
9. Horowitz, R.M. and Gentili, B., U.S. Patent 3,375,242, 1968.
10. Krbechek, L., Inglett, G., et al., J. Agr. Food Chem., 16, 108, 1968.
11. Robertson, G.H., Clark, J.P. and Lundin, R. Ind. Eng. Chem., Prod. Res. Develop., 13, 125, 1974.
12. Ueda, K. and Odawara, H. Japan Kokai, 75,149,635, 1975. Chem. Abstr. 84, 180577, 1976.
13. Japan Kokai, 75,154,261, 1975. Chem. Abstr. 84, 180574, 1976.
14. Westall, E.B. and Messing, A.W. Ger. Offen. 2,216,071, 1972. Chem. Abstr. 78, 43051, 1973.
15. Inglett, G.E., Krbechek, L., Dowling, B. and Wagner, R. J. Food Sci., 34, 101, 1969.
16. Gentili, B. and Horowitz, R.M. J. Chromat., 63, 467, 1971.
17. Linke, H.A.B. and Eveleigh, D.E. Z. Naturforsch., 30b, 940, 1975.
18. Schwarzenbach, R. J. Chromat., 129, 31, 1976.
19. Fisher, J.F. J. Agr. Food Chem., 25, 682, 1977.
20. Huber, U., Kossiakoff, N. and Vaterlaus, B. Ger. Offen. 2,445,385 Chem. Abstr. 83, 204966, 1975.
21. Guadagni, D.G., Maier, V.P. and Turnbaugh, J.H. J. Sci. Food Agric. 25, 1199, 1974.
22. Horowitz, R.M. and Gentili, B. U.S. Patent 3,890,296, 1975.
23. U.S. Patent 3,890,298, 1975.
24. Kamiya, S., Esaki, S. and Konishi, F. Agr. Biol. Chem. (Japan), 40, 1731, 1976.
25. Horowitz, R.M. in Biochemistry of Phenolic Compounds, edited by J.B. Harborne, Academic Press, New York, pp. 561, 562, 1964.
26. Farkas, L., Nogradi, M., Gottsegen, A. and Antus, S. Ger. Offen. 2,258,304 Chem. Abstr. 79, 78400, 1973.
27. Farkas, L., Nogradi, M., Pfliegel, T. and Antus, S.Hung. Teljes 10,931 Chem. Abstr. 85, 46201, 1976.
28. DuBois, G.E., Crosby, G.A. and Saffron, P. Science, 195, 397, 1977.
29. DuBois, G.E., Crosby, G.A., Stephenson, R.A. and Wingard, R.E., Jr. J. Agr. Food Chem. 25, 763, 1977.
30. Rajky-Medveczky, G., et al. Nahrung, 21, 131, 1977. Chem. Abstr. 87, 4147, 1977.

31. Warner Lambert Co. and Nutrilite Products, Inc., British Patent Specification 1,310,329, 1973.
32. Guadagni, D. Unpublished data.

Toxicity studies of neohesperidin dihydrochalcone*

M.R. Gumbmann, D.H. Gould, D.J. Robbins, and A.N. Booth

Toxicology and Biological Evaluation Research Unit, Western Regional Research Center, ARS, US Department of Agriculture, Berkeley, California 94710

SUMMARY

The sweetener, neohesperidin dihydrochalcone (NDHC), was fed to rats and dogs in a series of toxicity studies to evaluate the safety of the sweetener as a potential food additive and flavor agent. Rats fed 5% NDHC for 122 to 170 days developed no particular abnormalities with the exception of possibly slightly elevated thyroid weights. Such elevation was not observed in rats fed 5% NDHC for one year or in rats fed 10% NDHC for 11 months. Growth inhibition and increased testes weight relative to body weight were observed in rats fed 10% NDHC. A two-year carcinogenicity study in rats failed to detect any increase in tumor incidence that could be associated with ingestion of NDHC in the diet. In this study an apparent nutritional interaction between NDHC and a laboratory prepared diet was noted which resulted in a decreased growth rate in the rats. This could be corrected by a supplement of vitamins and minerals. No teratological abnormalities in rats were related to ingestion of NDHC.

Liver and thyroid weights were elevated in dogs fed daily doses of 2 g NDHC/kg body weight for two years. Mild thyroid hypertrophy and hyperplasia were confirmed in this dosage group. Marked testicular atrophy and degeneration occurred in one of three dogs in each of the two highest dose levels, 1 and 2 gm/kg.

It is not clear to what degree the biological alterations observed in these studies are to be regarded as toxic effects or whether they represent reversible adaptive responses associated with the metabolism of high doses of NDHC. The incidence of testicular atrophy appeared to exceed a normal expectation rate in beagle dogs and should be viewed as a suspect adverse effect.

Excretion studies showed that NDHC, at least the phenyl proprionic acid portion of the molecule, is readily absorbed and nearly completely (90%) excreted in 24 hours. Approximately 80% appears in the urine and the remainder in the feces.

INTRODUCTION

The synthetic sweetener, neohesperidin dihydrochalcone (NDHC), was developed in the early 1960's by Horowitz and Gentili, chemists with the USDA's Agricultural Research Service (1,2). NDHC is derived commercially from

---

* Reference to a company and/or product named by the Department is only for purposes of information and does not imply approval or recommendation of the product to the exclusion of others which may also be suitable.

naringin, a bitter substance naturally occurring in the rind of grapefruit and some other citrus varieties, and possesses a sweetness comparable to that of saccharin. NHDC has potential for use in citrus flavored beverages, chocolate, soft drinks, chewing gum, tooth paste and mouth wash.

Early toxicological studies at Western Regional Research Laboratory detected no adverse effects from NDHC fed to rats at a dietary level of 0.5%. It can be calculated that at 0.5%, NDHC in the diet of a rat is equivalent to approximately 100 times the per capita consumption of sucrose in the United States, based on relative sweetness. When larger quantities of the sweetener became available, longer term studies were conducted at levels up to 5 and 10% in the diet. Summarized here are the results of feeding studies in rats of NDHC at 5% of the diet for 170 days, one year, and two years, and at 10% of the diet for 11 months. Included are a three generation reproduction and teratogenesis study in rats, a two-year feeding in dogs, and excretion studies with carbon-14 labelled NDHC in rats.

METHODS AND RESULTS

Subacute Rat Feeding Studies. NDHC, containing less than 1% impurities, was fed to groups of 5 male and 6 female weanling Fisher inbred strain rats at 0 and 5% in the diet. The diet was Purina Laboratory Chow. No adverse effects were noted for weight gain, feed intake, and organ weights after 122 and 170 days for the males and females, respectively (Table 1).

Table 1. Short-term toxicity study of rats fed 5% NDHC in a commercial diet

	Control		5% NDHC	
	Male	Female	Male	Female
Weight gain at 90 days (gm)	272	137	262	148
Weight gain at autopsy (gm)[a]	309	202	300	205
Feed intake for 90 days (gm/rat/day)	15.0	10.2	15.3	10.7
Organ weights (% of body weight)				
Liver	3.38	3.30	3.33	3.17
Heart	.28	.31	.28	.33
Kidneys	.71	.73	.68	.72
Spleen	.17	.23	.18	.23
Pancreas	.42	.55	.48	.56
Testes	1.01	--	1.05	--
Adrenals (mg %)	14.9	26.7	15.5	27.7
Thyroids (mg %)	6.6	8.9	8.1[b]	9.4

[a]Males 122 days; females 170 days; [b]$p < .05$.

A possible exception was a diet-related increase in thyroid weight detected in the males and perhaps in the females. Gross and histological examination revealed no lesions which could be specifically associated with the ingestion of NDHC. After 90 days, a reproductive test was initiated. No impairment in reproductive performance was observed.

Six F_2 generation rats of each sex per dietary group were obtained as part of the three-generation reproduction study (see below) and maintained for one year. The dietary groups were 0, 0.5, 2.5 and 5.0% NDHC in a commercial rat diet (Wayne Lab - Blox). The results shown in Table 2 indicate that the rats grew at a normal rate during the 12-month period and that no highly significant differences in organ weights (p < .01) occurred between control and test groups.

Table 2. Body and organ weights of rats fed a commercial diet plus NDHC for one year[a]

NDHC in Diet %	Mean Weight (gm)	Organ Weights[b]			
		Liver	Kidneys	Testes	Thyroids
Males					
0.0	420	2.93	0.61	0.81	4.3
0.5	436	2.92	0.65	0.79	4.2
2.5	415	2.82	0.60	0.82	4.5
5.0	413	2.76[c]	0.60	0.82	4.4
Females					
0.0	231	3.15	0.66	--	6.3
0.5	236	3.30	0.68	--	6.2
2.5	220[c]	3.22	0.66	--	6.0
5.0	227	2.98	0.68	--	6.7

[a]No significant differences (p < .01); [b]Values expressed as gm/100 gm body weight, except thyroids in mg/100 gm; [c]p < .05

A group of weanling male Sprague-Dawley rats was fed the commercial (Wayne) diet to which was added 10% dextrose or 10% NDHC for a period of 11 months. At this concentration, the final body weights for the NDHC fed rats were significantly lower and their testes weights were significantly heavier than their control counter-parts (Table 3). None of the histological lesions that were observed in the F_2 generation rats fed 5% NDHC for one year or the group fed 10% NDHC for 11 months seemed to be associated with dietary regime. Life Span Rat Feeding Study. Groups of 24 male and 24 female Fischer strain rats, 33 days of age, were fed NDHC at dietary levels of 0, 0.5, 2.5 and 5.0% for their life-span or two years. The diet consisted of yellow corn meal, 72.8%; soybean meal, 10%; casein, 10%; brewer's dried yeast, 2%; corn oil, 3%; bone ash, 1.5%; sodium chloride, 0.5%; DL-methionine, 0.2%; and 2000 I.U. of vitamin A acetate and 200 I.U. of vitamin D_3 per 100 grams of diet.

Table 3. Body and organ weights of rats fed commercial diet containing 10% NDHC for 11 months

Diet Fed[a]	Mean Body Weight (gm)	Organ Weights[b]			
		Liver	Kidneys	Testes	Thyroids
Wayne + 10% Dextrose	623	2.27	0.61	0.59	3.8
Wayne + 10% NDHC	546[c]	2.45	0.60	0.73[c]	4.0

[a]Five weanling male rats per group, starting weight = 55 gm; [b]Values expressed as gm/100 gm body weight for liver, kidney and testes and mg/100 gm body weight for thyroids; [c]Significant difference (p <.01).

Inferior weight gain of both sexes fed the 5% NDHC diet became apparent within the first 10 weeks and was more pronounced in the females by the sixtieth week. At this time, the 5% NDHC diet for one-half of the rats of both sexes was supplemented with 3% USP Salts XIV (fortified with 40 ppm zinc and 2 ppm cobalt) and with 3% additional brewers dried yeast substituted at the expense of corn meal. These two supplements produced a dramatic response in growth, such that there was no significant difference at 100 weeks between the body weights of the rats fed the supplemented diet and those of the control group (Table 4).

Table 4. Mortality and body weights of rats fed NDHC for 100 weeks

NDHC in Diet	Number of Survivors[a]	Mean Body Weights (gm)
%	Male	
0.0	16	431
0.5	12	397
2.5	20	397
5.0	8	361[c]
5.0 + supplement[b]	11	431
	Female	
0.0	16	308
0.5	19	244[c]
2.5	17	213[c]
5.0	7	201[c]
5.0 + supplement[b]	11	301

[a]24 rats per group at start; [b]Supplement = 3% brewers dried yeast + 3% USP. XIV salt mixture; [c]p < .01.

A somewhat analogous situation has been reported by Ershoff (3) in which rats grew normally compared to controls when fed a commercial diet containing 5% sodium cyclamate, whereas rats fed a purified diet containing 5% sodium cyclamate were severely retarded in growth in relation to controls. In the present study, it is tentatively concluded that addition of NDHC to this laboratory prepared diet leads to a marginal deficiency in one or more nutrients, possibly due to interference in absorption. A specific mineral, vitamin, or other nutrient involved in this deficiency has not been identified.

Mortality was greatest in the control groups of both sexes at 100 weeks (66% survival), excluding the group of males fed 0.5% NDHC. The higher number of deaths, 50%, in this latter group is considered to be unrelated to the intake of NDHC since the mortality in the two groups of male rats fed 2.5 and 5.0 NDHC was quite low, namely 4/24 and 5/24, respectively.

Hematology and urinalysis data obtained at 6-month intervals were within normal limits regardless of the level of NDHC ingested. Plasma cholesterol, determined at autopsy, was lower in both sexes fed 5% NDHC than in the controls, whether or not the diet had been supplemented. Plasma cholesterol was also measured in the rats fed 10% NDHC in a commercial diet for 11 months and again found to be less than in the controls.

Organ weights of the rats surviving for two years are tabulated in Table 5.

Table 5. Organ weights of rats fed NDHC for two years

Percent NDHC in Diet	Organ Weights[a]					
	Liver	Spleen	Kidney	Heart	Adrenal	Thyroid
--- Male Rats ---						
0.0	3.36	0.28	0.68	0.31	14.5	20.8
0.5	3.60	0.25	0.71	0.31	13.5	16.5
2.5	3.22[b]	0.23	0.66[b]	0.30	13.5	27.1
5.0	4.21[b]	0.21[b]	0.82[b]	0.30	13.2	33.1
5.0 + supplement[c]	3.21	0.21[b]	0.64	0.26	13.0	8.6
-- Female Rats --						
0.0	3.06	0.18	0.61	0.27	16.4	21.0
0.5	3.22	0.17	0.62	0.27[b]	16.3	11.7
2.5	3.37[b]	0.20[b]	0.66[b]	0.33[b]	17.2[b]	20.9[b]
5.0	4.23[b]	0.24[b]	0.81[b]	0.36[b]	21.3[b]	41.0[b]
5.0 + supplement[c]	2.93	0.17	0.65	0.25	18.1	7.9

[a]Values in gm per 100 gm body weight for liver, spleen, kidney and heart, and mg per 100 gm body weight for adrenal and thyroid; [b]$p < .01$; [c]Supplement= 3% dried brewers yeast + 3% USP XIV salt mixture.

In rats of this age, greater variation in organ and body weight data is encountered as a result of increased incidence of emaciation and the presence of tumors. Thus, the value of such data as indicators of diet related toxicity is thereby reduced. On the diet containing 5% NDHC, without supplements, the liver and kidney weights of the males and all of the organ weights of the females were significantly heavier. These differences were eliminated in both sexes which received the supplements of yeast and minerals. This observation is particularly noteworthy in the case of the thyroid weights where even the control groups showed evidence of enlarged thyroids with respect to the thyroid weights of the rats receiving the supplementation.

Histological examination showed the incidence of tumors to be fairly uniformly distributed among the four groups and without any significant association with the presence of NDHC in the diet. The incidence of non-neoplastic lesions also showed no association to diet. A possible exception was the occurrence of focal renal cortical atrophy which tended to be higher in the 0.5, 2.5 and the non-supplemented 5% NDHC groups of females, as compared to the female controls and supplemented 5% NDHC group. Except for the animals in the supplemented 5% NDHC group, all animals exhibited diffuse thyroid follicular hyperplasia and hypertrophy consistent with a dietary iodine deficiency.

Three Generation Reproduction and Teratology Study. Weanling Fisher rats in groups of six of each sex were fed a commercial diet (Wayne Lab-Blox) containing 0, 0.5, 2.5 and 5.0% NDHC. This F_0 generation was mated at sexual maturity to produce an F_1 generation which, in turn, produced an F_2 generation. After producing an F_3 generation, the F_2 rats were continued on the same diets for one year and subjected to histological examination, as reported above. When the F_1 rats were weaned, the F_0 rats were used to produce second litters (F_{1b}), the embryos being removed by Caesarian section on the 20th day of gestation for teratological evaluation. No detrimental effects on reproduction due to the ingestion of NDHC were observed, except for slightly decreased survival of pups in the two highest level groups of the F_3 generation. The fetuses were examined grossly, after which one-third were prepared for visualization of the skeletal system and two-thirds were fixed for serial sectioning. Few abnormalities were detected, and no evidence was obtained for teratological effects related to NDHC.

Two-year Dog Feeding Study. Young beagle dogs, three of each sex per group, were fed NDHC in a commercial dog chow for two years. The daily dosages of NDHC were 0, 0.2, 1.0 and 2.0 g per kg body weight. For the high dose group, the average concentration of NDHC in the diet was estimated to be approximately 6% for the two-year period. At six month intervals, the dogs were given general physical examinations which included the collection and analysis of blood and urine samples. No effect related to feeding NDHC on growth or hematological values was demonstrated. Plasma alkaline phosphatase was consistently elevated in the 2 g/kg group of males at 12, 18 and 24 months,

but the elevation could not be associated with any other clinical or histological changes. In the females, plasma thyroxine concentration tended to
be decreased in the high level NDHC group from six months to the end of the
feeding trial.

A dose-related elevation of liver weight, expressed as percentage of
body weight, was suggested in both sexes (Table 6). Testes weights for one
dog in the 1 g/kg group and one in the 2 g/kg group were unusually low, 0.08
and 0.04% of body weight, respectively. Marked testicular atrophy and degeneration were confirmed histologically for these two dogs. Thyroid weight was
elevated in both sexes (significantly in the males) receiving 2 g/kg NDHC,
and mild thyroid hypertrophy and hyperplasia were detected in two of the
three dogs of each sex in this high dose group. Other histological lesions
could not be related to dietary groups.

Excretion Studies with NDHC-^{14}C in Rats. For excretion studies, NDHC was
labeled with ^{14}C, in the position indicated in Figure 1. The labeled NDHC
was dissolved in warm water and administered to rats by stomach tube in
doses of 1, 10 and 100 mg/kg body weight. The recovery of radioactivity in
respiratory carbon dioxide was 0.1% or less of the dose in 24 hours for
each of the levels tested. Approximately 90%, or more, of the administered

Table 6. Organ weight of dogs fed NDHC for two years[a]

Organ	NDHC Level (gm/kg body weight)							
	0		.2		1.0		2.0	
	Males	Females	Males	Females	Males	Females	Males	Females
Liver	2.81	3.26	3.12	3.14	3.21	3.51	3.69[b]	3.98
Heart	.74	.72	.84	.68	.81	.80	.81	.80
Kidneys	.43	.50	.51	.47	.54	.51	.53	.52
Brain	.67	.85	.80	.95	.78	.99	.77	.79
Spleen	.17	.19	.19	.20	.24	.29	.17	.24
Testes/Ovaries	.13	13	.16	13	.13	18	.13	10
Adrenals	10	13	12	16	12	17	10	14
Thyroids	7	8	11[b]	8	9	9	14[b]	15
Pituitary	1	.7	.6	1	.6	.8	.7	.8

[a] Values for liver, heart, kidneys, brain, spleen and testes are expressed as
1% body weight while values for ovaries, adrenals, thyroids and pituitary are
expressed as mg % body weight; [b] p <.05.

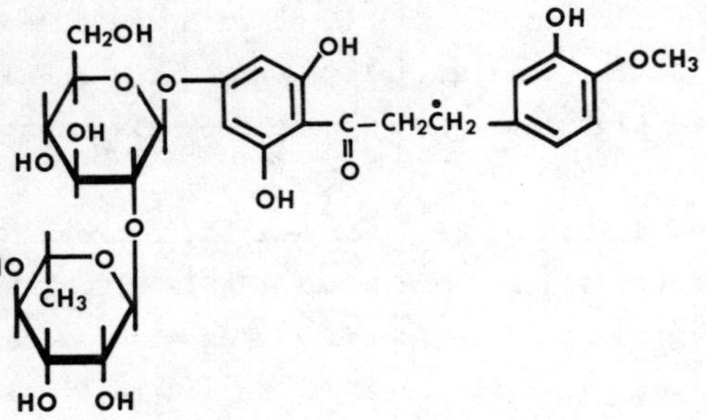

Figure 1. Formula of neohesperidin dihydrochalcone showing the carbon (C) labeled with ^{14}C

NDHC^{14}C was excreted in the first 24 hours, primarily in the urine. Increasing the dosage of NDHC increased the relative amount appearing in the feces. Only traces of radioactivity were found in various tissues after 24 hours. Most of the radioactivity remaining in the animal at this time was located in the gastrointestinal tract.

ACKNOWLEDGMENTS   Part of the NDHC used in these studies was a gift of Nutrilite Products, Inc., Buena Park, California. NDHC-^{14}C was a gift of the Givaudan Corporation, Clifton, New Jersey.
     The authors thank Glenda M. Dugan, Virgil V. Herring, Janie A. John and Lucille N. Rossi for their technical assistance.

REFERENCES

1. Horowitz, R.M., Relation Between the Taste and Structure of Some Phenolic Glycosides, in "Biochemistry of Phenolic Compounds," J.B. Harborne, ed., Academic Press, New York, pp. 545-71 (1964).

2. Horowitz, R.M., Gentili, B., U.S. Patent 3,087,821 (1963).

3. Ershoff, B.H. Comparative Effects of a Purified Diet and Stock Ration on Sodium Cyclamate Toxicity in Rats. Proc. Soc. Exp. Biol. Med. 141, 857 (1972).

DISCUSSION OF PAPERS BY DR. HOROWITZ AND DR. GUMBMANN

     DR. SHAW: When levels as high as 10% NDHC were included in the diet, did you keep any food consumption records to see whether or not, particularly in the rats, they increased their intake to take about 1.1 times as much food as the controls to compensate for the nutritionally inert material?

DR. GUMBMANN: I think consumption records would show a decrease in actual calories consumed. With 10% NDHC added to the commercial diet, I believe that too much of the diet is being replaced.

NDHC really isn't inert. It is metabolized in the liver, presumably with considerable conjugation, and is excreted in the kidneys. At 10%, it represents a real metabolic load.

DR. HILLER: Has the USDA or do you plan to submit a new drug application or food additive petition for dihydrochalcone?

DR. GUMBMANN: All these data have been submitted to the FDA in petitions, one by Nutrilite Products, Inc. of California and one by Research Organic/Inorganic Chemical Corporation of New Jersey.

DR. MOSKOWITZ: Dr. Horowitz, in the panel's data, were any other tastes observed beside the sweetness of the dihydrochalcones? When we worked with these compounds, we noted an odor and a persistent taste which kept interfering with the perception of sweetness. I can't describe the taste, but there was something beyond sweetness.

DR. HOROWITZ: When the compounds are pure they have no odor or off-taste. Some of the earlier commercially prepared samples did have a phenolic odor caused by an impurity. The taste of dihydrochalcones is long-lasting and, according to some tasters, has a cooling, menthol-like quality.

DR. MOSKOWITZ: Would that be a positive or a negative? Would this lasting positive taste which we noticed also be a potential drawback as a sweetener and could you get rid of it?

DR. HOROWITZ: The lingering effect seems to be inherent in these sweeteners and no one has yet been able to get rid of it. However, if the rhamnose is removed, as in HDG, the lingering effect is reduced.

DR. VAN DER WEL: Dr. Horowitz, may I ask you what is the influence of your compound on the bitterness of, for instance, quinine or coffee?

DR. HOROWITZ: I don't know whether the dihydrochalcones have been tried in the presence of quinine or in coffee.

DR. AMOS: I am not quite clear how much of the DHC gets into the body of the animal as evaluated by radioactive assay. Much apparently remains in the gut, but I wasn't sure of the distribution.

DR. GUMBMANN: Approximately 80% was absorbed from the gut and excreted by the urinary system within 24 hours.

DR. AMOS: Do simple sugars interfere with the uptake from the gut?

DR. GUMBMANN: We never evaluated that.

DR. AMOS: Even in tissue culture it would be interesting to see. There is a question about a number of drugs used for chemotherapy; could you improve their uptake by cells that you want to kill? If you put a sugar on the compounds, they would then home toward the sugar receptors with possibly increased intake.

DR. ROUSSOS: Could you tell us at present what potential commercial applications you can foresee for neohesperidin dihydrochalcone? Are you aware of current studies about applications?

DR. HOROWITZ: One of the applications is in chewing gum. Compounds such as neohesperidin dihydrochalcone are very good in chewing gum because of the long-lasting effect. They are said to be useful in pharmaceuticals for masking bitterness and they also work well in toothpastes and mouthwashes. There seem to be distinct possibilities of using the compounds in beverages, particularly those which are based on citrus flavors, but all the possibilities haven't been explored yet.

Monellin, a sweet-tasting protein

Robert H. Cagan, Joseph G. Brand, James A. Morris[+] and Robert W. Morris[++]

Veterans Administration Hospital and Monell Chemical Senses Center, University of Pennsylvania, Philadelphia, Pennsylvania 19104

SUMMARY AND PROSPECTS

Monellin is a sweet-tasting protein which is the sweet principle of the tropical fruit <u>Dioscoreophyllum cumminsii</u>. Monellin evokes an intense sweet sensation; when tasted at relatively high dosages there is a lingering sweetness but without a bitter taste. Monellin has a molecular weight of approximately 11,000 daltons and is comprised of two dissimilar polypeptide chains of known sequence. For a protein, the stability characteristics of monellin are reasonably good, with acid pH favoring retention of its sweetness.

Since monellin can be isolated in pure form, its preparation for commercial use would not appear to constitute a serious problem. However, a continuing supply of berries would need to be assured. The primary impediment to further testing at this time appears to be the regulatory climate that inhibits investment in the studies needed to establish required safety data before commercialization will be undertaken. Bringing monellin closer to the commercial stage is now largely a matter of the economic and regulatory aspects.

THE PROPERTIES OF MONELLIN

Monellin is an intensely-sweet-tasting protein[1,2] which is one of two chemostimulatory proteins[3] thus far described. It is itself a taste stimulus: when sampled at relatively high dosages it has a lingering sweetness without a bitter taste. This protein is the sweet principle of the fruit of the tropical plant <u>Dioscoreophyllum cumminsii</u>[4]. A number of its physicochemical characteristics are now well-defined, and some of these are summarized below.

Several methods were used to establish the molecular weight of monellin[5]. Gel filtration on Sephadex G-100 indicated a molecular weight near 10,000 daltons. In this case the eluate could be taste-tested, establishing that the protein of minimal molecular weight is the sweet-tasting species. Electrophoresis in polyacrylamide gels containing sodium dodecyl sulfate showed an apparent molecular weight of 10,500. The molecular weight

Table 1. Amino acid composition of monellin[**]

Amino Acid Residue	Residues/mol	
	Ref. (5)	Ref. (6)
tryptophan	1	1
lysine	8	9
histidine	0	–
arginine	7	6
aspartic acid/arparagine	10	10
threonine	4	4
serine	2	2
glutamic acid/glutamine	12	12
proline	6	5
glycine	8	8
alanine	3	3
cysteine	1	1
valine	4	4
methionine	1	1
isoleucine	6	8
leucine	6	6
tyrosine	7	7
phenylalanine	5	5

[**] Amino acid analyses are described in detail in references (5) and (6). In addition, an amino acid analysis in one additional study was published (2), but the results differed in several respects from the two studies summarized in this table.

calculated from the amino acid composition (Table 1) is approximately 10,700.

Monellin is composed of two dissimilar polypeptide chains that are non-covalently associated[6, 7, 8]. The amino acid sequence of monellin has been determined recently and is shown in Figure 1[6, 7]. A question asked frequently is whether there is any resemblance between the sweet-tasting dipeptide ester L-Asp-L-Phe-OMe[9] (Aspartame) and a dipeptide region within the monellin molecule. The dipeptide Asp-Phe does not occur in the sequence, thereby ruling out this simple hypothesis for monellin's sweetness[*]. To directly resolve detailed questions of the three-dimensional structure of monellin, X-ray crystallographic data[10] are necessary.

---

[*]It might prove interesting that Asp-(aa)$_n$-Phe occurs in two locations (where n is 2 or 3): residues 16-19 (subunit I) and 7-11 (subunit II). Models may show whether these regions could assume a conformation capable of bringing the relevant centers into juxtaposition to effect interaction with the taste receptor binding sites with which Aspartame interacts.

Pure monellin contains no carbohydrate and is therefore not a glyco-protein[1-3,5,7]. The isoelectric point was found by isoelectric focusing to be 9.3[5]. The spectral characteristics, absorption and fluorescence, are summarized in Table 2[5,11]. Absorbance in the ultraviolet is typical of a protein containing both tryptophan and tyrosine; the absorption characteristics of monellin (Figure 2) indicate a single tryptophan residue per molecule. The bathochromic shift in absorption maximum from 277 nm to 290 nm at alkaline pH (Table 2) is due, at least in part, to the ionization of tyrosine residues. Loss of sweetness also occurs at alkaline pH, but no direct correlation between these two observations is possible at this time because alkaline pH can also induce conformational changes in the protein. The fluorescence emission spectra (Figure 3) of native monellin are dominated by the tryptophan fluorescence, although significant tyrosine fluorescence can also be observed.

Recent preliminary studies suggest, as would be anticipated because monellin is a protein, that it is not cariogenic. The growth and acid

SUBUNIT I

```
1            5                  10
Arg-Glu-Ile-Lys-Gly-Tyr-Glu-Tyr-Gln-Leu-Tyr-Val-Tyr-Ala-Ser-

16           20                 25
Asp-Lys-Leu-Phe-Arg-Ala-Asn-Ile-Ser-Gln-Asn-Tyr-Lys-Thr-Arg-

31           35                 40
Gly-Arg-Lys-Leu-Leu-Arg-Phe-Asx-Gly-Pro-Val-Pro-Pro-Pro
```

SUBUNIT II

```
1            5                  10
Gly-Glu-Trp-Glu-Ile-Ile-Asp-Ile-Gly-Pro-Phe-Thr-Gln-Asn-Leu-

16           20                 25
Gly-Lys-Phe-Ala-Val-Asp-Glu-Glu-Asn-Lys-Ile-Gly-Gln-Tyr-Gly-

31           35                 40
Arg-Leu-Thr-Phe-Asn-Lys-Val-Ile-Arg-Pro-Cys-Met-Lys-Lys-Thr-

46
Ile-Tyr-Glu-Glu-Asn
```

Figure 1. Amino acid sequence of monellin. Details of procedures are in references (6) and (7), and additional details will be published.

Table 2.   Spectral characteristics of monellin *

Method	Characteristics			
Absorption	λmax (nm)	$E_{1\ cm}^{1\%}$	$\varepsilon(M^{-1}\ cm^{-1})$	
pH 7.20	277	13.7	1.47×10^4	
pH 12.8	290	17.1	1.83×10^4	
Fluorescence Emission	Excitation λ (nm)	Emission λ (nm)	Relative Quantum Yield	Halfwidth (nm)
0.1 M NaCl	260	340	0.072	60.0
	295	343	0.075	55.5
0.1 M acetate	260	340	0.069	58.5
	295	342	0.076	55.5

* The absorption spectra were described in reference (5) and the fluorescence analyses in reference (11).

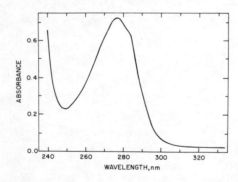

Figure 2.   Absorption spectrum of monellin.   Taken from reference (5) with permission.

production of cultures of <u>Streptococcus</u> <u>mutans</u> (strains GS5 and LM7) in vitro appeared to be unaffected by addition of monellin to the medium[12].

Additional research during the past two years has added greatly to our knowledge of the characteristics of monellin, and a number of these aspects are discussed below.

STABILITY AND ACTIVITY MEANS MAINTAINING SECONDARY, TERTIARY, AND QUARTERNARY STRUCTURE

The accumulated evidence points to the importance of maintaining the integrity of the secondary, tertiary and quaternary structure of the protein for retention of sweetness[1-3,5,6,8,11,13-15].   Based on knowledge from other

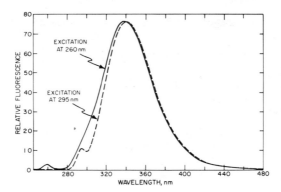

Figure 3.   Fluorescence emission spectra of native monellin at two
             excitation wavelengths.   Relative fluorescence was equalized at
             peak emission. Crossed polarizers were used to diminish the
             expression of excitation Rayleigh scatter. Monellin was in
             0.1 M NaCl.

biological systems of the mechanisms of interactions between small and large
molecules, it seems likely that only small regions of the protein monellin
actually interact directly (bind) with sweet receptor sites on taste bud
cells.   The major part of the monellin molecule would thus provide the
structural backbone for maintaining the correct spatial conformation for the
groups that actually bind to the taste receptor sites.   Recent studies have
yielded a methylated derivative of monellin with retention of sweetness[8].
This result has allowed us to synthesize a radioactively-labeled monellin
derivative, which we are now employing in binding studies.   Until the
details of the interaction between a taste stimulus molecule and its receptor
site are understood, the field will continue to rely for new sweeteners on
discoveries arising by accident, massive screening, or empirical structure-
activity studies.

     Experiments with common protein denaturants have yielded important
information about the pH stability of monellin.   The ability of this protein
to subsequently renature is dependent upon the initial pH that exists during
exposure to denaturants (Table 3), and is favored by acid pH.   For example,
monellin   exposed to urea at neutral pH regained less of its sweetness than
that exposed to urea at acid pH[13].   Although the pH at which the protein is
exposed to denaturant is important, the pH at which renaturation occurs
appears not to influence substantially the recovery of sweetness (Table 3).

     Following exposure to guanidine-HCl, recovery of sweetness was complete
in our experiments whether the exposure occurred at neutral or acid pH.   This
result was surprising because guanidine-HCl is commonly believed to be a more

Table 3. Effect of pH during denaturation-renaturation of monellin [*]

Denaturing Conditions	Dialysis pH	Sweetener Activity (%)
8 M urea in acetic acid, pH 4.2	2.9	107
8 M urea in acetic acid, pH 4.2	7.1	97
6 M guanidine-HCl in acetic acid, pH 2.8	2.9	93
6 M guanidine-HCl in acetic acid, pH 2.8	7.1	98
8 M urea in phosphate buffer, pH 7.0	2.9	56
8 M urea in phosphate buffer, pH 7.0	7.1	45
6 M guanidine-HCl in phosphate buffer, pH 7.0	2.9	87
6 M guanidine-HCl in phosphate buffer, pH 7.0	7.1	97

[*] Data are summarized from reference (13). Sweetener activity in percentage is derived from the sweetness expressed in units per milligram protein, and therefore does not show any actual loss of material during the procedures. See reference (13) for further details.

effective protein denaturant than is urea[16]. The conformational changes of monellin were examined recently using its fluorescence emission characteristics[11]. In an acidic medium, guanidine-HCl is in fact a more effective denaturant of monellin than urea. The better recovery of sweetness with guanidine may be related to the more rapid kinetics of denaturation with this reagent.

Monellin contains a single sulfhydryl (-SH) group. The evidence supporting this fact is now incontrovertible, based on amino acid analyses[5-7], on titration with -SH reagents such as PHMB or DTNB[6,14], and on spectral measurements[15]. In one report[2] it had been claimed that monellin does not contain any free -SH, but rather has a disulfide bridge that was suggested to be important in maintaining the tertiary structure. That they did not detect the free -SH can be readily explained by the finding[6,14] that the -SH is buried within the interior of the protein, and is titratable in native monellin only after prolonged exposure (hours) to -SH titrants. On the other hand, monellin in the presence of protein denaturants shows rapid titration with the -SH reagents. For example, native monellin when treated with iodoacetate (Table 4), in order to carboxymethylate the -SH group, showed only a 12% decrease in titratable -SH. However, when carboxymethylation was carried out in the presence of the protein denaturant guanidine-HCl, reaction of the -SH group was complete. These results demonstrated the inaccessibility of the -SH of native monellin and its reactivity when the

Table 4.   Carboxymethylation of monellin [*]

Treatment	% Sweetness	% -SH
none	100	100
carboxymethylated in tris/HCl buffer, pH 8.4	82	88
buffered 6 M guanidine-HCl, pH 8.4	52	60
carboxymethylated in buffered 6 M guanidine-HCl, pH 8.4	0	0

[*] Data are summarized from reference (14).

protein is unfolded.   The carboxymethylated monellin, like the monellin after titration with PHMB or DTNB, does not taste sweet.   Sweetness of the PHMB-monellin can be recovered, however, after appropriate dialysis against mercaptoethanol[14].

The near ultraviolet region of the circular dichroic (C.D.) spectrum[15] contains information from aromatic amino acids, that absorb light in this region.   Contrary to an earlier report[17], no dichromic bands assignable to disulfide bonds are observable[15].   This latter C.D. finding is in agreement with the results of the -SH titration studies discussed above.

It was concluded[14] that the -SH is buried within the interior of the protein so that it is relatively inaccessible to reaction, and furthermore that the -SH does not participate as a binding moiety with taste receptor sites.   The lack of oxidation of monellin and the lack of the need for special precautions against oxidation during preparation of monellin are readily explained by the free -SH being buried in this manner.

When the protein is unfolded, such as by denaturants, aggregation of monellin can occur[13] and the aggregated monellin no longer tastes sweet. The aggregated protein that precipitated had no titratable -SH, leading us to speculate that formation of aggregated monellin occurs initially through disulfide-linked species.

In the original publication[1], monellin solutions maintained in the cold (refrigerated at 4° C or frozen at -15° C) were reported to be quite stable. In the same paper, the stability at room temperature was greatly underestimated.   We had discussed[1] the loss of sweetness when dilute solutions of monellin were kept for one day at room temperature.   Based on studies since that time, we suggest that monellin in solution can be maintained for considerably longer periods, provided that the solutions do not become contaminated with microorganisms which may lead to putrefaction.   We also reported recently[18] on the difficulty of digesting monellin with proteolytic enzymes, even with time periods of 24 to 72 hours at room temperature. Studies of stability with respect to temperature show that the conformation of monellin, and its sweet taste, are stable up to 50 to 60° C (to be published). Between 70 and 80° C, a substantial loss of sweetness occurs, but heating solutions of monellin at 100° C for 15 minutes nevertheless resulted in retention of a small (25%) but definite residual sweetness. It is especially important to define the temperature stability characteristics of monellin in acid media.

The C.D. spectra of monellin have been reported for both the native (sweet-tasting) conformation and for various states of chemically-induced conformational transitions[15]. The C.D. spectrum of native monellin shows a high content of pleated sheet (β) conformation with a small amount (6 to 10%) of α-helix. The amide I band of the infrared spectrum of monellin is compatible with this assumption of a high content of β-structure, since it shows a distinct shoulder at 1640 cm^{-1} [19]. This content of β-structure may help to explain the relatively high stability of monellin.

Treatment of the protein with sodium dodecyl sulfate or increasing the concentration of ethanol results in C.D. changes[15] in both the near ultra-violet region, indicating local conformational changes about the aromatic nuclei, and changes in the 195 to 230 nm region that reveal an increase in the amount of α-helix in the amino acid backbone. Raising the pH of the medium to 10.9 or adding guanidine-HCl (2.5 M or above) results in complete loss of both β- and α-structure. Therefore the protein is effectively denatured. High pH also results in loss of sweetness[5,15]; subsequent neutralization or acidification causes recovery to a near normal C.D. spectrum and a sweet taste. Slight changes in the conformation of monellin therefore appear to be compatible with the retention of sweet taste. This conclusion is substantiated by our fluorescence measurements of the sweet-tasting methylated derivative of monellin[8] that showed a slight but definite increase in fluorescence emission bandwidth.

The fluorescence spectrum of monellin is dominated by the single trypto-phan (Figure 3, Table 2), which is not exposed to the solvent until considerable denaturation of the protein occurs. Denaturation with urea or guanidine-HCl lead to a broadening of the spectrum (by 55 to 60%) with an obvious appearance of tyrosine fluorescence at 304 nm (Figure 4). The tyrosine emission is the primary cause for the increase in bandwidth. When monellin is completely denatured the tryptophan fluorescence maximum is 350

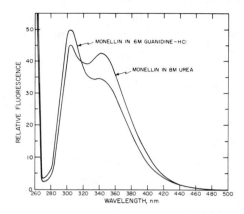

Figure 4. Fluoresence emission spectra of denatured monellin. Taken from reference (11) with permission.

nm, indicating that it is exposed to the solvent. Using the fluorescence emission characteristics, we found that guanidine-HCl exerts its maximal effects by 3M; with urea, even at 8M, denaturation may not be complete[11]. Denaturation here implies separation of the two chains of monellin as well as randomization of the conformation of each chain. Tryptophan is, of necessity, present in only one chain of the protein, while the seven tyrosines are distributed between the chains (Figure 1). Thus changes in the tryptophan emission reflect perturbations primarily in one of the two chains of monellin. The entire emission spectrum of monellin is, on the other hand, a composite of the aromatic amino acid fluorescence from both chains, and can be used as an indicator of the overall conformation of the molecule and hence its sweetness. The fluorescence emission bandwidth is a reasonable indicator of the conformation of monellin and predicts changes in sweetness with chemical[11], thermal and pH perturbants (to be published).

The possibility of producing a small fragment of monellin that would retain the intense sweet taste is, from a practical viewpoint, an attractive idea. To our knowledge this goal has not been achieved, but it could mean merely that the conditions were not appropriate. In a series of experiments, we attempted to degrade monellin with trypsin, α-chymotrypsin, or bromelain[18]; the protein is relatively resistant to proteolysis. Not one of the fragments from monellin was sweet. The sweetness remaining after proteolysis could be accounted for by the native monellin that remained in the digests. Its presence was indicated upon gel filtration chromatography. The individual subunits of monellin are not sweet[6]. It is important to note, in this context, that the information from a variety of studies indicates that the conformation of monellin dictates its sweetness (see above). This conclusion, therefore, argues against the possibility that a simple cleavage of this protein to produce a small peptide region will result in a sweet fragment.

Two preliminary experiments have addressed the stability question directly in food products. Neither study was exhaustive, since each was carried out only as an initial exploratory attempt to determine if monellin could survive normal processing conditions. In one study, it was desired to determine if monellin survives the processing temperatures during chocolate manufacture. Accordingly, chocolate bars were prepared[*] in which the sucrose was omitted and lactose was substituted to replace the bulk. Monellin was added at a level calculated using a sweetness conversion factor of 2000:1 (weight ratio, monellin:sucrose). From our taste-testing it was clear that the sweetness of monellin survived the processing. It was also apparent that the bulking problem caused by removing sucrose is considerable, with the lactose imparting a highly unpleasant texture. A preliminary experiment with chewing gum[**] was, on sensory grounds, highly successful. A sucrose-free gum was prepared using a commercial formulation for a sorbitol-saccharin gum. Samples were prepared with the standard formulation, with the saccharin omitted, and with monellin substituted for saccharin. Our observations suggested that the taste of the monellin-containing gum was highly acceptable compared with the regular product.

It is apparent from several studies that the conformation (secondary, tertiary and quaternary structure) of monellin is a requisite for the sweet taste of the molecule, although small changes in conformation are compatible with the retention of the sweetness. If it is assumed that the molecular structure in a simulus must be rigid in presenting itself to the taste receptor site in order for the sweetness of the stimulus to be recognized, then the regions of the molecule being probed by the spectral methods (primarily aromatic amino-acid side-chains and the amino acid backbone of the protein) may not be those directly involved in interacting (binding) with the receptor recognition site. On the other hand, since the molecule can undergo slight conformational changes yet still remain sweet, the receptor site may tolerate some degree of conformational deformability. Direct knowledge of the receptor site(s) for sweet taste would aid substantially in segregating the correct mechanism by which this chemostimulatory protein exerts its biological activity.

---

* Prepared by Dr. Barry Zoumas of Hershey Foods, Inc., Hershey, Pa.
** Prepared by Mr. Manoun Hussein of Life Savers, Inc., Port Chester, N. Y. .

THE PLANT GROWS IN AFRICA;   THE SWEETENER ISOLATED IS A PURE PROTEIN

Dioscoreophyllum cumminsii is a dicotyledonous plant of the family
Menispermaceae.   In English it is commonly called the guinea potato (also
the wild red berry or the serendipity berry).   Other names in African
dialects are ito-igbin and ayun-ita (Yoruba), moframofratwe (Twi), ekali-
bonte and kaligbonde (Timne) and nmimimi nwambe (Ibo)[3,4,20,21].

The plant grows wild in tropical West and Central Africa from Guinea to
Cameroon, and also in Gabon, Zaire, Sudan, southern Rhodesia, and Mozambique.
A number of individuals from West Africa have commented to us that the fruit
is eaten commonly by people.   Studies on the cultivation of the plant have
been in progress for several years in Ghana[22], and recent work has succeeded
in germination of seeds[23].   Nothing has been published about the biosynthe-
sis of monellin during development of the berries, but we observed in our
laboratory that the green immature berries do not taste sweet.   Small numbers
of these green berries sometimes were present in our shipments of fresh
berries from Ghana.   Based on the fact that monellin consists of two
dissimilar polypeptide chains, it is reasonable to postulate that a precursor
protein (called promonellin) might exist in the berries.   Monellin would be
produced by proteolytic cleavage of the hypothetical promonellin molecule to
form monellin and an additional inactive protein fragment.

Preparation of monellin in the laboratory employs a combination of
standard biochemical procedures[1].   Extraction of monellin from the fruit
with the skin removed is effected by leaching into water.   The proteins in
the extract are concentrated by salting out with ammonium sulfate, thereby
also purifying the monellin by approximately 1.8-fold.   Chromatography on a
column of DEAE-cellulose results in a broad elution peak which contains the
monellin; the dark brown contaminants are largely removed in this step.   The
material emerging is purified an additional 1.8-fold on this column.   Column
chromatography on CM-cellulose results in a product that is a homogeneous
protein.   In the original procedure[1], we described an initial CM-cellulose
column employing stepwise elution followed by a second column with gradient
elution to give further resolution and a very pure product.   It is suggested
that optimization of conditions could give an equally pure material by
employing only a single gradient elution from a CM-cellulose column.

In our more recent experiments, we have added a final chromatographic
step on Sephadex G-50[13] to remove a trace protein contaminant that is sometimes present.   Also, we have recently employed an ultrafiltration step
immediately following the ammonium sulfate fractionation.   The ammonium

sulfate fraction is filtered using an Amicon ultrafiltration apparatus equipped with an XM-100 filter, which retains molecules larger than approximately 100,000 daltons. By repeated washing of the ammonium sulfate fraction on the ultrafiltration apparatus, the monellin can thereby be recovered essentially quantitatively, while leaving behind a large portion of the highly viscous, and therefore troublesome, contaminants.

In view of our increased knowledge of the characteristics of the protein, it is suggested that the appropriate technology for purification of monellin on a larger scale would not be the limiting factor in its utilization as a sweetener. The original aims of our research were not towards quantitation of either the amount of monellin in the berries or the yield, hence only approximate values can be given. The yield of monellin can be estimated from the data on extractions and purifications that we have performed during the past few years in our laboratory investigations. Using data on extractions of kilogram batches of berries, we determined that 11 to 15 gm of total protein is extracted per kilogram of berries. We did not estimate the protein remaining in the berries following extraction, but merely subjected the berries to successive extractions until they yielded relatively little additional protein. If we assume that 10 to 15%[1] of the total protein extracted is ultimately obtained as pure monellin using our standard procedure, then the laboratory procedure yields approximately 1 to 2 gm of monellin per kilogram of berries. The most interesting calculation, however, is the following: if we use a conversion factor of 2000 (weight basis, monellin:sucrose) for the relative effectiveness of monellin compared with sucrose as a sweetener, then 1 to 2 gm of monellin is equivalent to 2 to 4 kilograms of equivalent sucrose sweetness per 1 kilogram of berries!

ACKNOWLEDGMENTS

The research from our laboratory was supported in part by NIH Research Contract No. N01-DE-22413 from the National Institute of Dental Research and by NIH Research Grant No. NS-08775 from the National Institute of Neurological and Communicative Diseases and Stroke.

+Present Address: International Flavors and Fragrances, Union Beach, New Jersey 07735

++Present Address: Widener College, Department of Biology, Chester, Pennsylvania 19013

REFERENCES

1. Morris, J.A., and Cagan, R.H. Biochim. Biophys. Acta 261, 114-122, 1972.
2. van der Wel, H., and Loeve, K. FEBS Lett. 29, 181-184, 1973.

3. Cagan, R.H. Science 181, 32-35, 1973.
4. Inglett, G.E., and May, J.F. Econ. Bot. 22, 326-331, 1968.
5. Morris, J.A., Martenson, R., Deibler, G., and Cagan, R.H. J. Biol. Chem. 248, 534-539, 1973.
6. Bohak, Z., and Li, S.-L. Biochim. Biophys. Acta 427, 153-170, 1976.
7. Hudson, G., and Biemann, K. Biochem. Biophys. Res. Comm. 71, 212-220, 1976.
8. Morris, R.W., Cagan, R.H., Martenson, R.E., and Deibler, G. Proc. Soc. Exp. Biol. Med., in press.
9. Mazur, R.H., Schlatter, J.M., and Goldkamp, A.H. J. Amer. Chem. Soc. 91, 2684-2691, 1969.
10. Wlodawer, A., and Hodgson, K.O. Proc. Nat. Acad. Sci. USA 72, 398-399, 1975.
11. Brand, J.G., and Cagan, R.H. Biochim. Biophys. Acta 493, 178-187, 1977.
12. Jacobsson-Hunt, U., Hammond, B.F. and Cagan, R.H. to be published.
13. Morris, J.A., and Cagan, R.H. Proc. Soc. Exp. Biol. Med. 150, 265-270, 1975.
14. Cagan, R.H., and Morris, J.A. Proc. Soc. Exp. Biol. Med. 152, 635-640, 1976.
15. Jirgensons, B. Biochim. Biophys. Acta 446, 255-261, 1976.
16. Tanford, C. Adv. Protein Chem. 23, 121-282, 1968.
17. van der Wel, H. in Biochemistry of Sensory Functions, edited by L. Jaenicke, Springer Verlag, New York, pp. 235-242, 1974.
18. Morris, J.A., and Cagan, R.H. J. Agr. Food Chem. 24, 1075-1077, 1976.
19. Cagan, R.H., Kare, M.R., and Morris, J.A. U.S. Patent No. 3,998,798, issued Dec. 21, 1976.
20. Dalziel, J.M. The Useful Plants of West Tropical Africa, Crown Agents for Oversea Governments and Administration, London, p. 14, 1937, reprinted 1955.
21. Irvine, F.R. Woody Plants of Ghana, Oxford University Press, London, p. 32, 1961.
22. Adansi, M.A. Ghana J. Agric. Sci. 3, 207-210, 1970.
23. Holloway, H.L.O. Econ. Bot. 31, 47-50, 1977.

DISCUSSION OF DR. CAGAN'S PAPER

DR. NEWBRUN: You mentioned some tests where you used monellin as a possible substrate for S. mutans to see if it was cariogenic. Quite frankly, I don't know of any single protein that is cariogenic.

DR. CAGAN: Although monellin can be theorized as unlikely to be cariogenic, a direct demonstration would be more convincing, and might also indicate any cariostatic activity, if present. Dr. Ulla Jacobsson-Hunt, in her studies on cariogenic bacteria, S. mutans, included a study with additions of monellin to the medium. Enough people have asked about that point so that it seemed well worth an initial survey. The results, while not extensive, are included in this review because this is a caries meeting.

DR. ROUSSOS: You have carried out at least two application trials, is this correct?

DR. CAGAN: Only two preliminary trials were conducted on applications of monellin because of the relatively limited amounts of material. We wanted to answer one question: could monellin survive the normal processing--was there any hope at all of using monellin in these products? Through the cooperation of two companies we answered that question. Hershey Foods, Inc. prepared chocolate bars with sucrose omitted and monellin incorporated to see if the sweetness of monellin survived, and it did. The texture was not acceptable, but the monellin sweetness was there.

Life Savers, Inc. prepared monellin-containing chewing gum, which had a very agreeable taste. The chewing gum contained both sorbitol and monellin. We may want to consider monellin in terms of combinations with other sweeteners.

DR. VAN DER WEL: The sweetness of monellin does not survive in coffee.

DR. CAGAN: The result actually depends upon how you make your coffee. If it's perked coffee, monellin will serve as a sweetener. If, however, it is freshly made instant coffee, then it may not be very effective as a sweetener. The answer, of course, is the difference in temperature.

Regarding temperature stability, the really surprising observation is that monellin heated in solution in a boiling water bath still retains some of its sweetness. I don't know if there is a conformation of monellin that is unusually stable, but that is an interesting speculation with many practical consequences. We don't know, however, whether we simply haven't heated sufficiently--long enough to completely denature the protein.

DR. VAN DER WEL: We heated monellin to 80 degrees at pH 2 and sweetness disappeared, and we put it in the refrigerator. After two days the sweetness came back. Do you have an explanation for this?

DR. CAGAN: Yes, it renatures.

DR. AMOS: Does monellin compete with sugars for receptors on the taste buds?

DR. CAGAN: We are using the labeled monellin, with _in vitro_ biochemical experiments, to answer that question.

DR. VAN DER WEL: Can you use cross-adaptation experiments?

DR. CAGAN: Cross-adaptation might be suggestive, but you cannot definitively distinguish, with cross-adaptation experiments, between receptor site events and later neural levels. In addition, with cross-adaptation between monellin and sugars, there would be difficulties experimentally in adjusting the concentration levels properly, because of the relative insensitivity of receptors for sugars.

DR. ARVIDSON: Together with Dr. Van der Wel we studied the stimulation of single human fungiform papillae with sucrose, monellin and thaumatin as well as miraculin followed by citric acid. We found cross-adaptation between sucrose, monellin and thaumatin and citric acid (pretreated with miraculin).

DR. BEIDLER: Two questions. Does gymnemic acid inhibit the sweetness of monellin? Second, what's your idea about how monellin works? Is it similar to our concept about how miracle fruit exerted its effect?

DR. CAGAN: First, the answer is yes. Second, you postulated that miraculin bound to the membrane and then exposure to acid induced a conformational change in the receptor membrane. We do not know which specific areas of the protein are directly binding to the receptor site. Our preparation of a radioactively labeled derivative of monellin now enables us to use it in an _in vitro_ system to answer that question. Undoubtedly, certain regions of the protein must interact rather specifically, probably in a multi-point attachment. We carried out degradative studies of monellin in which we were not able to produce a small fragment that was sweet.

DR. BIBBY: What is the FDA status of monellin? If approval were available, what additional uses would monellin have beside chewing gum?

DR. CAGAN: I don't know if any approach has been made to the FDA on behalf of monellin for commercial use. Other uses ought to be explored by companies making sweet-tasting products.

DR. SHAW: What is the potential supply of the berries from which monellin is prepared supposing that monellin is found to be useful in three or four different applications?

DR. CAGAN: Although I am not aware of cultivation of it in this country, the plant grows extensively in tropical West and Central Africa. The research institute in Ghana which has supplied us with berries has been working on cultivation of <u>Dioscoreophyllum</u> <u>cumminsii</u> because of its potential as a cash crop in Ghana. They recently reported success in propagating it from seeds.

DR. BEIDLER: Are you familiar with the fact that a Japanese scientist is studying organ culture as a means to produce some of the sweet proteins?

DR. CAGAN: I have heard about that long-term approach.

Potential intense sweeteners of natural origin

George E. Inglett

Cereal Science and Food Laboratory, Northern Regional Research Center, Agricultural Research Service, US Department of Agriculture, Peoria, Illinois 61604

SUMMARY

The need for a safe, nonnutritive sweetener for the diabetic and the diet conscious has spurred research in many diverse areas. Extensive evaluations of known sweet substances from both synthetic and natural materials are a continuing research effort. Many new analogs of both natural and synthetic sweet substances are continually being synthesized. Searching nature's storehouse of botanicals is a fascinating approach for finding a suitable intensely sweet, healthful sweetener.

This review covers those potentially promising, intensely sweet substances of natural origin excluding the dihydrochalcones.

GLYCYRRHIZIN

Licorice, well known for centuries and widely used, is obtained from the roots of Glycyrrhiza glabra, a small shrub, grown and hand-harvested in Europe and Central Asia. The roots contain from 6 to 14% glycyrrhizin. Licorice is the only botanical possessing significant amounts of glycyrrhizin.

Glycyrrhizic acid exists in licorice root as the calcium-potassium salt in association with other constituents, such as starch, gums, sugars, proteins, asparagine, flavonoids and resins[1]. Although this glycoside is difficult to free of nitrogen, minerals, color, and licorice flavor, it has been isolated in pure form. Glycyrrhizic acid is a glycoside of the triterpene, glycyrrhetic acid, which is condensed with O-β-D-glucuronosyl-(1'→2)-β-D-glucuronic acid. Colorless, crystalline glycyrrhizic acid was first isolated by Tschirch and Cederberg[2]. Although the empirical formula and optical inactivity that they reported were incorrect, the other properties given and their isolation procedures have been repeatedly reproduced[3-6]. A "very sweet" taste was ascribed to the free, tribasic acid. It was practically insoluble in cold water and soluble in hot water. The free acid and its ammonium and potassium salts formed pasty gels upon cooling the warm aqueous solutions; however, they could be recrystallized from glacial acetic acid or alcohol. An "intensely sweet" taste was ascribed to the water-soluble salts. Tschirch and Cederberg hydrolyzed potassium glycyrrhizinate in dilute sulfuric acid to obtain the crystalline aglycone, glycyrrhetic acid[2]. It was insoluble in water and tasteless. Glucuronic acid was identified in the mother liquor by Tschirch and Gauchmann[3].

Voss et al. first obtained the correct empirical formula for glycyrrhizin[7], and Voss and Pfirschke showed that the glucuronic acid found is linked to the hydroxyl group of glycyrrhetic acid as a disaccharide[8]. Lythgoe and Trippett proved the (1'→2)-interglycosidic linkage, also the pyranoside rings of the di-glucuronic acid moiety, by periodate oxidation and methylation analysis[4]. From optical rotation data, they could agree with Voss and Pfirschke's supposition that the interglycosidic linkage is beta. Both the interglycosidic linkage and the linkage to the aglycone were shown to be beta by Marsh and Levvy[5]. They hydrolyzed pure ammonium glycyrrhizinate with a β-glucuronidase isolated from mouse liver. This enzyme was shown to activate only β-glucuronides in pyranoside ring-form. Vovan and Dumazert have confirmed that the only sugar present in pure, crystalline glycyrrhizic acid is glucuronic acid[6,9]. Voss and Butter showed that the C-20 carboxyl group is not sterically hindered[10]. All three carboxyl groups are readily methylated[10A].

The absolute configuration of the aglycone, glycyrrhetic acid, is determined as the result of investigations too numerous to cite completely. Important contributions were made by Ruzicka et al.[11-15], Voss et al.[7,10], and Beaton and Spring[16]. Although two isomers (18-α and 18-β) have been isolated, Beaton and Spring have indicated that only the 18-β-glycyrrhetic acid is the natural isomer that occurs in glycyrrhizin[16].

The aglycone of glycyrrhizin is closely related to β-amyrin, a triterpene; however, it gives medicinal effects similar to those of deoxycorticosterone acetate. Both glycyrrhizin and glycyrrhetic acid, and their respective methyl esters, show a low hemolytic activity in comparison with some other triterpenoid saponins[17].

Countercurrent extraction process in which the most concentrated liquor passes over fresh root while fresh water scrubs the nearly spent root yields a variety of licorice products, including block licorice, licorice powder, liquid extracts, and glycyrrhizin. Ammoniated glycyrrhizin is manufactured by a special process in which only the glygyrrhizin is removed from the total extract[18].

The extracts are universally employed in the flavoring and sweetening of pipe, cigarette and chewing tobaccos[19]. They function as natural sweeteners, basic flavorants, and humectants. Tobacco products treated with licorice extracts show a marked reduction in harshness, better moisture retention, and improved burning qualities[20]. Licorice extracts are regularly used in confectionery manufacture. Other applications for refined licorice are found in pharmaceutical flavoring[21] where it is routinely employed for repressing the unpalatibility of many oral medicaments. Some segments of the flavor industry have long utilized these extracts in root beer, chocolate, vanilla, liqueur, and other flavors[22]. Ammonium glycyrrhizin (AG), the fully ammoniated salt of glycyrrhizic acid, is commercially available as a high-purity, spray-dried, brown powder. Further treatment and repeated crystallizations yield the more costly, colorless salt, monoammonium glycyrrhizinate (MAG). Both derivatives have the same degree of sweetness but they differ markedly from each other in solubility properties and sensitivity to pH.

AG is the sweetest substance on the FDA list of natural GRAS flavors. AG is 50 times sweeter than sucrose. In the presence of sucrose, AG is around 100 times sweeter than sucrose alone. This synergistic effect provided the basis for a U.S. patent[23].

In 1967, another patent relating to the application of AG in potentiating the flavor of cocoa and other cacao products[24] was obtained that showed the natural cocoa flavor is intensified to a point where a 25% reduction in cocoa content can be effected without changing acceptability.

MAG appears very effective in rounding out the harsh notes encountered in spice-mint blends which are regularly applied in flavoring toothpastes and mouthwashes. MAG, unlike AG, does not contribute any color to these products. The pH of aqueous MAG solutions is around 4.5.

Food applications with MAG have been broadened so that it can easily be applied to high-acid preparations such as beverages and gelatin desserts.

Many dentifrices which contain foaming agents such as sodium lauryl sulfate and hydrocolloid thickeners and stabilizers are sweetened exclusively with saccharin[25]. Because of its sweetening and nonfermentative properties, together with its excellent foaming, viscosity-building, and flavor-reinforcing action, MAG is well suited for dentifrice manufacture.

STEVIOSIDE

The sweet herb of Paraguay (Yerba dulce), called variously Caa-ehe, Azuca-caa, Kaa-he-e, and Ca-a-yupe by the Guarani, has long been the source of an intense sweetener. Natives use the leaves of this small shrub to sweeten their bitter drinks. The plant was first given the botanical name

Eupatorium rebaudianum, but this was later changed to Stevia rebaudiana Bertoni; recently, the name Stevia rebaudiana (Bert.) Hemsl has appeared. For the sweet, crystalline glycoside that has been extracted from the leaves of S. rebaudiana, the name stevioside was adopted by the Union Internationale de Chimie in 1921. Historical accounts of knowledge on stevioside and proposals for cultivation of S. rebaudiana for commercial use of stevioside as a sweetener have been reviewed by Bell[26], Fletcher[27], and Nieman[28]. Wood et al. reported a method for extracting stevioside in 7% yield from air-dried leaves of the Paraguayan plant[29].

Bridel and Lavieille reported stevioside as a white, crystalline, hygroscopic powder, approximately 300 times sweeter than cane sugar[30]. Very small amounts on the tongue gave a delectable sweetness, like the leaves of the plant; however, large amounts tasted sweet at first, then distinctly bitter. On the other hand, no bitter taste was attributed to stevioside by Nieman. By taste panel, Pilgrim and Schutz determined a relative sweetness of 280 for stevioside as against 306 for saccharin (sucrose = 1.00)[31].

Bridel and Lavieille showed that stevioside is rapidly hydrolyzed by enzymatic material extracted from the vineyard snail, Helix pomatia, giving rise to 3 moles of D-glucose and 1 mole of a tasteless, acidic aglycone which they named steviol[32]. Acidic hydrolysis gave the same percentage of D-glucose but a different aglycone named isosteviol. Their work on the constitution of stevioside was repeated, confirmed, and extended by Wood et al.[29] The identity and positions of the sugars linked to steviol were fixed by Wood et al[29] and by Vis and Fletcher[32A]. The absolute configuration of the diterpenoid aglycone was finally resolved by Mosettig et al.[33]

The disaccharide of stevioside is sophorose (2-O-β-D-glucopyranosyl-β-D-glucopyranose). It is linked to the tertiary α-hydroxyl at C-13 of steviol, whereas the monsaccharide, β-D-glucopyranose, is condensed with the sterically hindered α-carboxyl at C-4. The two sugars are, therefore, appended to the same side of the rigid aglycone at opposite ends. Alkaline splitting of the unusual D-glucoseto-carboxyl linkage produced levoglucosan (1,6-anhydro-β-D-glucopyranose) and the sophoroside of steviol, which was devoid of sweetness[29,32].

Pomaret and Lavieille reported that stevioside readily passes through human elimination channels in its original form[34]. It did not appear to be toxic to guinea pigs, rabbits or chickens. Furthermore, there are no recorded reports of ill effects in Paraguayan users of the leaves of S. rebaudiana. Nevertheless, the long-term effects of ingestion of stevioside would have to be investigated carefully before it could be considered for human use as a sweetener in the United States. The diterpenoid aglycone, steviol, has shown specific physiological activity[35] and weak antiandrogenic effects[36]. It remains to be proved that stevioside does not split to form any steviol in the human digestive tract.

Also from the leaves of Stevia rebaudiana, two new sweet glucosides, rebaudiosides A and B, were isolated besides the known glucosides, stevioside and steviolbioside. On the basis of IR, MS, IH, and ^{13}C NMR as well as chemical evidences, the structure of rebaudioside B was assigned as 13-O-[β-glucosyl(1-2)-β-glucosyl(1-3)]-β-glucosyl-steviol and rebaudioside A was formulated as its β-glucosyl ester[37]. In an attempt to produce stevioside as a substitute for synthetic sweeteners, Stevia rebaudiana is cultivated extensively in Japan.

MIRACLE FRUIT--SOUR →SWEET GLYCOPROTEIN

An important approach to sweet taste perception is the study of the strange properties of the miracle fruit (Synsepalum dulcificum). Although the miracle fruit's capacity to cause sour foods to taste sweet has been known in the literature since 1852[38], scientific investigations of the fruit were not made until Inglett and his associates found some experimental evidence that the active principle was macromolecular[39]. This berry possesses a taste-modifying substance that causes sour foods such as lemons, limes, grapefruit, rhubarb, and strawberries to taste delightfully sweet. Even dilute organic and mineral acids will induce a sweet taste. The berries are chewed by West Africans for their sweetening effect on some sour foods. The quality of the sweetness induced by this taste modifier is very similar to sucrose.

Preliminary studies on miracle fruit by Inglett et al. were intended to isolate, characterize and synthesize the taste-modifying substance[39]. The fruit was processed to give a stable concentrate. Since the active principle appeared polymeric, the material was incompatible with the sponsoring organization's mission, so research was discontinued.

Subsequently, Brouwer et al.[40] and Kurihara and Beidler[41] confirmed that the active principle was a glycoprotein with a molecular weight of around 42,000. That a taste-modifying substance could be a macromolecule was a new concept in sweet taste perception. Until this time, only small molecules were considered sweet-evoking substances. This was the first time that macromolecules were considered capable of participating in either taste perception or modification of taste[39].

With increasing pressure for an excellent nonnutritive sweetener in recent years, a corporation launched a venture on dietetic foods using miracle fruit concentrate as its source of latent sweetness. The FDA denied a petition for affirmation that miracle fruit, its extracts and concentrates, are generally recognized as safe[42]. They stated also that the substances cannot be approved as food additives because there is insufficient evidence of acceptable toxicological studies, insufficient history of consumption of the substances in the U.S. to demonstrate that no known hazards exist, and insufficient data to establish the conditions of safe use as food additives.

SERENDIPITY BERRIES --MONELLIN

While studying various natural sweeteners previously mentioned, the author discovered the intense sweetness of some red berries from West Africa. The fruit was called the serendipity berry, and its botanical name, Dioscoreophyllum cumminsii Diels, was established many months later[43-45].

Serendipity berries are indigenous to tropical West Africa. The Dioscoreophyllum cumminsii Diels plant grows from Guinea to the Cameroons and is also found in Gabon, the Congo, the Sudan, and southern Rhodesia. It grows in the rain forest during the rainy season from approximately July to October. The serendipity berries are borne by hairy climbing vines sometimes 15 ft. long and 1/8 to 3/16 in. in diameter. The berries are red in color, approximately 1/2 in. long, and grow in grapelike clusters with approximately 50 to 100 berries in each bunch. The tough outer skin of the berry encloses a white, semisolid, mucilaginous material surrounding a friable thorny seed. In spite of its intense sweetness, the fruit is not commonly cultivated or used by Nigerians[46].

Researchers at the Monell Senses Center and the Unilever Research Laboratorium, working independently, confirmed the protein nature of the serendipity sweetener[47-50].

Sweetness is lost on heating the protein at $50^{\circ}C$ (pH 3.2), $65^{\circ}C$ (pH 5.0) and $55^{\circ}C$ (pH 7.2). This unstable nature of the sweetener makes it unsuitable for many commercial applications.

KATEMFE--THAUMATIN I and II

Besides studies on miracle fruit and the serendipity berry, a large variety of plant materials were examined systematically by Inglett and May for intensity and quality of sweetness[46]. Another African fruit containing an intense sweetener was katemfe, or the miraculous fruit of the Sudan. Botanically the plant is Thaumatococcus danielli of the family Marantaceae.

Inside the fruit three large black seeds are surrounded by a transparent jelly and a light yellow aril at the base of each seed. The mucilaginous material around the seeds is intensely sweet and causes other foods to taste sweet. The seeds were observed to be present in trading canoes in West Africa as early as 1839, and were reported to be used by the native tribes to sweeten bread, fruits, palm wine and tea. Preliminary studies have indicated a substance similar to the serendipity berry sweetener[46]. Katemfe yields two sweet-tasting proteins[51-52] which they called thaumatin I and II.

Physical constants for the sweet proteins of katemfe are given in Table 1. Like Monellin, these protein sweeteners are heat sensitive and undergo irreversible heat denaturation. Because denaturation coincides with the loss of sweetness, the groups underlying the conformational change must also be responsible for generating their sweet taste. At least part of the intact tertiary protein structure must be required for the sweet taste.

A process for extraction of the thaumatins from the fruit was reported[53] and commercial interest in this sweetener is developing.

LO HAN FRUIT--TRITERPENOID GLYCOSIDE

Lo Han Kuo (Lo Han fruit), from _Momordica grosvenori_ Swingle, is a dried fruit from Southern China. The fruits are gourd-like, 6-11 cm long by 3-4 cm broad, dark brown, broadly ellipsoid, ovoid, or subglobose, with broadly rounded ends and a very thin rind (0.5-0.8 mm thick). It has 3 double locules (compartments or cavities), each with 2 rows of seeds (about 10-12 in a row). The brownish-gray pulp dries to a light fibrous mass. The seeds are light brownish-gray, flattened (15-18 mm long, 10-12 mm broad, and 3-4 mm at

Table 1. Physical Properties of the Sweet-Tasting Proteins from Katemfe[51]

Criteria	Thaumatin I	Thaumatin II
Isoelectric point	12	12
Molecular weight	$21\ 000 \pm 600$	$20\ 400 \pm 600$
$A_{1cm}^{1\%}$ (pH 5-6, 278 nm)	7.69	7.53
Sweetness intensity (times sweeter than sucrose):		
on a molar basis	1×10^5	1×10^5
on a weight basis	1600	1600
Temperature (^{o}C) above which sweetness disappears at:		
pH 3.2	55	55
5.0	75	75
7.2	65	65

edges) with a depressed area in the center of each side. Characterization of the tissue, as well as the taxonomic status of the plant, was described by Swingle who reported that the species was introduced to the United States through the Division of Plant Exploration and Introduction, Bureau of Plant Industry[54]. Swingle also reported that 1000 tons of the green fruits were delivered every year to the drying sheds at Kweilin (Kwangsi Province). The fruits lost much weight in drying and were then carefully packed in boxes and shipped to Canton, but large amounts were also exported to Chinese communities outside China. The dried fruit is a valued folk medicine used for colds, sore throats, and minor stomach and intestinal troubles. Lee found that the sweet principle could be extracted by water from either the fibrous pulps or from the thin rinds of Lo Han Kuo[55]; 50% ethanol was also found to be a good extractant. Rinds afforded a more easily purified extract. Sweetness of Lo Han sweetener was accompanied by a lingering taste described as licorice-like, somewhat similar to that of stevioside, glycyrrhizin, and the dihydro-chalcones. Structural studies indicate the sweetener to be a triterpenoid glycoside with 5 or 6 glycose units[56]. The purified sweetener has a more pleasant sweet taste than the impure material. The purest sample is about 400 times sweeter than sucrose.

PHYLLODULCIN

A sweet tea, Amacha, is served at Hanamatsuri, the flower festival celebrating the birth of Buddha. Amacha is the dried leaves of Hydranga macrophylla Seringe var. Thunbergii Makino. The sweet principle, phyllo-dulcin, was isolated and the structure determined by Asahina and Asano[57-59]. The absolute configuration of phyllodulcin was shown to be the 3R configuration at the asymmetric center at C(3) by identification of a malic acid from ozonized phyllodulcin[60]. Phyllodulcin is the first representative of a natural isocoumarin. More information is needed about the intensity and sweetness quality of phyllodulcin. Some 3, 4-dihydroisocoumarins were reported to have a sweet taste[61].

OSLADIN

The sweet taste of rhizomes of the widely distributed fern, Polypodium vulgare L., has attracted the interest of many chemists and pharmacists. Van der Vijver and Uffelie[62] and others have shown that the sweet substance is not glycyrrhizin, as was once proposed. Many constituents of the rhizomes have been isolated, but the substance that resembles saccharin in sweetness was isolated and identified only recently[63-64]. The name osladin is based on the Czech name for polypody, osladic. Osladin comprised only 0.03% of the dry weight of the rhizomes. Its chemical structure was revealed as a bis-glycoside of a new type of steroidal saponin.

The glycoside that results by replacement of the monosaccharide radical with hydrogen was isolated separately and named polypodosaponin. Its absolute configuration was determined by Jizba et al[65]. They made no comment on its taste.

The disaccharide of osladin was shown to be neohesperidose, 2-O-α-L-rhamnopyranosyl-β-D-glucopyranose. The glycosidic linkage was shown to be beta by cleavage of neohesperidin, with a specific β-glucosidase from Aspergillus wentii. Therefore, neohesperidose is in the same glycosidic configuration in osladin as is found in the intensely sweet neohesperidin dihydrochalcone. Neohesperidose is only slightly sweet, if at all[66-67]. The configuration of the glycosidic linkage of the monosaccharide, L-rhamnose, has not been determined. It is probably α-L-, as shown, because this corresponds to the β-D-linkage usually found in natural glycosides. The molecular structure of osladin resembles that of stevioside; it shows a (1'→2)-linked disaccharide at one end of the aglycone and a monosaccharide at the other.

The C-26-O-methyl polypodosaponin shows hemolytic activity and a strong inhibition of fungal growth[65]. Even without regard to biological activity, the very low concentration of osladin found in polypody rhizomes dims the prospects for developing osladin as a commercial sweetener.

REFERENCES

1. Nieman, C. in Adv. in Food Res. Vol. 7, Academic Press, New York, 1957.
2. Tschirch, A. and Cederberg, H., Arch. der Pharm. 245, 97-111, 1907; Chem. Zentr., 1799, 1907(I).
3. Tschirch, A. and Gauchmann, S., Arch. der Pharm. 246, 545, 1908; Chem. Zentr., 1604, 1908(II).
4. Lythgoe, B. and Trippett, S., J. Chem. Soc. 1983, 1950.
5. Marsh, C.A. and Levvy, G.A., Biochem. J. 63, 9, 1956.
6. Vovan, L. and Dumazert, C., Bull. Soc. Pharm. Marseille 19, 45, 1970.
7. Voss, W., Klein, P. and Sauer, H., Ber. 70B, 1212, 1937.
8. Voss, W. and Pfirschke, J., Ber. 70B, 132, 1937.
9. Vovan, L. and Dumazert, C., Bull. Soc. Pharm. Marseille 19, 45, 1970.
10. Voss, W. and Butter, G., Ber. 70B, 1212, 1937.
10A. Brieskorn, C.H. and Sax, H., Arch. Pharm. (Weinheim, Ger.) 303, 905, 1970.
11. Ruzicka, L., Furter, M. and Leuenberger, H., Helv. Chim. Acta 20, 312, 1937.
12. Ruzicka, L., Jeger, O. and Ingold, W., Helv. Chim. Acta 26, 2278, 1943.
13. Ruzicka, L. and Leuenberger, H., Helv. Chim. Acta 19, 1402, 1936.
14. Ruzicka, L., Leuenberger, H. and Schellenberg, H., Helv. Chim. Acta 20, 1271, 1937.
15. Ruzicka, L. and Marxer, A., Helv. Chim, Acta 22, 195, 1939.
16. Beaton, J.M. and Spring, F.S., J. Chem. Soc., 3126, 1955.
17. Schloesser, E. and Wulff, G., Z. Naturforsch. Teil B 24, 1284, 1969.
18. MacAndrews and Forbes Company, Camden, N.J., 1970.

19. Cook, M.K., Flavour Ind. 1, No. 12, 831, 1970.
20. Cook, M.K., U.S. Pat. 3,342,186. September 19, 1967.
21. Cook, M.K., Drug Cosmet. Ind. 76,#5, 624, 1955.
22. Cook, M.K., Flavour Ind. 2,#3, 155, 1971.
23. Muller, R.E., U.S. Pat. 3,282,706. November 1, 1966.
24. Morris, J., U.S. Pat. 3,356,505. December 5, 1967.
25. Cook, M.K., Drug. Cosmet Ind., 82,#3, 314, 1958.
26. Bell, F., Chem. Ind. (London) 897, 1954.
27. Fletcher, H.G., Jr., Chem. Dig. 14, 7, 18 (July-August), 1955.
28. Nieman, C., Suesswaren-Wirtsch, 11, 124-126, 236, 1958.
29. Wood, H.B., Allerton, R., Diehl, H.W. and Fletcher, H.G., Jr., J. Org. Chem. 20, 875, 1955.
30. Bridel, M. and Lavieille, R., Compt. Rend. 192, 1123; J. Pharm. Chim. 14(3), 99,; 14(4), 154, 1931.
31. Pilgrim, F.J. and Schutz, H.G., Nature (London) 183, 1469, 1959.
32. Bridel, M. and Lavieille, R., Compt. Rend. 193, 72-74, 1931; Bull. Soc. Chim. Biol. 13, 636, 1931.
32A.Vis, E. and Fletcher, H.G., Jr., J. Am. Chem. Soc. 78, 4709, 1956.
33. Mosettig, E. et al. J.Am. Chem. Soc. 85, 2305, 1963.
34. Pomaret, M. and Lavieille, R., Bull. Soc. Chim. Biol. 13, 1248, 1931.
35. Vignais, P.V., Duee, E.D., Vignais, P.M. and Huet, J. Biochim. Biophys. Acta 118, 465, 1966.
36. Dorfman, R.I. and Nes, W.R. Endocrinology 67, 282, 1960.
37. Kohda, H. et al. Phytochemistry 15, 981, 1976.
38. Daniell, W.F. Pharm. J. 11, 445, 1852.
39. Inglett, G.E., Dowling, B., Albrecht, J.J. and Hoglan, F.A. J. Agric. Food Chem. 13, 284, 1965.
40. Brouwer, J.N., Van der Wel, H., Francke, A. and Henning, G.J. Nature (London) 220, 373, 1968.
41. Kurihara, K. and Beidler, L.M. Science 161, 1241, 1968.
42. Anon. Federal Register, May 24, 1977.
43. Inglett, G.E. and Findlay, J.C. Abstr. Pap. 75A presented to the Div. of Agriculture and Food Chemistry, 154th Am. Chem. Soc. Meeting, Chicago, IL., 1967.
44. Inglett, G.E.and May, J.F., J. Food Res. 34, 408, 1969.
45. Inglett, G.E., In Symposium: Sweeteners, Avi Publishing, Westport, CT, 1974.
46. Inglett, G.E. and May, J.F. Econ. Bot. 22, 326, 1968.
47. Morris, J.A. and Cagan, R.H. Biochim. Biophys. Acta 261, 114, 1972.
48. Morris, J.A., Martenson, R., Deibler, G. and Cagan, R.H. J. Biol. Chem. 248, 534, 1973.
49. Van der Wel, H. FEBS Lett. 21, 88, 1972.
50. Van der Wel, H. and Loeve, K. FEBS lett. 29, 181, 1973.
51. Van der Wel, H. and Loeve, K. Eur. J. Biochem. 31, 221, 1972.
52. Van der Wel, H. In Symposium: Sweeteners, Inglett (Editor), Avi Publishing Co., Westport, CT., 1974.
53. Higginbotham, J.D., U.S. Pat. 4,011,206, March 8, 1977.
54. Swingle, W.T., J. Arnold Arbor. Harv. Univ. 22, 198, 1941.
55. Lee, C.H., Experientia 31(5), 533, 1975.
56. Lee, C.H., General Foods Corporation, Tarrytown, N.Y., 1977.
57. Asahina, Y. and Asano, J., Ber. Dtsch. Chem. Ges. 62, 171, 1929.
58. Asahina, Y. and Asano, J., Ber. Dtsch. Chem. Ges. 63, 429, 1930.
59. Asahina, Y. and Asano, J., Ber. Dtsch. Chem. Ges. 64, 1252, 1931.
60. Arakawa, H. and Nakazaki, M., Chem. Ind. 671, 1959.
61. Yamato, M., et al. J. Pharm. Soc. Jpn. (Yakagaku Zasshi) 92, I. 367; II. 535; III. 850, 1972.

62. Van der Vijver, L.M. and Uffelie, O.F., Pharm. Weekbl. 101, 1137-1139, 1966; Chem. Abstr. 66, 52936, 1966.
63. Jizba, J. and Herout, V. Collect. Czech. Chem. Commun. 32, 2867, 1967.
64. Jizba, J., Dolejs, L., Herout, V. and Sorm, F. Tetrahedron Lett. 18, 1329, 1971.
65. Jizba, J. et al. Chem. Ber. 104, 837, 1971.
66. Koeppen, B.H. Tetrahedron 24, 4963, 1968.
67. Horowitz, R.M. and Gentili, B. J. Agric. Food Chem. 17, 696, 1969.

DISCUSSION OF DR. INGLETT'S PAPER

DR. HERBERT: Does monellin also have a hydrophobic buoyance?

DR. INGLETT: Yes. Several of the amino acids in monellin, phenylalanine and tryosine, for example, are hydrophobic. Distribution of the hydrophobic and hydrophilic centers may be related to sweetness. It's not going to be a simple matter because the tertiary structure may play an important role in sweetness. Once you denature monellin, the sweetness is lost, but the amino acids are still there, including the hydrophobic ones.

DR. HERBERT: Does the hydrophobic section have anything to do with attaching to the taste bud cell?

DR. INGLETT: The intense sweeteners have a hydrophobic section to them that could theoretically be very much involved in producing the intense sweetness by binding with taste bud cells as opposed to the ordinary level sweetness in sucrose and ordinary sugars.

DR. CAGAN: Just a general comment in terms of the distribution of hydrophobic residues in proteins that are as soluble as monellin. The likelihood is that the hydrophylic residues are on the outside and that the center of the protein has a greater preponderance of the hydrophobic residue. The only way to really answer that question is with the x-ray structures.

We have done some fluorescent studies that pointed to a fair degree of hydrophobicity in the interior of the protein. However, there could be some small segment of the external region that was relatively hydrophobic and interact with the distribution of the membrane. It's all very speculative.

DR. SHAW: Dr. Inglett, you and Dr. Beidler are persons that I think of with breadth in the subject of natural sweeteners. Of all these available compounds which one or two do you think are particularly exciting and hopeful for extensive research?

DR. INGLETT: Of the materials I discussed, glycyrrhizin is being used and is on the GRAS list. As long as people don't consume large amounts of this material, it's likely that this can be used as a flavoring agent for years ahead. Stevioside is second in potential importance. The Japanese apparently are considering it and are at the present time growing the plant for the production of stevioside. Of course, the dihydrochalcones that Dr. Horowitz discussed are very much in the forefront because they have a lot of toxicological studies behind them.

DR. SHAW: Dr. Horowitz particularly mentioned the availability of the hydrochalcones from the citrus industry. What amounts of these materials that you speak of could become available if safe and commercially usable?

DR. INGLETT: Generally if there is a market for a plant or a product from it, efforts are made to expand their cultivation. I mentioned the plantation of Katemfe in Africa at which the plants were grown to provide the leaves for wrapping purposes.

DR. VAN DER WEL: There is a company with a plantation for cultivation of Katemfe. They are purifying and extracting thaumatin on a pilot plant scale. Perhaps thaumatin will be one of the sweeteners in the future.

DR. SHAW: Dr. Beidler, if you had a certain amount of money to invest in the development of a natural sweetener, which one or two would you go after?

DR. BEIDLER: I was involved in a company that did invest seven million in miracle fruit (miraculin). Many thousand plants were grown and an abundant supply of miracle fruit was available. I would like to see development of several of the sweet proteins. Miracle fruit has a head start since much development and testing has already taken place. Perhaps we need several new sweeteners instead of just one. All of the sweet proteins appear to have limited applicability. Certainly miracle fruit was shown to be acceptable to diabetics.

DR. HERBERT: We should not be caught in a common, current, public misconception that because a material is natural it is non-toxic. That is not necessarily true. One need only refer to the book "Toxicants Naturally Occurring in Foods," published by the National Academy of Sciences in 1973 to be aware that many human foods contain substances which are toxic. One African food is a typical example; cassava, eaten by hundreds of thousands of Africans, contains cyanide. It is harmful to those people in many thousands of cases in producing nerve damage and producing blindness even though they soak it overnight and beat it in order to volatilize as much as possible of the cyanide as hydrocyanic acid.

DR. CAGAN: I'd like to make a seconding comment to that. If you notice how many toxic alkaloids and other categories of compounds in nature are intensely bitter, a very sensitive sense of bitter taste must have had survival value for us when we were out foraging for roots and herbs.

DR. INGLETT: I didn't mean to imply that all naturally occurring sweeteners are healthful. I think most of us are aware that digitalis and other alkaloids are toxic and occur in nature. But on the other hand, the survey that we made was looking for things that people consumed. Birds won't eat some berries, and people won't eat some foods. However, a lot has to do with the concentration of various poisons in the diet.

SESSION V.

FUTURE OPTIONS

Moderator:

Juan Navia
School of Dentistry
The Medical Center
The University of Alabama in Birmingham
University Station, Birmingham
Alabama 35294

New dimensions in synthetic sweeteners, anatomical compartmentalization by adjustment of molecular size

Guy A. Crosby, Grant E. Dubois, Ron L. Hale, Steven C. Halladay, Rebecca A. Stephenson, Patricia C. Wang, and Robert E. Wingard, Jr.

Dynapol, 1454 Page Mill Road, Palo Alto, California 94304

SUMMARY

A new approach to the development of synthetic sweeteners is presented. This approach recognizes and effectively deals with the stringent biological standards needed for such compounds by applying the concept of anatomical compartmentalization. The concept can be realized by increasing molecular dimensions of the synthetic material to a level that defeats intestinal absorption. However, the leashing of simple, low-molecular-weight sweeteners to macromolecules has generally failed to provide materials exhibiting significant sweet taste characteristics. This failure is explained as being the result of too drastic a change in those molecular parameters necessary for proper interaction with the taste receptor. Successful implementation of the compartmentalization effect has been achieved by use of the 4-0 substituted hesperetin DHC nucleus in conjunction with the core concept, which simply involves linking two or more of the sweetener nuclei together through a common bridge. In this manner sweeteners have been obtained which affords $\leq 1.0\%$ intestinal absorption as determined by studies with ^{14}C-labeled analogs in rats. A major problem remains the development of a sweetener of this type possessing a sucrose-like taste.

INTRODUCTION

The worldwide need for new and nontraditional food sources is paralleled by a similar need for food additives to render these materials palatable and acceptable to the consumer. The terms palatable and acceptable have, in the past, included a wide range of organoleptic and aesthetic properties. However, safety is an important aspect of food additives as evidenced by public opinion studies and recent FDA actions[1]. New approaches to food additives must recognize this fact and deal with it effectively.

Food additives were originally picked from available industrial chemicals. Food dyes, for example, were chosen from the myriad of organic and inorganic compounds first intended for fabric and paper coloration. Even the additives most recently approved for use in food, TBHQ as an antioxidant and FD&C Red #40 as a color, were originally designed for other purposes. A new approach to the development of essential food additives, which recognizes, understands, and accepts stringent biological standards for such materials, must be taken.

The concept of <u>anatomical compartmentalization</u> of substances can be achieved by appropriately adjusting molecular dimensions and has previously been applied to pharmacological agents. We are pursuing this approach, which calls for increasing molecular volume or size to a point which prohibits intestinal absorption, for the development of food colors[2-5], food antioxidants[6-8], and synthetic sweeteners[9]. If the process of intestinal absorption is considered analogous to simple dialysis, it is apparent that there is some size above which essentially no transport will take place via passive diffusion. The additive will therefore pass through the intestinal tract and be excreted. The nonabsorption principle is illustrated schematically in Figure 1.

CONTROL OF ABSORPTION

One of the first studies of intestinal transport versus molecular size was made by Höber and Höber[10]. Transport rates of polyhydric alcohols ranging from ethylene glycol to mannitol, and of amides ranging from acetamide to succinimide, were measured using isolated rat intestinal loops in situ. The results showed a decrease in absorption with molecular weight. However, because the compounds were all of small size, only a trend with molecular weight was obtained rather than an upper limit for measurable transport.

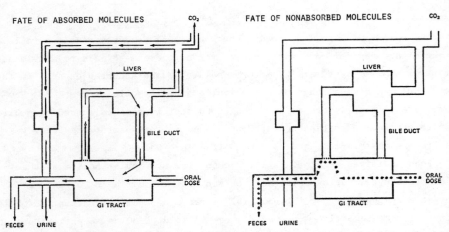

Figure 1. The biological fate of absorbable versus non-absorbable molecules is shown in this schematic representation. Reproduced with permission from T.E. Furia, Food Technol. (Chicago), 31(5), 34 (1977).

Subsequent studies showed that increasing molecular size into the polymer range afforded a marked reduction in intestinal absorption[11-13].
Polyethylene glycol is the most studied polymer in this regard since it is a
commonly used marker for monitoring fluid movement in the intestine.
Biological fate studies with this material at 4,000 peak molecular weight
showed 98% excretion in the feces for both man[12] and rats[13].

Additional insight can be provided into size control by examining
published data on the intestinal absorption of a number of mono-, di- and
polysaccharides[14]. Table 1 presents the effect of molecular size on the
rate of absorption for a series of sugars. Compounds with a molecular weight
from 200 to 400. are just able to penetrate the cell while larger carbohydrates such as insulin and starch are not appreciably transported across
the membrane barrier.

A detailed rat and mouse absorption study of ^{14}C-labeled polymeric
antioxidant was recently carried out[15]. Data were obtained that indicate a
molecular weight limit for transport of about 600-700 for these materials.
Intestinai absorption studies performed at Dynapol with radiolabeled polymeric dye precursor 1[2] provide additional evidence that the molecular
weight of synthetic polymers controls transport. In this work a sample of 1,
which was free of polymer below 900 molecular weight, was essentially
nonabsorbed. The full biological details of this study are available[16].

EXTRACELLULAR NATURE OF SWEET TASTE RESPONSE

An exhaustive discussion of the sweet taste response, including a
proposed model for a sweet taste receptor, has been described by Crosby

Table 1. Effect of Molecular Size on Rate of Absorption of Sugars.

Carbohydrate	Molecular weight	Radius of Equivalent sphere($\mathring{A}$)[a]	Rate of movement from mucosal to serosal sides (μmole/100 mg wet wt/h)[b]
Polysaccharide (starch)	50,000±		0
Inulin	5,000	14.8	0
Disaccharide (lactose)	342	4.4	0.5
Hexose (mannose, sorbose)	180	3.6	1.9
Pentose (ribose, arabinose)	150		2.2
Triose (glyceraldehyde)	90		4.5

[a]The radius of a particle that would show the same diffusion coefficient in
solution as that actually observed for the particular molecule (S.L. Palay
and L.J. Karlin, J. Biophys. Biochem. Cytol., 5, 373, 1959. [b]The initial
concentration gradient for the four sugars was 0.3M while that for inulin
and starch was 1%. (Wilson and T.N. Vincent, J. Biol. Chem., 216, 851, 1955)

$\underset{\sim}{1}$

et al[9]. It can be argued that a nonabsorbable sweetener would, by definition, be a nonexistent entity were the sweet taste response not extracellular in nature. At the present, it is generally accepted that the sweet taste response involves the very precise complexation of a molecule of appropriate size, shape, functionality, and hydrophobic-hydrophilic balance with a proteinaceous component of the plasma membrane of a receptor cell[17-18]. The protein component undergoes a conformational change upon complexation, which presumably results in a change in the electrical potential of the cell membrane (ion flux, depolarization)[19]. The precise chain of events leading from cell depolarization to neurotransmitter release are not well-understood at this time, although some meaningful results have been reported by Cagan and co-workers[20-22]. The evidence that the complexation is extra-cellular includes the findings that:

(a) The sweet taste response is both rapid and reversible[23].

(b) The proteins monellin[24-27] and thaumatin[28], which for reasons of size seem unable to penetrate cells, are intensely sweet.

(c) The taste-modifying glycoprotein miraculin (mol wt 44,000) also appears to be too large to penetrate the taste cell[29-31].

Points (b) and (c) strongly suggest that macromolecules can elicit a sweet taste. It is instructive to amplify the theme derived from these points. The different classes of reportedly sweet molecules exhibit a truly astonishing array of size, shape, and functionality. Consider for example just the differences in molecular weight of the various sweeteners listed in Table 2. The low end of the scale is represented by the simple amino acids, such as glycine and D-phenylalanine, cyclamate, saccharin, nitroanilines such as P-4000, and acetosulfam. Most of these materials have molecular weights considerably below 200. A tremendous number of dihydro-chalcone (DHC) and dipeptide analogs comprise the moderate (roughly 300-600)

Table 2. Relative molecular weights of synthetic and natural sweeteners

Compound	Molecular weight (free acid)
Lysine	75
Acetosulfam	163
D-phenylalanine	165
Cyclamate	179
Saccharin	183
P-4000	196
Aspartame	280
4-O-carboxymethylhesperetin DHC	362
Neohesperidin DHC	612
Stevioside	805
Glycyrrhizin	823
Osladin	884
Monellin	10,700
Thaumatin I and II	22,000

molecular weight range. The list also includes very large (mw > 800) sweet molecules, such as the stercidal saponin osladin, the diterpene glucoside stevioside, and the triterpenoid flavoring agent glycyrrhizin. The structures of these three large, naturally occurring sweeteners are shown in Figure 2.

Monellin. The tropical plant Dioscoreophyllum cumminsii produces a light red fruit, the pulp of which is intensely sweet[32-33]. Morris and Cagan isolated a pure substance, designated monellin, from the fruit and determined it to be a protein of molecular weight 10,700[24-26]. This finding has been independently confirmed by Bohak and Li[27], by Zuber[34], and by van der Wel and Loeve[35].

Bohak and Li found the sweet protein to consist of two noncovalently bound subunits of 50 and 42 amino acid residues[27]. The two chains could be separated chromatographically. Automatic sequential Edman degradation gave the complete sequence of the longer subunit, and a partial sequence of the shorter one. The sweetness of monellin requires the undissociated molecule. The individual subunits were not sweet, nor did they block the sweet sensation of sucrose or monellin. Solutions of the two separated chains could be mixed and seen by electrophoresis to return slowly to monellin, accompanied by a return of sweet taste.

Thaumatins I and II. The fruit of the tropical plant Thaumatococcus danielli Benth has been used as a source of sweetness for many years by the inhabitants of certain sections of Western Africa[32-33]. In 1972, van der Wel

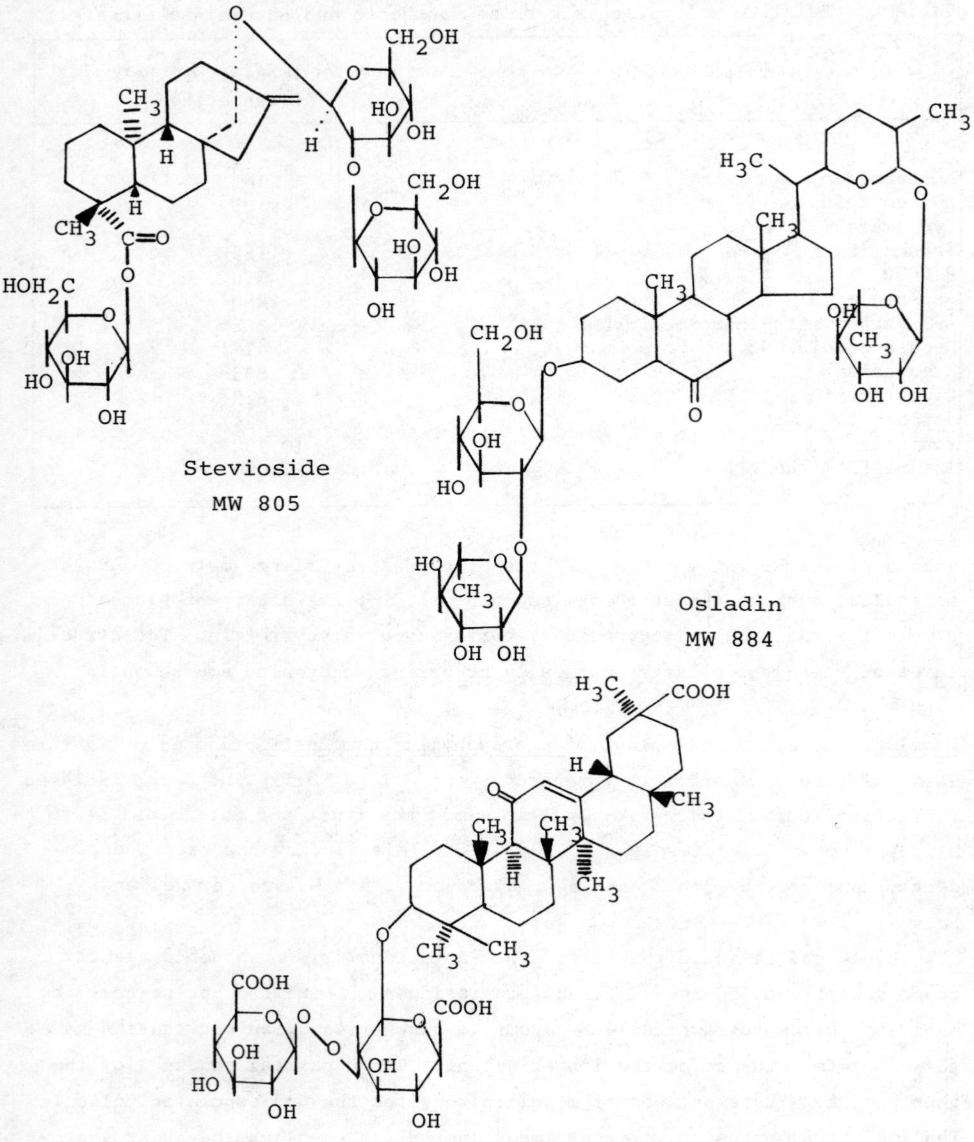

Stevioside
MW 805

Osladin
MW 884

Glycyrrhizin
MW 823

Figure 2. The structures and molecular weights of several large natural sweeteners.

and Loeve isolated two very similar proteins (named thaumatins I and II) from an aqueous extract of the fruit[28]. Both proteins possess molecular weights of approximately 21,000 and have nearly identical sensory properties. The taste intensity of the thaumatins has been estimated at approximately 1600X sucrose (threshold level); however, the materials exhibit a persistent licorice-like aftertaste. The finding that cleavage of the disulfide linkages present results in loss of sweet taste suggests that the taste sensation is a result of the intact protein, rather than a fragment[28].

Miraculin. Miraculin is a glycoprotein that is not sweet per se. After the tongue is exposed to the material for a short period, both mineral and organic acids taste sweet. The research groups of Beidler[29-31] and van der Wel[36] were both successful in isolating the glycoprotein. The molecular weight of miraculin was estimated by both research teams to be in the range of 4.2 to 4.4 x 10^4, while the carbohydrate content was reported by the Beidler group to be a combination of glucose, ribose, arabinose, galactose, and rhamnose units.

Beidler suggested a mechanism of action for the taste-modifying protein involving the binding of the protein portion of the molecule to an appropriate location on the taste cell membrane adjacent to a receptor site[30]. Exposure to acid was postulated to result in a change in conformation of the protein such that the carbohydrate portion(s) of the molecule interacts with a sweet receptor site.

Implications for Taste Mechanism. The ability of macromolecules to elicit taste sensations provides strong, although indirect, evidence supporting the idea that the sweet taste response is extracellular in nature. The initial interaction of the macromolecules certainly occurs at the surface of the plasma membrane of the receptor cell. It is most unlikely that these materials, all of which have molecular weights above 1 x 10^4, must first pass through the membrane of a receptor cell before initiating depolarization.

Polymer-Leashed Sweeteners. The straightforward idea of leashing low-molecular-weight sweeteners to water-soluble polymers is one approach to non-absorbable sweeteners. This early plan was patterned after the successful design utilized for affinity chromatography supports[37] and polymer-bound hormones[38]. In each case the active substrate must be tethered to the polymer backbone by a long leash. Examples of polymer-leashed sweeteners that were prepared are illustrated in Figures 3 and 4. The leashing of alkoxy nitroaniline sweeteners to soluble dextran derivatives (Figure 3) produced polymers with no taste. The leashing of an analog of a sweetener developed by Wyeth Laboratories[39-41] to a copolymer of acrylic acid and

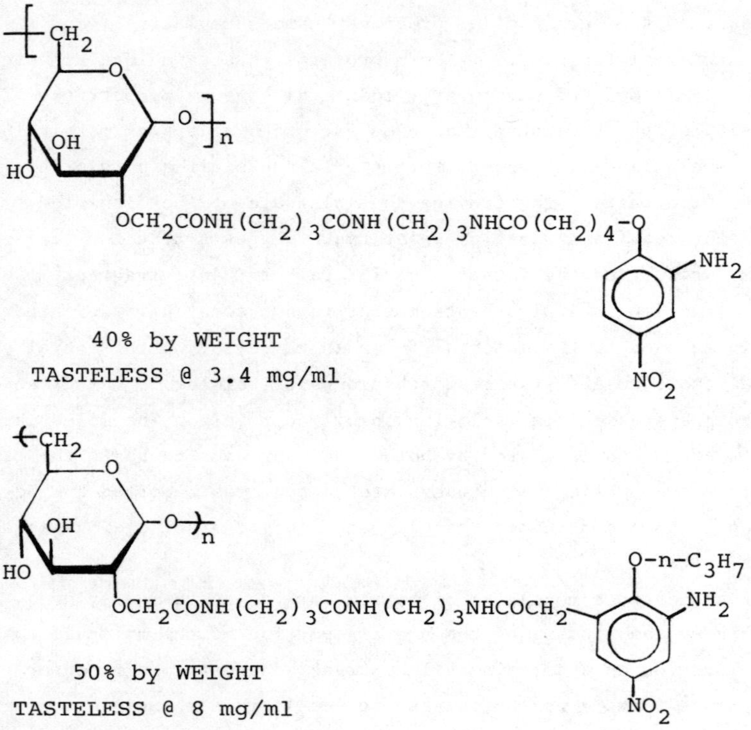

40% by WEIGHT
TASTELESS @ 3.4 mg/ml

50% by WEIGHT
TASTELESS @ 8 mg/ml

Figure 3. Polymer leashed alkoxy nitroanilines. The polymer backbone is a soluble dextran derivative.

N-vinylpyrrolidone provided a polymeric derivative which, although possessing a strong taste, was not significantly sweet (Figure 4).

Other results with polymer-leashed, low-molecular-weight sweeteners were similarly disappointing. Extensive studies, in both our laboratories and elsewhere, have shown that the low-molecular-weight sweeteners, such as alkoxy nitroanilines, saccharins, and sulfamic acid salts, generally cannot be structurally modified without detrimental effects on taste quality. The task of finding a chemically modifiable site in a sweetener molecule that does not result in the loss of sweet taste is a substantial problem that is addressed in some detail as described in the following sections.

NONABSORBABLE DIHYDROCHALCONE (DHC) SWEETENERS

Fourteen years ago Horowitz and Gentili reported that citrus peels contained glycosidic flavononids which could, by simple chemical modification,

WYETH SWEETENER

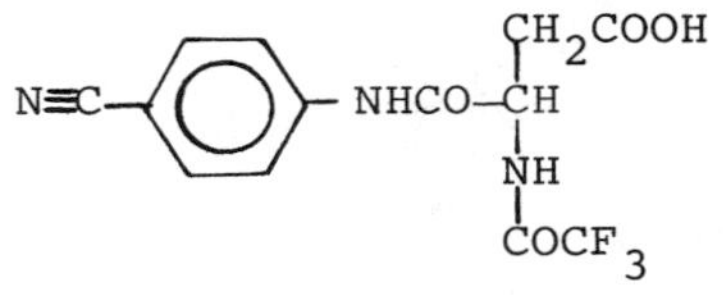

3000X Sucrose

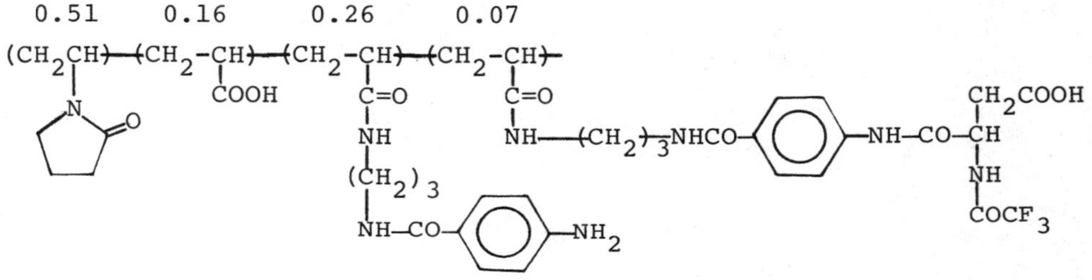

Strong Taste But Not Significantly Sweet At 18 mg/ml

Figure 4. A polymer leashed derivative of the Wyeth sweetener. The polymer backbone consisted originally of a 1:1 acrylic acid/N-vinylpyrrolidone copolymer.

be converted into a new class of sweet compounds[42-46]. For example, neohesperidin (2), a bitter flavanone 7- β-neohesperidoside found in the Seville orange, provided an intensely sweet DHC (3) upon alkaline hydrogenation. Derivatives of the hesperetin DHC molecule (4) possess a number of strong advantages for the development of nonabsorbable derivatives. These inherent advantages include:

(a) The molecular size ofthe hesperetin DHC nucleus (MW 304) is large enough to make this material a rather attractive starting point.

(b) These materials are natural products and extensive testing has indicated the relative biological safety of this class. For example, extended feeding studies of the sweetener 3 in a variety of animals have provided no evidence of toxicologically significant ill effects[47].

(c) The taste intensity of these materials is high. Sweetener 3, for example, is 600X sucrose on a weight basis and 1100X sucrose on a molar basis[48].

(d) Various structural features of the hesperetin DHC nucleus may be readily manipulated synthetically with retention of sweet taste.

(e) The structural features may be changed in such a way as to insure metabolic stability in the intestine.

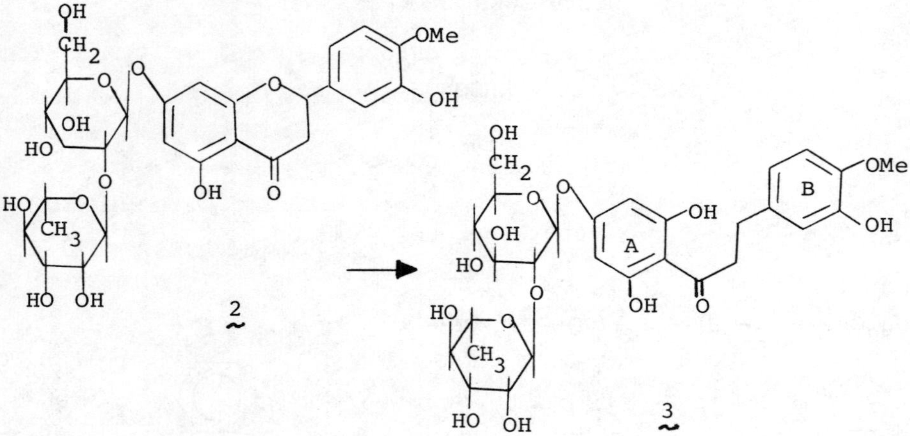

Structural Manipulatability. Studies in our laboratories have now conclu-
sively established that the structural elements of dihydrochalcones responsible for inducing the sweet taste response reside wholly on the aromatic
nucleus[48-51]. The key hydrogen donor-acceptor features based on the concept
developed by Shallenberger and Acree[52] are speculated in Figure 5.

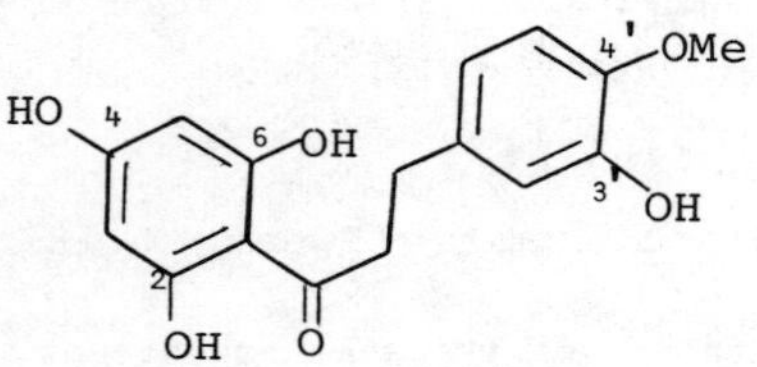

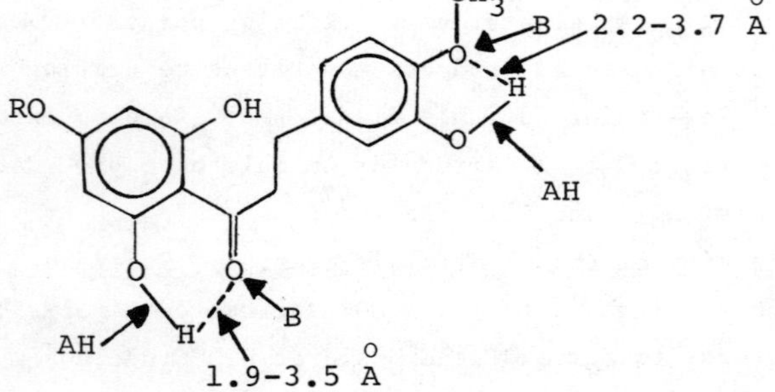

Figure 5. The structural features of hesperetin DHC responsible for
producing the sweet taste response.

We have found it possible to prepare simple, nonglycosidic DHC's which are intensely sweet. Hesperetin DHC derivatives bearing simple 4-0-carboxyalkyl[49] and 4-0-sulfoalkyl[48] substituents have been found to have excellent water solubility and to display taste properties which compare favorably with neohesperidin DHC (see Table 3). The function of the A-ring C_4-0-substituent is primarily to provide water solubility while maintaining proper hydrophobic-hydrophilic balance of the molecule.

Several straightforward and generally applicable means for the preparation of these simplified sweeteners have been developed[48,50]. Figure 6 depicts the most general route which involves the regioselective alkylation and direct alkaline hydrogenation of the flavanone hesperetin (6). This aglycone is economically obtainable in quantity and in high state of purity by the sulfuric acid-catalyzed methanolysis[53] of the rutinoside hesperidin (5), which is the main flavonoid constituent of lemons and oranges. Differences in the acidity of the phenolic hydroxyls of hesperetin account for the high regioselectivity observed in the alkylation.

Table 3. Selected Hesperetin Dihydrochalcone Sweeteners Possessing Non-glycosidic Solubilizing Groups[48-49]

R	Taste intensity[a]	Flavor judgment[b]
β-neohesperidose	612	80/12
CH_2CO_2Na	504	82/7
$(CH_2)_3CO_2K$	308	74/11
CH_2SO_3Na	432	84/10
$(CH_2)_2SO_3K$	705	83/7
$(CH_2)_3SO_3Na$	496	80/9
$(CH_2)_4SO_3K$	226	50/19

[a] These values were determined on a weight basis with a trained taste panel utilizing the magnitude estimation technique which consists of ranking the total taste intensity (i.e., all flavors combined) of a test solution relative to a sucrose standard. Full details of the sensory evaluation methodology employed may be found in M.L. Swartz and T.E. Furia, Food Technol. (Chicago), in press. [b] The ratios listed are percent taste identified as sweet/percent identified as bitter.

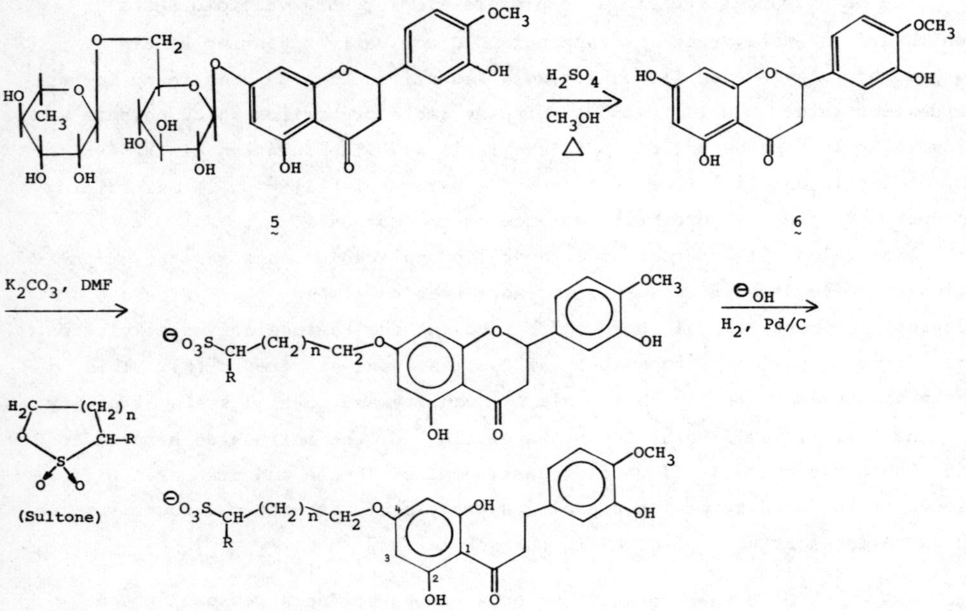

Figure 6. The preparation of nonglycosidic DHC sweeteners via the regio-selective alkylation (as pictured with a generalized sultone) and direct alkaline hydrogenation of the flavanone hesperetin.

Thus we were encouraged by the observation that the C_4-O-substituent in hesperetin DHC can vary dramatically in size and shape without resulting in the loss of sweetness. This position appeared to present a logical site for attachment to a polymer backbone.

<u>Metabolic Stability and Absorption Studies</u>. Hesperetin-3-C^{14} was prepared by a five-step reaction sequence from acetonitrile-2-^{14}C (Figure 7). The metabolic fate of this flavanone in vivo after oral and intraperitoneal administration to intact control, bile duct ligated, and bile duct cannulated rats and in vitro with rat cecal microflora was studied[54]. Forty percent of the radioactivity orally administered to intact rats was expired as $^{14}CO_2$, while virtually no $^{14}CO_2$ was produced upon incubation with cecal microflora. The major labeled metabolites found in both the urine of orally dosed animals and in vitro incubations were 3-phenylpropanoic acids. The results indicated that bacterial enzymes were responsible for the metabolism of hesperetin-3-^{14}C to labeled phenylpropanoic acids, while mammalian hepatic enzymes mediated their further breakdown to benzoic acids and $^{14}CO_2$.

Figure 8 presents the results of a USDA study on the metabolic fate of the sweetener neohesperidin DHC (3) in rats[47,55]. At feeding levels of

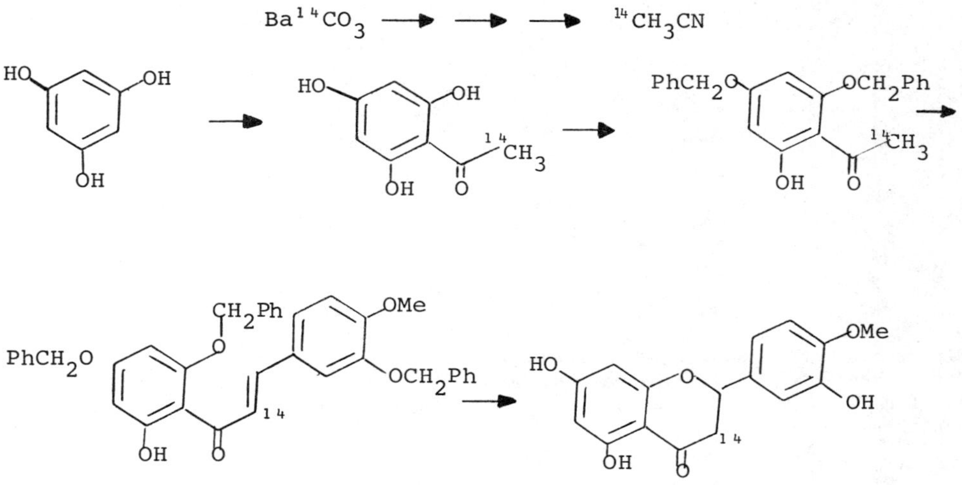

$$Ba^{14}CO_3 \longrightarrow \longrightarrow \longrightarrow {}^{14}CH_3CN$$

HESPERETIN-3-^{14}C (A FLAVANONE) TO RATS AT 0.15-0.30 mg/RAT
>90% ABSORBED, NO INTACT HESPERETIN DETECTED, COMPLETE METABOLISM

43% RADIOACTIVITY AS $^{14}CO_2$, THREE LABELED METABOLITES

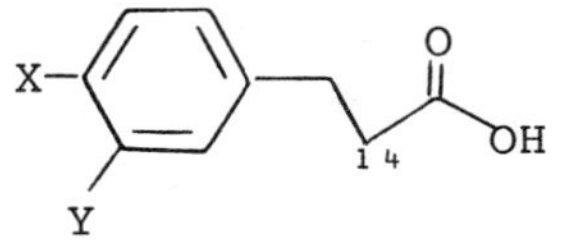

X = OCH_3, Y = OH
X = OH, Y = OH
X = H, Y = OH

BACTERIAL ENZYMES: Metabolism to Phenylpropanoic acids
MAMMALIAN HEPATIC ENZYMES: Further breakdown to benzoic
acids and $^{14}CO_2$

Figure 7. The synthesis and metabolic fate of hesperetin-3-C^{14} in rats.
Data taken from T. Honohan, R.L. Hale, J.P. Brown, and R.E. Wingard, Jr.,
J. Agric. Food Chem., 24, 906, 1976.

1, 10 and 100 mg/kg this ^{14}C-labeled sweetener was completely metabolized.
Urinary excretion of radioactivity varied from 76% at the highest feeding
level to 87% at the lowest. Only one labeled metabolite of significance was
detected. Even though its identity has not been reported, we can safely
speculate that the metabolic pathway involves first the enzymatic removal of
the sugar residues followed by a degradation of the DHC nucleus similar to
that observed with hesperetin-3-$^{14}C^{54}$. No $^{14}CO_2$ was involved in this case
because of the position of the label.

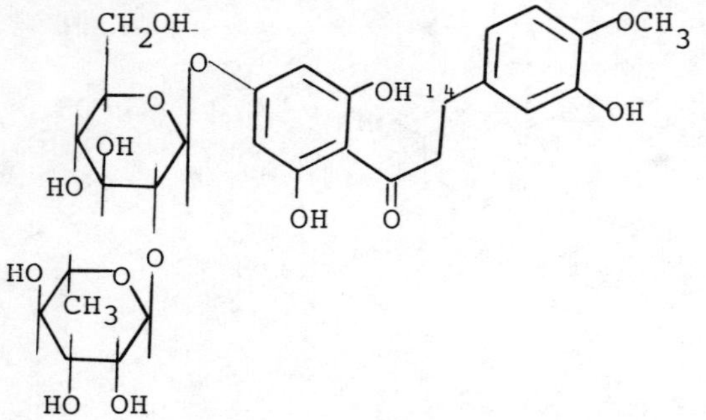

Figure 8. The metabolic fate of radiolabeled hesperidin DHC (3) in rats.
Data taken from M.R. Gumbmann, D.H. Gould, D.J. Robbins and A.Ñ. Booth, 1976,
and from R.M. Horowitz and B. Gentili, both presented at the Citrus Research
Conference, Pasadena, Calif., December 8, 1976.

Axiomatic, of course, to the principle of nonabsorption is that the
sweetener must be stable to the biological degradation processes found in
the human intestinal tract. We have found that the attachment of a metabolically stable substituent to the 4-hydroxyl of hesperetin DHC or to the
7-hydroxyl of the flavanone hesperetin blocks metabolic degradation. Figure
9 presents the results of radiolabeled metabolic fate/absorption studies for
7-O-sulfopropylhesperetin, 4-O-sulfopropylhesperetin DHC, and 2,4-bis-O-(sulfopropyl)hesperetin DHC in rats[56]. The three materials underwent essentially
no metabolism and were recovered intact in both the urine and feces. Absorption drops from 12.3 to 4.43% in proceeding from the monosulfopropyl DHC (mol
wt 426) to the 2,4-bis(sulfopropyl) analog (MW 549). This results from both
the increase in molecular weight (a difference of 123), and the presence of a
second negative charge. The extent of transport within the 300 to 800 molecular weight range generally decreases with increased charge and polarity[57].

Ignoring biological aspects, the metabolic degradation of the DHC
molecule may be pictured as consisting of an acyl cleavage generating
phloroglucinol and phenylpropanoic acid followed by a series of oxidative
decarboxylations involving the latter to produce CO_2 and benzoic acid. The
aryl substituent modifications, such as those shown in Figure 7, are the
result of well-known enzymatic processes[58]. Although the reasons are not
well understood at this time, substitution of the 4-hydroxyl on the DHC
nucleus appears to effectively defeat the initial step (acyl cleavage) of
the metabolic breakdown.

The Core Concept. Initial efforts toward a non-absorbable DHC sweetener
involved polymer leashing. A 2:1 copolymer (gel permeation chromatography

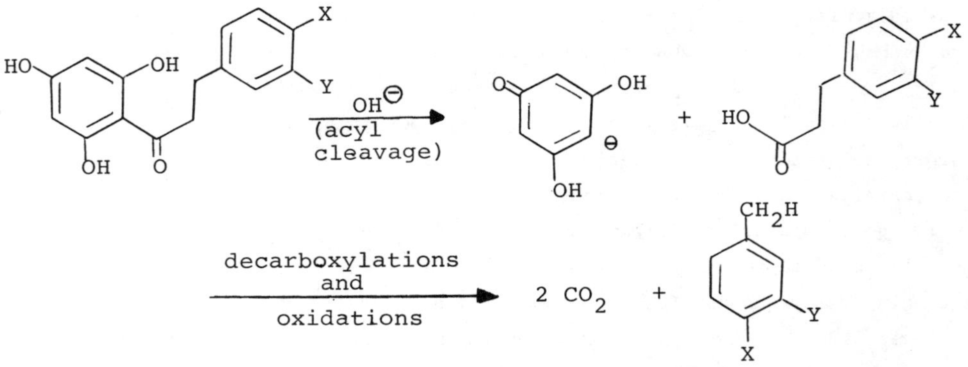

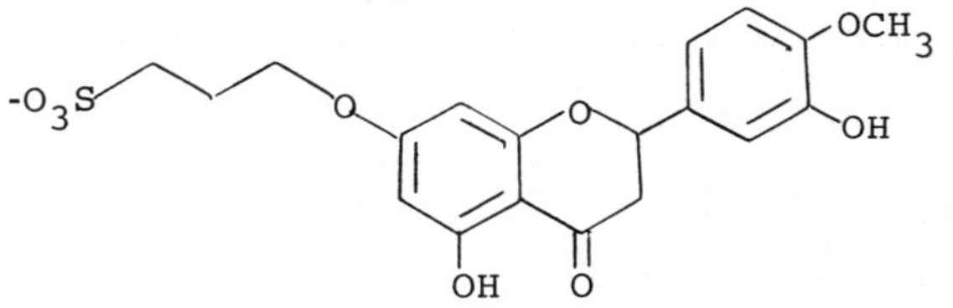

MW 424   22.8% ABSORBED   (urinary excretion)

NO PHENYLPROPANOIC ACIDS DETECTED   0.33% $^{14}CO_2$

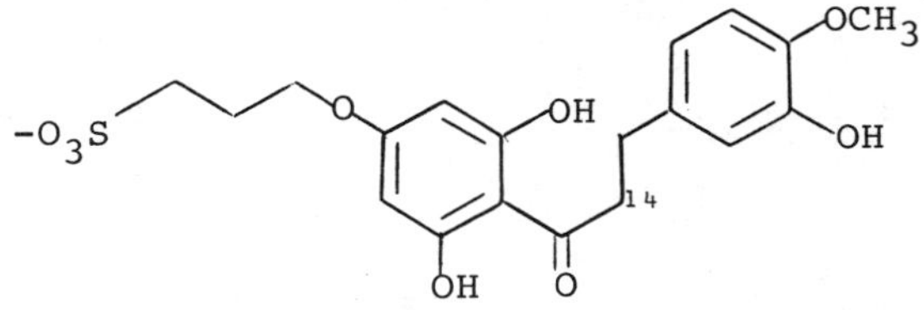

MW 426   12.3% ABSORBED (urinary excretion)

NO PHENYLPROPANOIC ACIDS DETECTED   0.00% $^{14}CO_2$

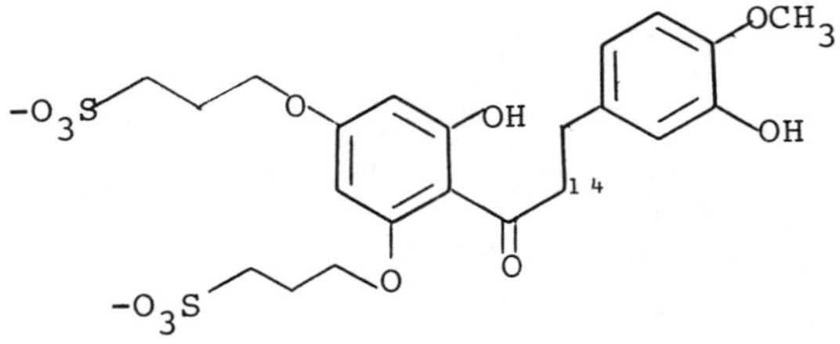

MW 549   4.43% ABSORBED (urinary excretion)

NO PHENYLPROPANOIC ACIDS DETECTED   0.07% $^{14}CO_2$

Figure 9. Metabolic fate/absorption data for three moderate molecular weight sulfopropylated flavonoids. Molecular weights are for the free acids.

peak molecular weight 1.5×10^4) was prepared from N-vinylpyrrolidone and propene sultone (9)[59], and the hesperetin DHC molecule was attached (Figure 10). Polymer 11 was completely tasteless as a 0.20% aqueous solution. Equally disappointing results were obtained upon incorporation of the hesperetin DHC molecule into other copolymeric sultones. These macromolecules are tasteless most probably because the DHC portion is sterically prevented, by the polymer backbone, from interacting with the receptor sites.

A second approach to nonabsorbable DHC sweeteners involved simply attaching a large substituent to the 4-hydroxyl. For example, the use of Carbowax-350 for this purpose provided a mixture of DHC's (18) with molecular weights of 640 to 728 (Figure 11). This mixture exhibited only an intense bitter taste that results from the overall molecular hydrophobic-hydrophilic balance being so shifted by the large relatively hydrophobic substituent, that partitioning onto the sweet receptor site is retarded.

A straightforward strategy for obtaining less sterically encumbered high molecular weight DHC sweeteners with suitable hydrophobic-hydrophilic balance

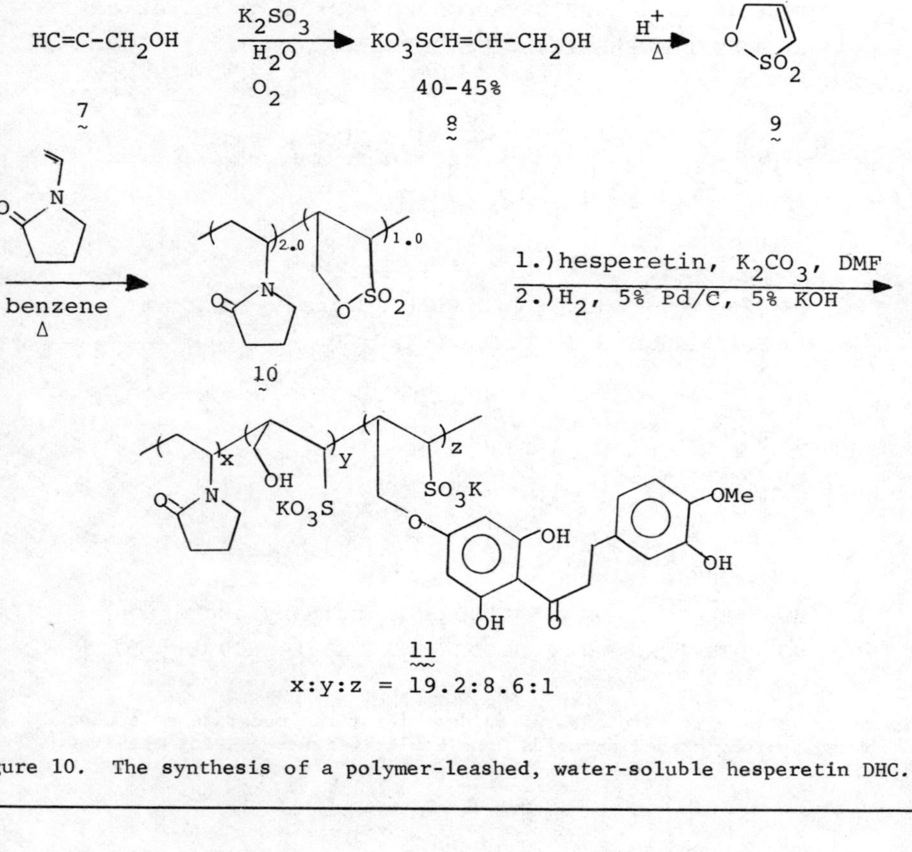

Figure 10. The synthesis of a polymer-leashed, water-soluble hesperetin DHC.

involves linking two or more molecules through a core, while providing water solubility with sulfonates attached either to the core or to the DHC molecules. This strategy is called the <u>core concept</u>. Figure 12 schematically represents the preparation of a typical core-concept dimer, which in this case was prepared from propane sultone (19), trans-1,4-dibromo-2-butene, and hesperetin. DHC dimer 22, as well as the similarly prepared materials in Figure 13, exhibited moderately intense sweetness, at < 0.1 wt.%, followed

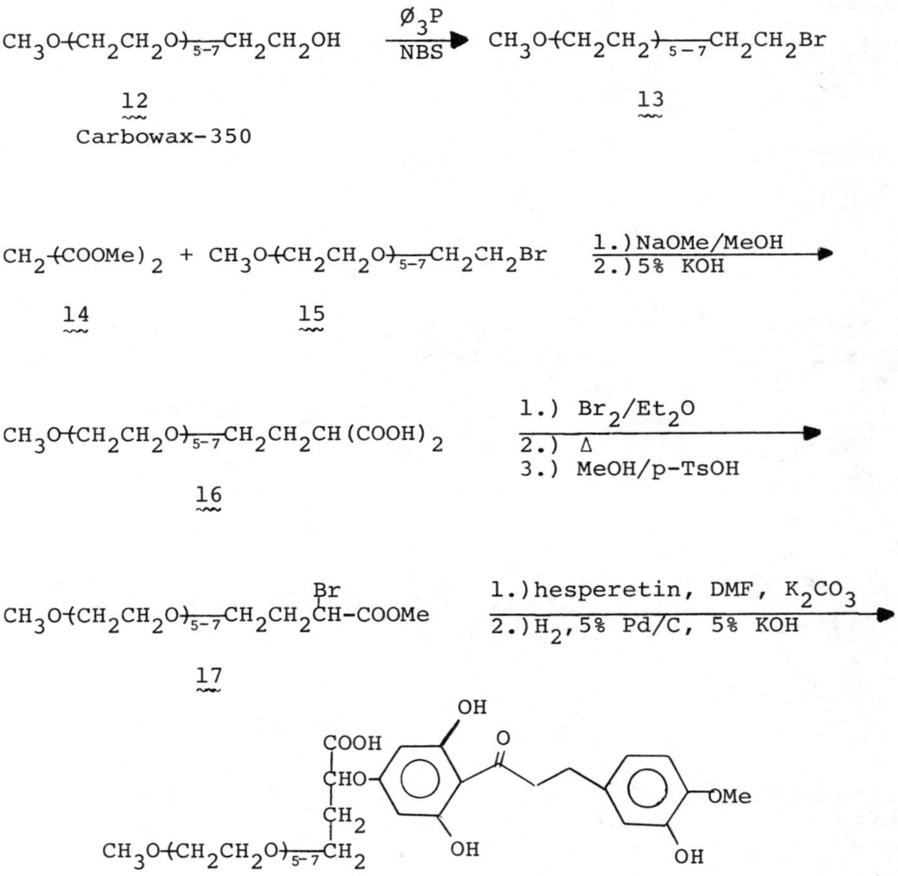

$$CH_3O(CH_2CH_2O)_{5-7}CH_2CH_2OH \xrightarrow[\text{NBS}]{\emptyset_3P} CH_3O(CH_2CH_2)_{5-7}CH_2CH_2Br$$

<u>12</u>

Carbowax-350

<u>13</u>

$$CH_2(COOMe)_2 + CH_3O(CH_2CH_2O)_{5-7}CH_2CH_2Br \xrightarrow[\text{2.) 5\% KOH}]{\text{1.) NaOMe/MeOH}}$$

<u>14</u>   <u>15</u>

$$CH_3O(CH_2CH_2O)_{5-7}CH_2CH_2CH(COOH)_2 \xrightarrow[\text{3.) MeOH/p-TsOH}]{\text{1.) } Br_2/Et_2O \quad \text{2.) } \Delta}$$

<u>16</u>

$$CH_3O(CH_2CH_2O)_{5-7}CH_2CH_2\overset{Br}{\underset{}{CH}}-COOMe \xrightarrow[\text{2.) } H_2, 5\% \text{ Pd/C, 5\% KOH}]{\text{1.) hesperetin, DMF, } K_2CO_3}$$

<u>17</u>

<u>18</u>

(MW 640-728)

Figure 11. The synthetic scheme used for the preparation of a hesperetin DHC analog substituted at the 4-hydroxyl with a high-molecular-weight substituent.

THE CORE CONCEPT

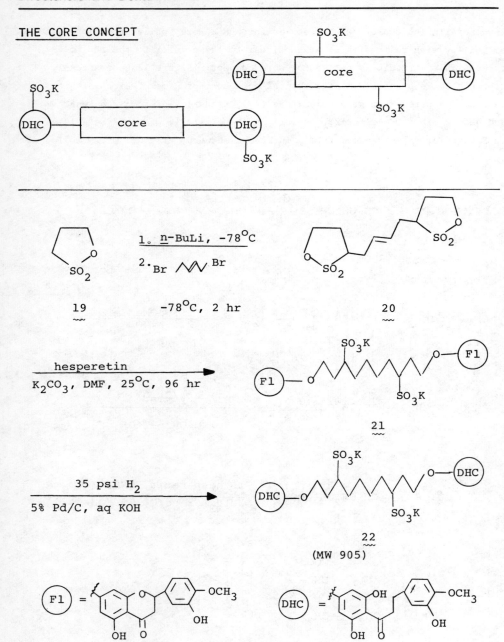

Figure 12. The synthetic scheme employed for the preparation of a simple, sweet, core-concept dimer. The molecular weight of the product is based on the free diacid.

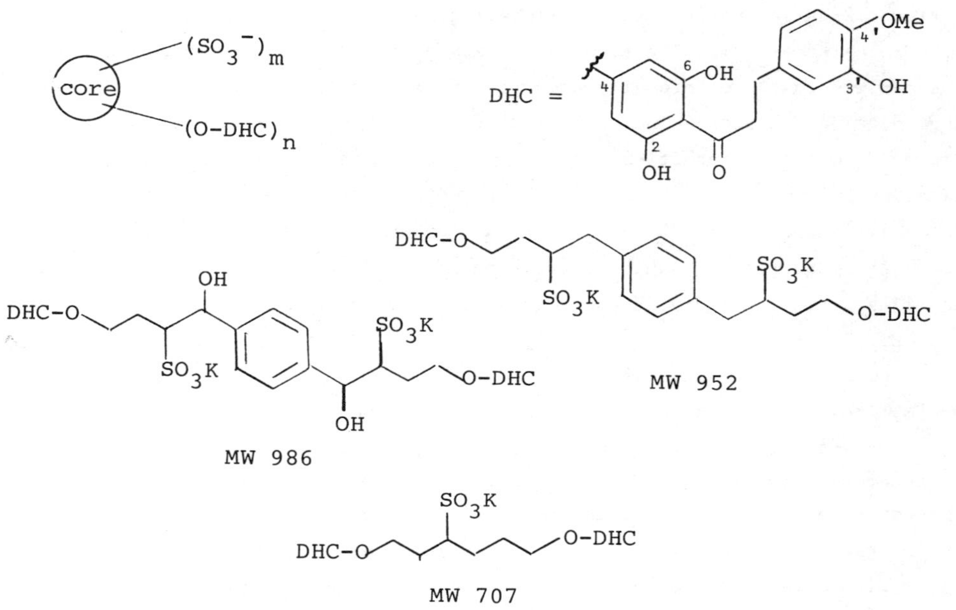

Figure 13. Examples of sweet, core-concept dimers. Molecular weights are based on the free acids.

by a strong lingering methanol-cooling-like aftertaste. A core-concept dimer prepared by linking via the C_2-hydroxyl groups was found to be tasteless (Figure 14), providing additional support for the importance of selecting the proper site for modification.

Biological fate studies were carried out in rats with a radiolabeled sweet core-concept dimer of molecular weight 986[56]. No metabolic breakdown was revealed and intestinal absorption was at the 1.0% level (Figure 15). A plot of molecular weight versus intestinal absorption is presented in Figure 16 for several radiolabeled flavonoids and saccharin. Saccharin (MW 183) is, of course, almost totally absorbed[60].

DHC Sensory Properties. The fact that sweetener research is aimed at a sucrose-like substance cannot be overemphasized. The DHC dimer 22 is un-attractive as a commercial candidate for this reason. Although the taste of a prospective sweetener can generally be described as some composite of the four primary tastes (sweet, sour, salty, bitter), some classes possess taste components usually described as a lingering aftertaste. The aftertaste varies from the tongue-numbing anesthetic effect of the nitroanilines, to the

bitter of the saccharins, to the lingering menthol-like sweetness of the
dihydrochalcones. The lingering sweetness of dihydrochalcones results in
the maximum in taste intensity occurring at a later time than sucrose, and
then dissipating over a longer period of time. The difference in taste-
timing (temporal) properties between sweeteners has been recognized for some
time. Even glucose has been reported to develop its sweet taste somewhat
more slowly than sucrose. Sweeteners are capable of exhibiting a range of
temporal properties, of which sucrose and dihydrochalcones represent extremes,
as illustrated in Figure 17 where sucrose exhibits Type A behavior and DHC's
Type B.

The slow onset-lingering taste behavior renders the presently known
DHC's, including the dimers, incompatible with the flavor components of many
food systems. This problem can be readily appreciated if coffee or cola base
is sweetened with a DHC. The intrinsic bitter and other flavors inherent in
the beverage are experienced immediately, and, as they decay, are replaced by
an intense sweetness which fades slowly. The overall gustatory response
bears little resemblance to a sucrose-sweetened counterpart.

Dihydrochalcones, such as those in Table 3, are not larger or more

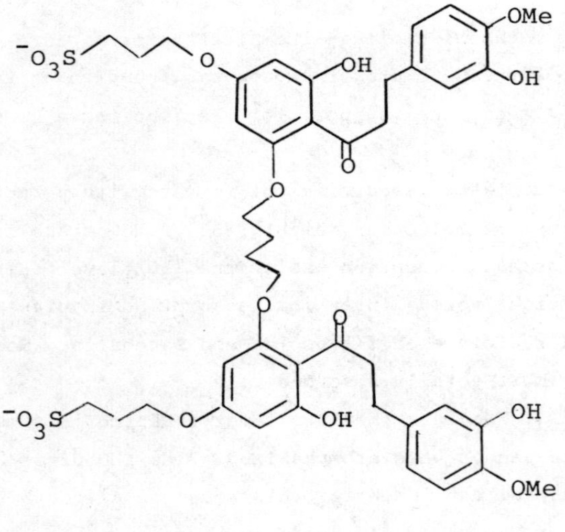

MW 804

Figure 14. A core-concept dimer with linking through the C_2-hydroxyl
groups. This material is tasteless.

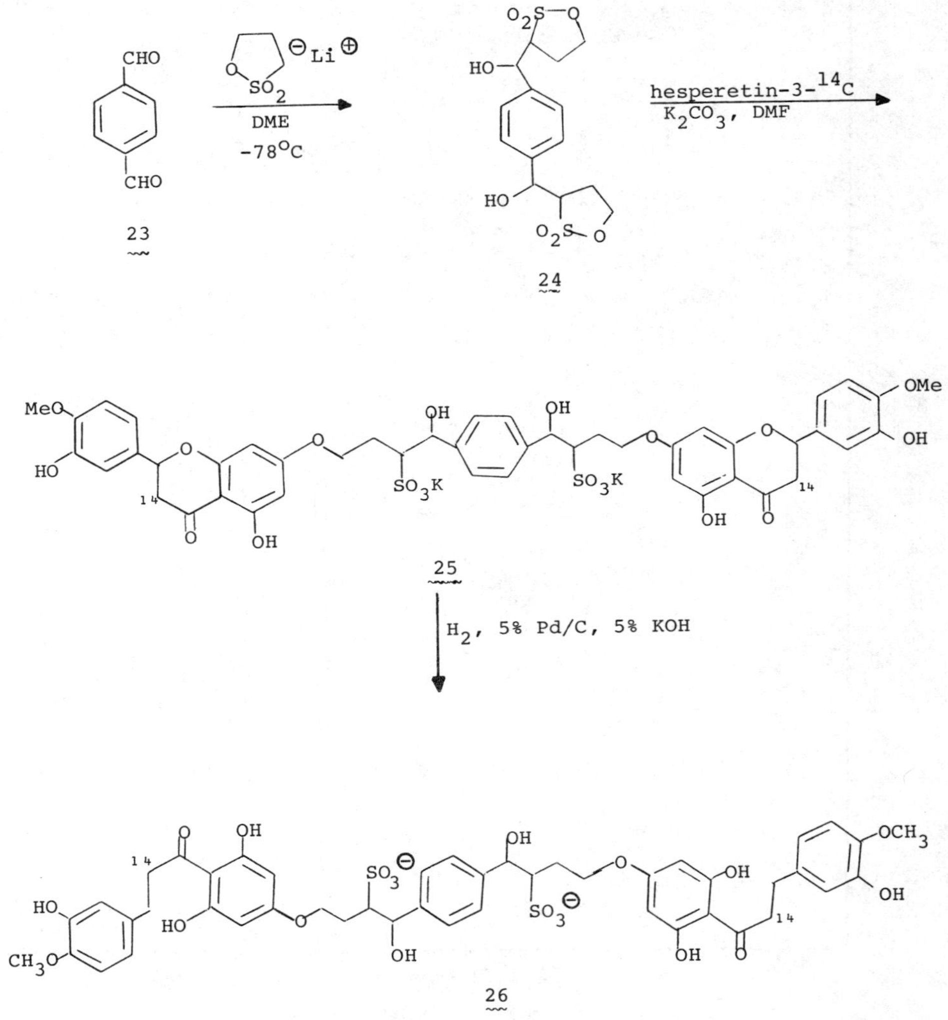

MW 986 (free acid) 1.00% ABSORBED (urinary excretion)
NO PHENYLPROPANOIC ACIDS DETECTED 0.04% $^{14}CO_2$

Figure 15. The synthesis and biological fate of a radiolabeled version of a sweet, core-concept DHC dimer.

hydrophobic than many dipeptide esters that display excellent temporal qualities, suggesting the lagging and lingering taste characteristics are not the result of these molecular parameters. One explanation lies in the presence

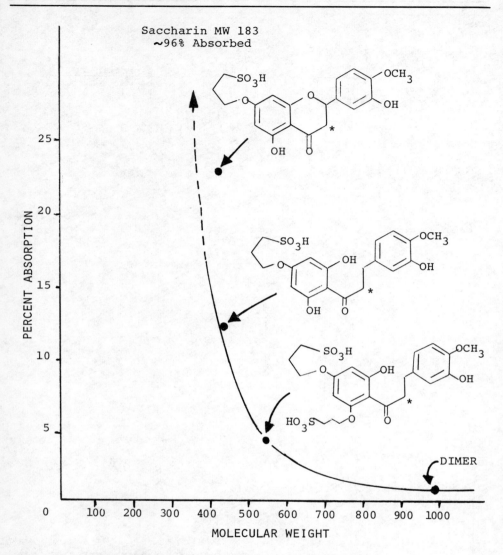

Figure 16. A plot of molecular weight vs. intestinal absorption for saccharin and a series of ^{14}C-labeled flavonoids bearing sulfonate groups.

of two identifiable hydrogen donor-acceptor systems (see Figure 5)[48]. It can, for example, be shown that the second equilibrium constant (K_2) for the process depicted in Eq. (1) is much larger than the equilibrium constant (K_1) for the first step. Thus, formation of the first noncovalent bond makes formation of the second energetically more favorable (just as an intramolecular reaction is much more favorable than a corresponding intermolecular

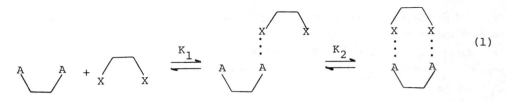

reaction). It follows that the duration of the binding, which is likely related to the lingering taste sensation, will be much greater for a stimulant molecule possessing two binding units than for a similar molecule equipped with a single binding unit.

The presence of two donor-acceptor units within a single DHC nucleus raises interesting comparisons with molecules containing a solitary unit (e.g., cyclamate, saccharin, and the amino acids). For example, it may be an indication of more than one type of receptor for sweet substances in humans. In this regard, DHC's elicit a sweet response not only on the tongue but also on the roof and back of the mouth, while cyclamate and most other sweet substances appear to produce a reaction predominantly on the tongue. Alternatively, a single receptor may contain two hydrogen donor-acceptor units. Accordingly, two molecules of saccharin or cyclamate would be required to elicit the same response provided by a single DHC. The unusual lingering

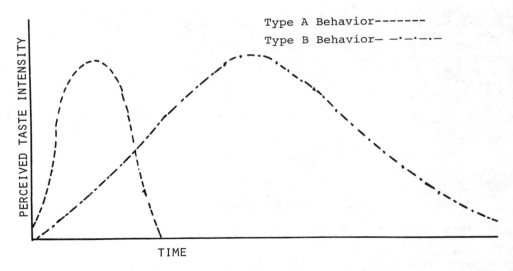

Figure 17. Perceived taste intensity vs. time for a Type A sweetener (rapid taste onset and cutoff), compared to a Type B sweetener (slow taste onset, lingering aftertaste). Reproduced with permission from G.E. DuBois, G.A. Crosby, R.A. Stephenson and R.E. Wingard, Jr., J. Agric. Food Chem., 25, 763, (1977).

taste of dihydrochalcones indirectly implicates this mechanism. Additional support comes from the studies of Morita and Shiraishi[62]. These workers investigated the stimulation of the labellar sugar receptor of the fleshfly by mono- and disaccharides and determined that a 1:1 complex is formed between the receptor and sucrose, while glucose or fructose form a 2:1 complex. It was concluded that the sugar receptor of the fleshfly is comprised of two subunits; for excitation the receptor must be simultaneously occupied at the two subunits. Similar observations have recently been made with the gerbil by Jakinovich and Goldstein[63].

CONCLUSION

The anatomical compartmentalization approach has now provided synthetic sweeteners which afford $\leq 1.0\%$ intestinal absorption in the rat. Successful implementation was based upon the use of the hesperetin DHC nucleus in conjunction with the core concept, which involves simply linking two or more sweetener nuclei together with a common bridge. The remaining major hurdle is the development of a sweetener of this type possessing the molecular parameters which will afford a sucrose-like taste. Significant progress has been made toward this goal during the past year.

Acknowledgment. The authors wish to acknowledge all those members of the Dynapol research staff involved with synthetic sweeteners for their dedicated and innovative pursuit of the goals of this program. We are especially indebted to P. Saffron and P. Frank for synthetic assistance during the early phases of this research, to F.E. Enderlin, B.A. Ryerson, T. Honohan, and Dr. T.M. Parkinson for the metabolic fate/intestinal absorption studies, to C.T. Seitz for analytical support, and to M. Swartz for sensory evaluation.

REFERENCES

1. Weinshenker, N.M. Polym. Prepr., Am. Chem. Soc., Div. Polym. Chem. 18(1), 531, 1977.
2. Dawson, D.J., Gless, R.D., and Wingard, R.E., Jr. J. Am. Chem. Soc. 98, 5996, 1976.
3. Dawson, D.J., Gless, R.D., and Wingard, R.E., Jr. Chem. Technol. 6, 724, 1976.
4. Furia, T.E. Food Technol. (Chicago) 31(5), 34, 1977.
5. Dawson, D.J., Otteson, K.M., Wang, P.C., and Wingard, R.E., Jr. Macromolecules submitted.
6. Furia, T.E. and Bellanca,N. J. Am.Oil Chem. Soc. 53, 132, 1976.
7. Furia, T.E. and Bellanca, N. J. Am. Oil Chem. Soc. 54, 239, 1977.
8. Dale, J.A., Ng, S., and Wang, P.C. manuscript in preparation.
9. Crosby, G.A., DuBois, G.E., and Wingard, R.E., Jr., in Drug Design, Vol. VIII, edited by E.J. Ariens,, Academic Press, New York, N.Y., in press.

10. Höber, R. and Höber, J. J. Cell. Comp. Physiol. 10, 401, 1937.
11. Soergel, K.H. and Hogan, W.J. Gastroenterology 52, 1056, 1967.
12. Shields, R., Harris, J., and Davies, M.W. Gastroenterology 54, 211, 1968.
13. Miller, D.L. and Schedl, H.P. ibid. 58, 40, 1970.
14. Wilson, T.H. and Vincent, T.N. J. Biol. Chem. 216, 851, 1955.
15. Parkinson, T.M., Honohan, T., Enderlin, F.E., Halladay, S.C., Hale, R.L., deKeczer, S.A., Dubin, P.L., Ryerson, B.A. and Read, A.R. Food Cosmet. Toxicol. submitted.
16. Honohan, T., Enderlin, F.E., Ryerson, B.A. and Parkinson, T.M. Xenobiotica , in press.
17. Kikuchi, T. and Shimada, I. Kagaku No Ryoiki 30, 510, 1976
18. Price, S. and DeSimone, J.A. Chem. Senses Flavor, in press.
19. Beidler, L.M. and Gross, G.W. Contrib. Sens. Physiol. 5, 97, 1971.
20. Cagan, R.H. in Sugars in Nutrition, edited by H. Sipple and K. McNutt, Academic Press, New York, N.Y., pp. 19-36, 1974.
21. Brand, J.G. and Cagan, R.H. J. Neurobiol. 7, 205, 1976.
22. Cagan, R.H. J. Neurosci. Res. 2, 363, 1976.
23. Beidler, L.M. Prog. Biophys. Biophys. Chem. 12, 107, 1962.
24. Morris, J.A. and Cagan, R.H. Biochim. Biophys. Acta 261, 114, 1972.
25. Cagan, R.H. Science 181, 32, 1973.
26. Cagan, R.H. and Morris, J.A. Proc. Soc. Exp. Biol. Med. 152, 635, 1976.
27. Bohak, Z. and S.-L. Li, Biochim. Biophys. Acta 427, 153, 1976.
28. van der Wel, H. and Loeve, K. Eur. J. Biochem. 31, 221, 1972.
29. Kurihara, K. and Beidler, L.M. Science 161, 1241, 1968.
30. Kurihara, K. and Beidler, L.M. Nature (London) 222, 1176, 1969.
31. Kurihara, K., Kurihara, Y. and Beidler, L.M. in Olfaction and Taste.III, C. Pfaffman, Ed. Rockefeller Univ. Press, New York, N.Y., 1969.
32. Inglett, G.E. and May, J.F. Econ. Bot. 22, 326, 1968.
33. Crosby, G.A. Crit. Rev. Food Sci. Nutr. 7, 297, 1976.
34. Zuber, H. Eidgenossiche Technische Hochschule, Zurich, private communication, 1974.
35. van der Wel, H. and Loeve, K. FEBS Lett. 29, 181, 1973.
36. Brouwer, J.N., van der Wel, H., Franke, A. and Henning, G.J. Nature (London) 220, 373, 1968.
37. Lapidus, M. and Sweeney, M. J. Med. Chem. 16, 163, 1973.
38. Cuatrecasas, P. and Anfinsen, C.B. Annu. Rev. Biochem. 40, 259, 1971.
39. Venter, J.C., Dixon, J.E., Maroko, P.R. and Kaplan, N.O. Proc. Nat. Acad. Sci. U.S.A. 69, 1141, 1972.
40. Suzuki, F., Daikuhara, Y., Ono, M. and Takeda, Y. Endocrinology 90, 1220, 1972.
41. Venter, J.C. and Kaplan, N.O. Science 185, 459, 1974.
42. Horowitz, R.M. and Gentili, B. U.S. Patent 3087821, 1963 Chem. Abstr. 59, 11650c, 1963.
43. Horowitz, R.M. in Biochemistry of Phenolic Compounds, J.B. Harborne, Ed., Academic Press, New York, N.Y.,pp. 545-571, 1964.
44. Horowitz, R.M. and Gentili, B. J. Agric. Food Chem. 17, 696, 1969.
45. Horowitz, R.M. and Gentili, B. in Sweetness and Sweeteners, G.G. Birch, L.F. Green and C.B. Coulson, Ed., Applied Science, London, pp. 69-80, 1971.
46. Horowitz, R.M. and Gentili, B. in Symposium: Sweeteners, G.E. Inglett, Ed., Ari Publishing, Westport, CT, pp. 182-193, 1974.
47. Gumbmann, M.R., Gould, D.H., Robbins, D.J., and Booth, A.N., presented at the Citrus Research Conference, Pasadena, CA December 8, 1976, and earlier presentations by this group at this annual conference.
48. DuBois, G.E., Crosby, G.A., Stephenson, R.A. and Wingard, R.E., Jr. J. Agri. Food Chem. 25, 763, 1977.

49. DuBois, G.E., Crosby, G.A. and Saffron, P. Science 195, 397, 1977.
50. DuBois, G.E., Crosby, G.A. and Saffron, P. Synth. Commun. 7, 44, 1977.
51. Crosby, G.A., DuBois, G.E. and Wingard, R.E., Jr. in Flavor: Its Chemical, Behavioral, and Commercial Aspects, C. Apt, Ed., Westview Press, Boston, MA, in press.
52. Shallenberger, R.S. and Acree, T.E. Nature (London) 216, 480, 1967.
53. Seitz, C.T. and Wingard, R.E., Jr. J. Agric. Food Chem. in press.
54. Honohan, T., Hale, R.L., Brown, J.P. and Wingard, R.E., Jr. J. Agric. Food Chem. 24, 906, 1976.
55. Horowitz, R.M. and Gentili, B., presented at the Citrus Research Conference, Pasadena, CA December 8, 1976.
56. Crosby, G.A., DuBois, G.E., Hale, R.L., Honohan, T. and Wingard, R.E., Jr. unpublished results.
57. Magee, D.F. Gastro-Intestinal Physiology C.C. Thomas & Co., Springfield, IL 1962.
58. Hathway, D.E. Adv. Food Res. 15, 1, 1966.
59. Helberger, J.H. and Müller, G. German Patent 1146870, 1963. Chem. Abstr. 59, 11259d, 1963.
60. Couch, M.W., Das, N.P., Scott, K.N., Williams, C.M. and Foltz, R.L. Biochem. Med. 8, 362, 1973.
61. Dahlberg, A.C. and Penczek, E.S. Agric. Exp. Stn., Ithaca Bull. 258, 1, 1941.
62. Morita, H. and Shiraishi, A. J. Gen. Physiol. 52, 559, 1968.
63. Jakinovich, W. and Goldstein, I.J. Brain Res. 110, 419, 1976.

DISCUSSION AFTER DR. CROSBY'S PAPER

DR. BOLLENBACK: You say that your molecules are not absorbed and not metabolized. What happens to the molecules? Are they excreted in the feces?

DR. CROSBY: That's right, that's where we detect the labelling.

DR. BOLLENBACK: Do you know any particular effect that the material might have on the lower bowel?

DR. CROSBY: We have not noticed any effect yet. We have not done a great deal of work of this sort with the sweetener molecules, but with the polymeric dyes that we have made, we have not seen adverse effects.

DR. BOLLENBACK: You don't see any promotion of diarrhea, for instance?

DR. CROSBY: No, although perhaps at very high feeding levels one might expect to see some.

DR. CAGAN: Guy, the possibility of leashing a molecule to a polymer so that it is not absorbed is interesting in terms of function. In terms of commercial realities, how would FDA approval of one material affect the cost and ease of getting later molecules approved by the FDA?

DR. CROSBY: I don't think it will reduce it one bit. I think it will cost as much to carry out all the safety testing for any type of material, whether it's absorbed or not absorbed. What we have done initially is at least guarantee to ourselves that, once we start out with a product with desirable functionality, we have a very good chance of bringing it through all the sensitive and lengthy safety testing without seeing any signs of toxicity.

At this early stage in our program, I do not think that we can justify to the FDA any less testing than is required of all food additives now.

That's our general philosophy. We expect the same amount of time and money to be spent.

DR. CAGAN: Do you think that this will be true even for subsequent compounds?

DR. CROSBY: Perhaps in the future if the safety of such unabsorbed and unmetabolized compounds can be demonstrated a number of times, there may be a change in philosophy. However, initially for our first few products, we are not counting on this at all.

DR. SHAW: What is the potential cost, if you come up with a compound with desirable functionality and safety? For example, how would it compare with saccharin?

DR. CROSBY: The problem, of course, is that this is a fairly expensive approach. As the price of saccharin goes, we would never expect to be in a region to compete with it. We have been attempting to look at this realistically in terms of the cost on the basis of sucrose sweetness. We hope to be within the overall cost of sugar because so much less of these intense sweeteners is needed. We are talking about substances that need to be 100 to 500 times the potency of sugar so that our manufacturing cost can be perhaps several hundred times the cost of sugar.

DR. NOZNICK: In the phenomena called pinocytosis, large molecules of the size of egg albumin can pass into the cells lining the intestinal villi. If something that large can pass into the villi, I would imagine that something smaller than that, such as your molecules, might also be able to move across the intestinal barrier by pinocytosis. I think that's an area that you should investigate.

DR. CROSBY: We have done that with the polymeric dyes, compounds that are real polymers with a peak molecular weight of about 60,000. We purify them by ultrafiltration, so that we remove essentially everything below molecular weight, 2,000. These compounds can be radiolabelled; when we purify these polymers carefully, only a few hundredths of a percent of the label is absorbed. This amount of absorption probably indicates that the purification process leaves tiny traces of low molecular weight materials. We have done studies where we constantly decrease the amount of low molecular weight impurities and in turn decrease the amount of absorption. It appears as though we are looking at essentially a diffusion controlled transport of what small amount of material does get across this membrane.

DR. AMOS: Isn't it surprising, though, if that's the case. I think on your last slide when you simply opened the ring and put on a hydroxyl group, the absorption dropped by a very large factor.

DR. CROSBY: Yes, absorption dropped to one half. Of course, the size, the molecular weight of that molecule is in a region where there is great sensitivity to the size of the molecule versus the amount of transport. The flavanone is a closed molecule, so its molecular size is smaller than one which is open and extended and probably solvated in the aqueous system. We feel the apparent molecular size of the open molecule, as seen by pores in the membrane, is much larger than the flavanone. That's our explanation for the difference in absorption between the flavanone and the dihydrochalcone.

DR. AMOS: We are speaking of two different processes. With pinocytosis, the larger molecules up to a certain size, especially positively charged molecules at pH 7, are taken up much more rapidly. There's been a lot of work on that, on both naturally occurring histones, and some synthetic polylysines. Their molecular size up to 100,000 or so daltons is a positive attribute.

DR. CROSBY: Even visible beads of polystyrene can be absorbed by this mechanism into isolated cell systems. So far we have not seen what happens with our particular dyes.

DR. INGLETT: Concerning the taste quality of your polymeric dihydrochalcone sweeteners, can you comment on the quality of the sweetness of some of these polymeric materials?

DR. CROSBY: Less than desirable from a commercial point of view! The closest so far is a material that is perceived much like neohesperidin dihydrochalcone. In other words, we still have some lag and a certain amount of lingering taste. The potency of this material has been reduced; it's only about 70 times sucrose for the most recent material. From our point of view, we still hope to have compounds that are better overall in their properties than what we have now. A special problem seems to be when these compounds are evaluated in water solution, panelists find them quite acceptable because they have no time reference to compare them with. They are just evaluating total taste response.

The real problem comes about when you evaluate food systems and you then have a comparison of the sweet taste versus other tastes in the system. The difference then becomes drastic. This is where you really run into the problems of the acceptability of sweeteners like hydrochalcones when you are comparing a number of different tastes. The taste intensities are quite different with time. We have been studying time/intensity of several of these types of compounds for some time.

DR. TOWNSLEY: Only about 15 percent of the estrogen administered to humans is excreted in the urine; most of it ends up in the feces, but it is extensively metabolized and conjugated in the liver. To detect that, you have to do biocanulation. Could you use biocanulation studies to verify whether your criteria for absorption is correct?

DR. CROSBY: Yes. We have performed bile duct ligation experiments in vivo in rats. For example, for dihydrochalcone, about 12% of the ^{14}C label occurred in the urine. Actually a great portion of this has come in through the bile system. A number of these compounds go into the liver to be excreted back in the feces through the enterohepatic system.

DR. MACKAY: Is it fair to say that this idea of leashing the sweetener with a polymer is in effect being abandoned for the idea of making a polymeric sweetener? Have you ever found a leash so tenuous it hasn't seriously affected the sweetness of the molecule? If I understand your approach, you decided that it's better to try to get a number of sweetening centers together to minimize the effect of the leash.

DR. CROSBY: We got away from the real polymeric system because they were just too difficult and time consuming to make, and because we really hadn't isolated products that we considered exhibited a satisfactory sweetness property. We thought of moving towards these directions of simpler molecules that were single molecular weight sweeteners. We were concerned about having a polymer with a distribution of many molecular weight species. When the product contains a wide array of different molecular weight compounds, each one may actually behave a little differently, and interact with a receptor a little differently. Under those conditions we may not be able to develop a clean sweet taste with a substance that's a mixture of materials rather than a single molecular weight sweetener.

DR. MACKAY: Is it also fair to say as you succeed in increasing the polymeric weight of your system, you are increasing its colloidal nature and the likelihood of interaction with food systems? Is the colloid going to be disrupted so that it would not act in a food system as it would in a so-called colloidal suspension? To put it another way, have you looked at the colloidal nature of the material to find out if it's the type of colloid that you expect to persist in a food system?

DR. CROSBY: We have looked at the polymeric food dyes because they are much larger but we have not really examined the sweeteners in that fashion.

DR. JONES: You apparently recovered the label in the feces. Did you do any recovery studies to see whether the material got through the gastro-intestinal tract intact? How can you be sure it was the active material itself?

DR. CROSBY: Do you mean to actually identify the material?

DR. JONES: Yes. How do we know some of the side chains didn't get split off and absorbed?

DR. CROSBY: The only way we have of knowing that is with the in vitro studies where we incubated the material with the rat fecal contents and examined the mixture by high pressure liquid chromatography. We looked for any breakdown material. Since we knew basically what the breakdown products were from at least the dihydrochalcone portion, we looked for these materials, and in fact didn't see anything in the mixture other than the intact sweetener molecule.

The result of these in vitro studies and their correspondence with recovering over 98 percent of the label in the in vivo system made us pretty confident the material was not being metabolized or degraded in any way. I really do not feel that my answer to your question was that satisfactory because I did not do any of the work on the biological study. That was all conducted by the biological sciences department at Dynapol.

Industrial potential of sweeteners other than sucrose and simple carbohydrates

William A. Hoskins

Research and Development Center, Foremost Foods Company, 6363 Clark Avenue, Dublin, California 94566

SUMMARY

The use of sucrose and other carbohydrate sweeteners by the food industry is extensive, six million tons per year and 59 billion dollars worth of products manufactured annually. Development of snack foods using other sweeteners appears to be technically feasible. In order to make the task of reducing sucrose in foods more manageable, the development of such foods should probably be centered around snack type products. If possible, those products should be sold primarily to that age group most susceptible to caries. These products could have three potential markets--calorie reduction, diabetic and caries reduction. Further studies appear to be needed to define and substantiate the caries reduction benefit to the consumer when such products are used.

CHANGING USES OF SWEETENERS

To provide a framework from which to discuss the potential of sweeteners other than sucrose and simple carbohydrates in the food industry, I would first like to review the present status of sucrose and other sweeteners in that industry.

Sucrose is used in several segments of the food industry, but primarily in bakery products, beverages, confections, dairy products and processed food such as fruits, vegetables and meats. The USDA data shown in Table 1 indicate the quantity of sugar used in manufactured food and beverage products and also nonindustrial food use from 1970 to 1976. Several items are worth noting in this table: the overall use of sucrose has fluctuated over the years and is declining; approximately 2/3 of the sucrose used in the United States is by industry as opposed to household use; the major users of sucrose are the bakery and beverage industries, with the beverage use of 30+% being by far the largest. If we reviewed consumption figures for the 1920-1930's, we would see a different pattern. At that time, approximately 25% of the sugars sold in the United States were delivered to industrial users with the remaining 75% being used and controlled in individual households[1]. The de-

Type of buyer	Calendar year						
	1970	1971	1972	1973	1974	1975	1976
	1,000 tons, refined sugar						
Industrial users:							
Food use							
Bakery and cereal products	1,420	1,361	1,449	1,454	1,443	1,241	1,313
Confectionery products	1,098	1,056	1,057	1,035	1,018	795	911
Processed foods	969	1,027	987	1,025	949	743	737
Dairy products	537	561	599	595	570	511	553
Other	433	493	508	502	514	486	520
Total	4,456	4,498	4,600	4,611	4,494	3,776	4,034
Beverage use	2,356	2,365	2,437	2,469	2,350	2,074	2,253
Total industrial users	6,812	6,863	7,037	7,080	6,844	5,850	6,287
Non-industrial users:							
Institutions							
Eating and drinking	90	79	85	94	91	72	64
Other[1]	99	95	88	106	121	85	135
Total institutions	189	174	173	200	212	157	199
Wholesale and retail							
Wholesalers, jobbers, and sugar dealers	2,206	2,156	2,103	2,064	2,002	1,919	2,144
Retail grocers, chain stores, and supermarkets	1,330	1,324	1,316	1,316	1,353	1,261	1,310
Total wholesale and retail	3,536	3,480	3,419	3,380	3,355	3,180	3,454
Minus consumer size packages[2]	2,544	2,610	2,557	2,530	2,581	2,409	2,440
Redistributed to industrial and other users[3]	992	870	862	849	774	771	1,014
Total non-industrial users	3,725	3,654	3,592	3,580	3,567	3,337	3,653
Total food use	10,538	10,517	10,629	10,669	10,411	9,187	9,940
Non-food use[4]	84	93	91	111	128	86	103
Total food and non-food use	10,621	10,610	10,720	10,771	10,539	9,273	10,043

[1] Includes deliveries to government agencies and the military. [2] Less than 50 pounds. [3] Includes some deliveries to eating and drinking places and institutions. [4] Used largely for pharmaceuticals and some tobacco.

Source: Fruit and Vegetable Division, AMS, USDA.

Table 1: U.S. sugar deliveries to industrial and non-industrial users, calendar years, 1970-1976

velopment of convenience foods by industry and their acceptance by the housewife accounts for this change in use pattern. Prepared foods of all varieties are now available, using sugar that in prior years was used by the housewife.

Table 2 gives the per capita U.S. consumption of caloric and noncaloric sweeteners from 1960 to 1976. A decline in the per capita consumption of sucrose began in 1974, but a fairly steady increase in the overall per capita consumption of sweeteners as a whole has been occurring for the last 16 years.

Table 3 indicates the dollar value of the products manufactured from sucrose since 1970. Two items of interest here are: the size of that portion of the food industry that we are discussing is currently over 59 billion dollars a year; some of the largest users of sucrose (Table 1) generate products which have a low dollar value. The products which use the most sucrose also cost the least.

From the figures presented in Tables 1 to 3 we can get an idea as to the

Table 2 [3]

Caloric and noncaloric sweeteners: Per capita U.S. consumption, 1960-76

Calendar year	Refined cane and beet sugar						Corn sweeteners[1]				Minor caloric[1]			Total caloric	Noncaloric sweeteners[2]		
	U.S. grown sugar			Cane sugar		Total	Corn sirup		Dextrose	Total	Honey	Edible sirups	Total		Saccharin	Cyclamate	Total noncaloric
	Beet sugar	Cane sugar	Total	Imported	Total		High-fructose	Other									
	Pounds	Pounds	Pounds	Pounds	Pounds	Pounds	Pounds	Pounds	Pounds	Pounds	Pounds	Pounds	Pounds	Pounds	Pounds	Pounds	Pounds
1960	25.2	28.1	53.3	44.3	72.4	97.6	—	8.2	3.4	11.6	1.2	0.8	2.0	111.2	1.9	0.3	2.2
1961	26.1	28.7	54.8	43.0	71.7	97.8	—	8.6	3.4	12.0	1.1	.8	1.9	111.7	2.1	.4	2.5
1962	23.9	28.0	51.9	45.4	73.4	97.3	—	9.3	3.6	12.9	1.1	.9	2.0	112.2	2.5	.4	2.9
1963	27.2	27.8	55.0	41.7	69.5	96.7	—	9.9	4.3	14.2	1.0	.7	1.8	112.7	3.0	.7	3.7
1964	28.5	30.3	58.8	37.9	68.2	96.7	—	10.9	4.1	15.0	1.1	.7	1.7	113.4	3.5	1.3	4.8
1965	29.4	30.3	59.7	37.1	67.4	96.8	—	11.0	4.1	15.1	1.0	.7	1.8	113.7	4.0	1.7	5.7
1966	28.3	28.6	56.9	40.3	68.9	97.2	—	11.2	4.2	15.4	1.0	.7	1.7	114.3	4.5	1.9	6.4
1967	26.6	29.9	56.5	41.8	71.7	98.3	—	11.9	4.2	16.1	.9	.5	1.4	115.8	4.8	2.1	6.9
1968	27.8	26.5	54.3	44.7	71.2	99.0	—	12.6	4.3	16.9	.9	.7	1.6	117.5	5.0	2.2	7.2
1969	30.1	25.2	55.3	45.4	70.6	100.7	—	13.2	4.5	17.7	1.0	.6	1.6	120.0	5.3	1.6	6.9
1970	31.4	25.0	56.4	45.5	70.5	101.9	—	14.0	4.6	18.6	1.0	.5	1.5	122.0	6.2	(}	6.2
1971	31.1	22.8	53.9	48.5	71.3	102.4	—	15.0	5.0	20.0	.9	.5	1.4	123.8	5.7	(}	5.7
1972	30.4	25.4	55.8	47.0	72.4	102.8	0.9	15.6	4.4	20.9	1.0	.5	1.5	125.2	5.7	(}	5.7
1973	30.4	24.9	55.3	46.2	71.1	101.5	1.4	16.7	4.8	22.9	.9	.5	1.4	125.8	5.7	(}	5.7
1974	26.1	21.0	47.1	49.5	70.5	96.6	2.3	17.4	4.9	24.6	.8	.4	1.2	122.4	7.0	(}	7.0
1975	30.5	24.9	55.4	34.8	59.7	90.2	4.7	17.7	5.1	27.5	.9	.4	1.3	119.0	7.0	(}	7.0
1976[4]	32.2	22.7	54.9	39.8	62.5	94.7	7.1	17.7	5.1	29.9	1.0	.4	1.4	126.0	8.0	(}	8.0

[1] Dry basis. Recent corn sweetener consumption may be under stated due to incomplete data. [2] Sugar sweetness equivalent—assumes saccharin is 300 times as sweet as sugar, and cyclamate is 30 times as sweet as sugar. [3] Cyclamate food use was banned by the Food and Drug Administration, effective in 1970. [4] Preliminary.

Sweeteners and sweetener-containing products: Estimated manufacturer value of annual shipments, calendar years, 1970-77[1]

Industry or product	Value of shipments					Projection estimates		
	1970	1971	1972	1973	1974	1975	1976	1977
	Million dollars	Million dollars	Million dollars	Million dollars	Million dollars	Million dollars	Million dollars	Million dollars
Sweetener industries								
Sweetening sirups and molasses ...	149	145	167	174	254	274	295	320
Wet corn milling (corn refining) ...	728	753	787	1,083	1,772	1,642	1,828	2,023
Cane sugar, except refining (raw cane sugar)	364	408	440	559	1,321	1,416	1,062	1,083
Cane sugar, refining	1,589	1,640	1,747	1,925	4,424	4,305	2,833	2,890
Beet sugar	727	809	867	922	1,668	1,512	1,391	876
Total sugar	2,680	2,857	3,054	3,406	7,413	7,233	5,286	4,849
Total sweetener industries	3,557	3,755	4,008	4,663	9,439	9,149	7,409	7,192
Sweetener-containing products industries								
Cereal and bakery								
Cereal breakfast foods	822	867	935	1,073	1,364	1,579	1,800	2,029
Flour mixes and refrigerated doughs								
Made in flour mills	167	152	144	156	240	252	277	306
Not made in flour mills	536	578	652	696	943	925	950	969
Bread and other bakery products	4,721	4,830	5,181	5,642	6,903	9,083	9,779	10,580
Crackers and cookies	1,501	1,578	1,713	1,890	2,268	2,902	3,080	3,270
Total flour and bakery	6,925	7,138	7,690	8,384	10,354	13,162	14,086	15,125
Total cereal and bakery	7,747	8,005	8,625	9,457	11,718	14,741	15,886	17,154
Confectionery products								
Candy and other confectionery products	2,285	2,358	2,335	2,524	3,074	2,830	3,031	3,156
Cocoa and chocolate products	601	655	724	814	1,075	1,324	1,452	1,601
Chewing gum	407	429	385	405	439	639	656	727
Total confectionery products ...	3,293	3,442	3,444	3,743	4,588	4,793	5,139	5,484
Processed foods								
Canned specialities	1,448	1,581	1,630	1,837	2,092	2,565	2,700	2,803
Canned fruits and vegetables; jams, jellies and preserves	3,379	3,569	3,923	4,480	5,286	5,784	6,002	6,242
Pickled fruits and vegetables, vegetable sauces, seasonings, and salad dressings	994	1,070	1,165	1,342	1,688	1,425	1,495	1,555
Frozen fruits, juices, and vegetables	1,151	1,308	1,649	1,824	2,220	2,145	2,334	2,575
Frozen specialities	1,477	1,484	1,742	2,107	2,373	2,610	2,871	3,221
Total processed foods	8,449	9,012	10,109	11,591	13,659	14,529	15,402	16,396
Dairy products								
Condensed and evaporated milk ...	1,475	1,591	1,706	1,916	2,315	1,979	2,158	2,343
Ice cream and other frozen desserts	1,343	1,389	1,520	1,639	1,788	1,502	1,671	1,857
Flavored milks	311	349	423	450	512	550	580	618
Total dairy products	3,129	3,329	3,649	4,005	4,615	4,031	4,409	4,818
Miscellaneous sweetened food preparations, not included elsewhere	1,677	1,815	2,030	2,353	3,626	4,092	4,477	4,951
Total sweetener-containing foods	23,473	25,603	27,857	30,076	38,206	42,186	45,313	48,803
Beverages and flavorings								
Bottled and canned soft drinks	4,178	4,323	4,807	4,961	5,901	7,290	7,829	8,337
Flavoring extracts and sirups	1,347	1,380	1,453	1,590	2,195	1,844	1,992	2,151
Total beverages and flavorings ...	5,525	5,703	6,260	6,551	8,096	9,134	9,821	10,488
Total sweetener-containing products	28,998	31,306	34,117	36,627	46,302	51,320	55,134	59,291
Animal feed industries								
Dog, cat, and other pet food	1,240	1,343	1,451	1,891	2,137	2,885	3,365	3,846
Other animals and fowl	3,907	4,325	4,658	6,334	7,134	10,001	10,631	11,384
Total animal feed	5,147	5,668	6,109	8,225	9,271	12,886	13,996	15,230
Total sweetener-containing foods, beverages, and animal feeds	34,145	36,974	40,226	44,852	55,573	64,206	69,130	75,521
GRAND TOTAL	37,702	40,729	44,234	49,515	65,012	73,355	76,539	82,713

[1] The value of shipments of sweetener and sweetener-containing products shown here was not adjusted for double counting.

Source: Bureau of Census, U.S. Dept. of Commerce, 1970-74. Projections estimates for 1975-77. Domestic and International Business Administration, U.S. Dept. of Commerce.

Table 3 [4]

profile of the industry we are talking about -- it manufactures over 59 billion dollars a year in products, which account for six million-plus tons a year of sucrose consumption and when utilized accounts for approximately 126 lbs. per capita consumption of sweeteners.

The use of sweeteners in the food industry is in a dynamic state. Since

1973 the use of corn sweeteners, particularly the fructose type, has been increasing rapidly; the use and role of sucrose in foods declining. Also the use of non-nutritive sweeteners has been steadily increasing since the 1960's. There are probably several reasons for these changes. One reason is the sudden increase in sucrose prices in 1973, coupled with the introduction of new higher-fructose corn syrups and the food industry's ability to blend these newly available sweeteners to optimize functionality and cost. A second reason could be the continuing increase in demand from portions of the consumer market for lower calorie food and beverage products. This demand comes from the estimated 50% of the adult United States population that is overweight and is continually seeking ways to reduce calorie consumption; from the more than ten million diabetics in the United States who would like to have nutritional products which are sufficiently sweet to be both orally acceptable and yet satisfactory for use in their restricted diets; and most recently the demand from various consumer groups to remove foods which contain a high percentage of sugar -- "junk snack foods" and non-nutritious foods--from the items served to children in federally supported school feeding programs.

To meet these changing consumer needs the food industry has used a combination of sweetener products such as cyclamates, saccharin, sorbitol, mannitol, xylitol, and high fructose corn syrups.

ROLES OF SUCROSE IN FOOD PRODUCTS

The task of using other items to reduce or replace the amount of sucrose in those products which normally contain a large amount of sucrose varies in difficulty with the type of product being made. To help understand the problem that faces a food scientist when he is asked to perform such a task, let us first take a look at the various roles sucrose can play as an ingredient in a food product.

Sweetening Agent Sweetness and the level of sweetness is very important to the acceptance of certain foods. Sucrose is still the most commonly used ingredient by the food industry to provide the sweet taste. Other factors such as pH, concentration, salt, temperature and other ingredients may also enter into determining the overall perception of sweetness in a food product.

Studies indicate that the age of the consumer significantly affects the preference for sweetness in foods, with the young and elderly preferring sweeter foods than those age groups in between[5].

Flavor Blender or Modifier Sucrose in some uses such as canned vegetables, fruits, mayonnaise, soups and catsup seems to act as a blender of flavor

notes and helps provide a well-rounded flavor profile for the product. In
acid foods such as pickles and some soft drinks, sucrose is sometimes used
to reduce the acid bite and sour taste.

Texture or Bodying Agent Many beverage products and canned fruits, where
sucrose is one of the major ingredients, rely on sucrose to provide a body or
"mouthfeel" to the product and thereby enhance its acceptance. In some baked
goods such as cakes, sucrose acts together with other ingredients as a
"tenderizer" and contributes to the texture or structure of that product.

Dispersing/Lubricating Agent Many dry mix products such as cakes, flavored
drink mixes, gelatin desserts, etc. utilize sucrose not only as a flavoring
agent but as a dispersing agent to keep other ingredients apart and thereby
provide more uniform mixtures. In cake or cookie batters sucrose acts as a
lubricant for other food ingredients in the mixture and aids in their mixing
and baking. The mouthfeel of many products can also be attributed to the
lubricating property of sucrose.

Caramelization and Color The caramelization of sucrose during a baking or
cooking process produces a brown color which contributes to the acceptance
of many baked goods. Flavor by-products are also produced during this caramelization process which contribute to the food product's characteristic
flavor and aroma.

Bulking Agent Compared to some sweetener compounds sucrose has a low sweetening strength to weight ratio. Therefore, in some applications where sucrose
is replaced with sweeter but less bulky compounds, other ingredients must
be added to replace the lost sucrose "bulk" in order to maintain the product's
normal appearance and consistency.

In summary, sucrose performs a variety of functions in foods in addition
to providing a sweet taste.

CHARACTERISTICS OF MARKET FOR SWEETENERS

To completely replace the approximate 95 lbs. of sucrose and 30 lbs. of
corn sweeteners per year that the American consumer presently uses would be
a large task and one I am not sure we would ever want to consider. However,
as the purpose of this discussion is to target such a replacement relative to
reducing the incidence of caries in the United States consumer, a more reasonable approach might be to identify that group of consumers most susceptible
to caries, such as children from the age of 8 to 19. Then, if we studied
the eating habits of this group, we would probably find a high and increasing
incidence of snacking, eating in between meals and a high consumption of
dessert-type foods, usually containing a high percent of sucrose or other
caloric sweeteners.

Snack Foods The trend of increased consumption of snack foods was noted in a 1970 convenience food study by the Arthur D. Little Company, where it was noted that "Many persons have only two formal meals a day, with the third or even a fourth meal being a continuous snacking event throughout the day."[6] Actually, snack foods are not new, since most snack items - crackers, candies, cookies, ice creams, soft drinks, etc. - have been around a long time. The big change seems to be in consumer attitudes toward snacks and the manufacturers' marketing strategies to meet the consumers' demand created by: prosperity, a more informal lifestyle, increased leisure time, plus time for more home entertainment. Foods included in this category usually contain substantial amounts of sucrose and are consumed as both snacks and desserts. Today, with increasing frequency, the dessert habit is considered by many an indulgence and something that, as much as possible, should be unlearned. Consumption patterns further indicate that the use of desserts or dessert-like products are more acceptable when used not as desserts, but rather as snacks. This trend was indicated in a 1969 Marketing and Economic Report on Frozen Desserts[7] where 4,000 households were surveyed about their usage of fourteen food items during the 1962 to 1963 and 1967 to 1968 periods. The results, as indicated in Table 4, indicate a shift from dessert to snack usage in most of the food product categories about which we are concerned.

There also seems to be a correlation between the consumption pattern of snacks and desserts and their caries potential. That is, that when consumed, neither food is followed immediately by other foods which would serve to scrub or remove remaining particles from the tooth surface. Since both snacks and desserts contain foods with substantial sucrose or other caloric sweeteners, both categories could contribute significantly to caries production.

YOUNGER POPULATION

For our next step we should look at the size of our target market. As our population expansion continues, we are at the same time becoming a younger nation. Today, nearly half of the nation's people fall into the under 27-year-old category. According to Department of Commerce figures, by 1982 we will have increased in number by 22%, with 47% of our population in the under 24-year-old category. A majority will be 14 to 24 years of age[8]. This trend toward a younger nation with more individuals in the age groups with a high caries susceptibility coupled with a changing lifestyle which favors an increased consumption of snack items, many containing substantial amounts of sucrose and other caloric sweeteners, combines to project a situation which may contribute to an increase in caries production.

Table 4. Index of Household Servings of Selected Food Product Categories[1]
Total Servings, Servings as Dessert, Servings as Snack - July, 1962-
June, 1963 and July, 1967 - June, 1968

Type of Food	Total No. of Servings (Index, 1962-63 = 100)		Total Dessert Servings (Index, 1962-63 = 100)		Total Snack Servings (Index 1962-63 = 100)	
	1967-68	% change	1967-68	% change	1967-68	% change
1. Juices, Drinks,Ades						
Ades	96.3	-3.7	182.4	82.4	100.3	0.3
Single noncitrus fruit	87.4	-12.6	107.7	7.7	96.2	-3.8
2. Soft Drinks	132.5	32.5	214.3	114.3	126.3	26.3
3. Cakes	90.2	-9.8	67.2	-32.8	170.4	70.4
4. Cookies	88.1	-11.9	57.4	-42.6	139.9	39.9
5. Pies	98.5	-1.5	79.7	-20.3	124.0	24.0
6. Other Baked Desserts	99.1	-0.9	75.5	-24.5	174.8	74.8
7. Gelatin	116.1	16.1	104.9	4.9	154.1	54.1
8. Pudding Desserts	97.5	-2.5	92.2	-7.8	142.6	42.6
9. Ice Cream & Related Prod.	93.1	-6.9	82.7	-17.3	103.0	3.0
10. Fruit	89.1	-10.9	61.5	-38.5	156.1	56.1
11. Quick Bread,Toasted Prod.	133.6	33.6	84.0	-16.0	135.8	35.8
Coffee Cake	89.9	-10.1	75.9	-24.1	112.8	12.8
Donuts	73.4	-26.6	57.7	-42.3	103.6	3.6
12. Snacks,Curls,Chips,Nuts	128.1	28.1	101.0	1.0	163.2	63.2
13. Candy (Chocolate)	129.2	29.2	62.8	-37.2	146.5	46.5
14. Other Candy	141.0	41.0	91.7	-8.3	148.0	48.0

[1]Selected product categories are those that have at least 20 percent of their
total servings either as a dessert or as a snack serving.

Thus, if possible, we might want to go one step further in defining our
replacement target and classify those food products which are most frequently
consumed by this age group, which due to their composition, structure or other
characteristics appear to be a prime substrate for caries production. Now we
have a more manageable and defined target from a product and consumer use
standpoint to discuss the potential of "non-cariogenic sweeteners" in food
systems.

POSSIBILITIES FOR INCREASED USE OF NON-CARIOGENIC SUBSTANCES

When considering the technical feasibility of manufacturing products
with a majority of the sucrose and/or other simple carbohydrates replaced,
such commercial products already exist for dietetic reasons in the soft drink,
jams, jellies and frozen dessert areas. A no-sucrose bread product has also
been developed recently but has yet to go into large scale commercial production and use.

Pies, cakes, cookies and confectionery items appear to be food categories where successful low sucrose commercial products have yet to appear.

It appears technically feasible to produce consumer-acceptable pies and
cookies without sucrose or other simple carbohydrates. Using the variety of
gums, stabilizers, modified starches and flavoring systems that a food tech-
nologist currently has available to provide body, viscosity and texture,
coupled with polyols and/or a non-nutritive sweetener to provide the sweet
taste, the feasibility to produce such items appears good. Technically the
most difficult items to produce at present appear to be the cake-type products and confection items. In both of these types of product, sucrose represents a substantial portion of the total composition and functions as a
structural and/or texture agent as well as a sweetener. It appears technically feasible to produce products in these categories using non-nutritive sweeteners; however, chances are that their sensory properties (appearance, texture and taste) would be noticeably different from their current commercial
counterparts. Therefore, a major portion of their consumer acceptability and
use would rely upon establishing that a caries benefit does exist and commu-
nicating such a benefit to the consumer in order that products of different
nature would become acceptable.

That leads us to the next area of consideration when determining a commercial potential for sucrose replacers for caries reduction purposes, that
of consumer acceptance. In products such as are being considered in this
discussion, the most important benefit or claim for the products to provide
would be caries reduction. Reduced calories and benefits to diabetics may
also be obtained from some of the products.

A caries reduction benefit from the use of such products is one of the
major areas yet to be substantiated. Questions yet to be answered include
areas such as "What are the quantities of cariogenic sweeteners that must be
replaced in the diet to achieve significant caries reduction?" "What is the
relationship of a food's total composition, form and texture to its cariogenicity?" and "Are there other nutritional effects to be considered if a substantial replacement of sucrose and simple carbohydrates in the diet is made?"

The question of whether an extensive dietary change is the most cost-
effective and acceptable method of controlling caries can best be addressed
once such an effect is established and answers to questions such as those
previously mentioned are obtained. Once this information is established and
accepted by consumer and regulatory groups, then assessing the commercial
potential of foods made with non-cariogenic sweeteners becomes one with which
the food industry is familiar: that of designing products with a balance of
sensory acceptability, cost and benefit to achieve consumer acceptability.

Studies which will provide answers to questions such as those discussed are complex, require a broad range of capabilities and can probably best be accomplished by a consortium of university, industry and government working together. Once benefits from model product systems or specified diets have been established, these benefits made available to and accepted by the public, then industry could develop and produce the products to meet those requirements specified by the model system.

REFERENCES

1. Junk, W.R. and Pancoast, H.M. Handbook of Sugars, The Avi Publishing Company, Inc., Westport, CT, pp. 1-6, 1973.
2. USDA, Sugar and Sweetener Report 2, (5),27, 1977.
3. Ibid., 2,(5) 31, 1977.
4. Ibid, 2,(5) 35, 1977.
5. Simone, et al., Food Technology 10, 279-282, 1956.
6. Anon., Snack Food 59(12), 39-41, 1970.
7. Anon., An Economic & Marketing Report on Frozen Desserts, International Association of Ice Cream Manufacturers, October, 1969.
8. Anon., Snack Food 59 (12), op. cit.

DISCUSSION AFTER MR. HOSKINS' PAPER

DR. MOSKOWITZ: In talking about the nature of product acceptability, Bill, you mentioned that special products should be targeted to a certain group, e.e., the 8 to 19-year-olds, who may have limited incomes. Typically these individuals are going to be faced with a product that is probably not one that they had before. Also we know that teenagers and young kids are finicky about foods to the point of occasional irrationality if it has an off taste. Finally, their food habits are very hard for someone else to change.

We are almost faced with a dilemma. In order to give this age group the health benefits that they need, we have got to give them a more expensive and possibly a different tasting food. But their own behavior, economically as well as in terms of eating patterns, will militate against acceptance. Is there a way around that or are we stuck with that dilemma?

MR. HOSKINS: You are right, it's a tough task, Howard. The approach that I proposed that might make it a little bit more feasible is to determine if a caries benefit could be obtained from substituting a small number or type of products in their diets. If you could demonstrate that there is a benefit with regard to dental caries by providing these kinds of foods to that particular market, then you might influence the prime purchaser of those items who is probably a parent providing the income to buy the non-cariogenic foods.

And hopefully, although NIDR hasn't emphasized public information, I think there would have to be some overall nutritional education coming down to the people in that age group to make them understand that proper nutrition and use of food is going to benefit them.

DR. MOSKOWITZ: One corollary to what you said became clear when the announcement of the saccharin ban was made. We found a weak tradeoff between taste benefits and health benefits. It seems that the health benefits for a large number of people do not outweigh the sensory benefits. This fact raises serious problems in changing food habits.

MR. HOSKINS: Yes, you are correct. The food industry has seen similar responses in other areas, such as nutritional labeling. To obtain a health related change in eating habits will be a long-term process and will require good and consistent education, starting at an early age.

MR. CHOATE: I must rise to debate Mr. Hoskins, wherein he says that the replacement of some of the sucrose by corn sweeteners would be a large task and one I'm not sure we would ever want to consider.

I also would point out to the Chairman of this meeting, Dr. Navia, that his statement that we are looking at all angles of the sweetness problem is not true. It would probably be wise for this audience to reflect upon the fact that dozens of meetings are going on all over the country today on what can be done about the seeming obsession of the American public with sweet products.

You here have been addressing the cavity issue. Other people in the room are interested in the diabetic problems of today's youth and adults. There has been some reference to the problem of obesity. There has been relatively little mention about the behavior modification that is taking place daily both through the printed page and through television and radio commercials urging us to eat sweets.

Sweetness is being advocated a tremendous number of times per year to the young public. Our studies indicate that a moderate television watching child is exposed to between 8,500 and 13,000 food and beverage commercials each year. I do not know what percent of those is for a high sucrose product. I would be very interested in the study Mr. Hoskins mentioned to find out what is the relative sucrose content of the products that are seemingly being used more and more as snacks.

We do need to look into this area. I think the problem is not only what is the degree of sweetness in the products that are being consumed by the public, but what is the degree of sweetness in the messages that now are being broadcast to the American audience and are starting to be addressed to an international audience by many of these same food companies.

The Federal Trade Commission has paid some attention to food advertising in the last two years. You probably know that there are some 8,000 pages of transcript and maybe another 5,000 pages of records of the 1976-1977 food hearings; you probably know that there will be another phase of those food hearings starting in 1978. Some of you may have heard the rumor that the Federal Trade Commission will be addressing the problem of sweetness promotion particularly to children in rule making hearings some time in the next six months.

Some of you may realize that the National Science Foundation has just concluded a lengthy study of what is known about the communication to children of food messages and what impact it has on their diet. I bring this to your attention only to point out that the discussion here has been confined to "So we have a great many sweet products; how do we handle the cavities and perhaps the diabetes and obesity that those sweet products bring to the American diet?"

Not much attention has been paid in this forum to changing the attitudes towards sweetness in the American public; a number of people are generating meetings just like this in other disciplines trying to bring about a change in behavior rather than trying to change the product.

One additional point I think would be an interesting interaction with

Dr. Hoskins' last table. Let me read to you the sequence of products which received the highest television advertising budgets in the country in 1975. These are network figures for the first nine months, not for the twelve months, so you can multiply it by four-thirds to get the approximate total. Local spot commercials are not included.

The most advertised food product category in the United States over television is cereals, which in 1975 in nine months had 8,166 separate commercials. The second most advertised food product category is candies and gums, which had 4,083 in nine months. Third is shortening and oils, fourth is cookies and crackers, fifth is desserts, sixth is non-carbonated soft drinks, seventh is carbonated soft drinks, eighth is meats and poultry, ninth is macaroni and spaghetti, and way down the list are vegetables, citrus and cheese.

I think this supports the table that you showed up there as to the shift occurring toward certain types of products. The shift reflects where the food companies are putting their advertising.

DR. CAGAN: If you wish to initiate a program of reeducation, such as you are suggesting, then you must do it in view of the biological realities. For example, it is now well established by a number of studies that newborn human infants prefer the sweet taste of sugars.

MR. CHOATE: I understand that.

DR. CAGAN: If advertising is to be made a chief component of a program, then it must be done with the realization that advertising may not be establishing a new pattern, but simply playing on a biological drive that is already present and reinforcing it. This means a very different kind of approach is needed than simply saying if you stop the advertising, then consumption of sweets will stop.

MR. CHOATE: Accepted. I think advertising is a reinforcing mechanism; I think it brings about a great deal of peer pressure on those who do not see a television set. I think also that one would be foolish to say that one could not advertise sweet products over the air. That's not the private enterprise way.

I do think we might give some consideration to the need for either including in advertising messages some information about the caries-producing danger of certain products. Or we might adopt a policy where more public service announcements are made on that same medium and educate the public about the very matters were are talking about here.

DR. CAGAN: It is unfortunate that nutritionally well planned meals, such as in the school lunch programs, are not eaten consistently by the children. If the flavor is unappealing, then the nutritional value is zero, because the food is not eaten. Flavor (taste, smell, texture, etc.) should be made a component of good nutrition.

MR. CHOATE: True, but good taste doesn't have to be sweet.

DR. CAGAN: Taste can be used in the service of good nutrition rather than to regard good taste as unworthy of good nutrition.

MR. CHOATE: Accepted.

DR. ARVIDSON: I have a question about your TV programs. Don't you have advertisements about brushing your teeth to remove all the plaque around the teeth? They are required regularly in Sweden.

MR. CHOATE: Public service announcements are what those types of 30 and 60 second messages are called in the United States. They have a very weak standing on our airwaves today. Every three years a broadcaster licensee must apply for a new license and therein promise how much time that licensee will devote to such things as public service announcements. The only standard by which the broadcaster is judged three years later, when the time for renewal comes up, is, "Did they live up to their previous promise?" A broadcaster can promise two messages a day or 200 public service messages a day and that is the only standard by which they are judged.

It is interesting that in Holland a little logo of a toothbrush with some toothpaste coming on it appears, I believe, for 1-1/2 seconds at the end of every candy advertisement, sweet beverage advertisement, and I believe some sugared product advertisements. That had some impact in the beginning. In our conversation of the other day, our guest from Holland said it had now been on the air about two years and was starting to be disregarded because there weren't any public service announcements explaining the logo to the public.

I think it is possible in this country to increase the number of public service announcements relating to dental health. We presented a petition to the American Dental Association to that effect in early October.

DR. BRIN: I'd like to make two comments. One: Mr. Choate is to be commended for stimulating the improvement of the nutritional quality of the cereals which are advertised. Perhaps it is appropriate that these cereals get a major portion of the advertising. If children consume half a glass of milk with those cereals, this would comprise a good breakfast. The school lunch program is holding open hearings all next week in all parts of this country to get input from parents, teachers and others to determine how the school lunch or school breakfast program can be modified to be more nutritious. Secondly, as a consequence of the concept of empty calories, so-called, such as in cookies and crackers, the Food and Nutrition Board has made recommendations for the broadened vitamin and mineral enrichment of cereal grains, wheat, corn and rice at the mill rather than at the bakery.

DR. MOSKOWITZ: I'd like to address myself to Mr. Choate's statement about advertising and the implicit message of sweeteners. If you look at some of the advertising approaches, they embody what David Ogilvy called a unique selling proposition (USP). The USP is what differentiates one product from another. In the case of advertising to children, if you look at children's behavior--what they recall, what they perceive, why they buy a cereal or any other product--you will probably find that young children do not usually turn on to flavor per se, but rather to "Tony the Tiger" or to some picture of the individual on a box or to the reward inside.

This USP generates awareness of the product and purchase interest, so that a child nudges his mother to buy it. What maintains the child's interest is the flavor of the product and sweetness. But this is not the initial or even overall critical motivating force for purchasing it or for sparking the child's interest in it.

I think what has happened in the food industry is that the advertising agencies have realized you can't advertise flavor or sweetness per se. These sensory perceptions lead to repeat use and eventually support product use. Rather, an image has to be advertised. The same thing applies to adult advertisements for cigarettes. When they talk about flavor, it's a very nebulous kind of attribute. They have to talk about something that can be pictorialized. In that way they can interest the person to buy. Afterwards it's taste and flavor (in cereals it's also the sweetness) that continues to maintain the interest.

So you have to differentiate what the advertising message is conveying and what's really helping maintain the product.

DR. BIBBY: I want to turn the discussion back to the sucrose question. The point should be made that it's a mistake to think there is a direct relationship between the amount of sucrose in the food and its caries causing potential. There are many modifiers of the cariogenicity of a food or confectionery item besides the sugar content. I think that should be on the record. This has been shown in vitro in animal studies and findings in man.

DR. HERBERT: I gather that sticky sugar is cariogenic. Are high sucrose cereals cariogenic?

DR. BIBBY: I know of no evidence to show that high sugar cereals are uniformly more cariogenic than low sugar cereals. There's no definitive study at the clinical level to show that high sugar cereals cause more caries than low sugar cereals. There is a study by Dr. Glass, with which I am not overly impressed, that failed to find such evidence. Neither has it been shown in rats that all high sugar cereals cause more caries than all low sugar cereals. We have one experiment in which we find that a high sugar cereal containing 40% sucrose caused less caries than one containing 8%. Further, neither measurements of enamel demineralization nor the speed of food removal from the mouth are adversely affected by high sucrose content. Indeed, the reverse seems to be the case.

I'm not saying high sugar cereals are good, but I do want to emphasize that we do not have the evidence to condemn them. I dislike them, I think they are bad on many counts, but if we are trying to make a decision on a scientific rather than on an emotional basis, then we have to face the facts as we know them today.

DR. NEWBRUN: I have to respond to some comments of Dr. Bibby's. First of all, he mentioned the human studies which have been extensively publicized by some of the cereal manufacturers as evidence that sugar containing cereals are harmless in terms of cariogenicity. We have to realize in this conference that we cannot consider one food item alone because we are omnivorous. In those cereal studies conducted by Glass at Forsyth (J. ADA 88:807, 1974) and by Rowe at Michigan (J. Dent. Res. 53:33, 1974) they varied only one food product. We are consuming about 96 pounds of sugar a year in many food items. If you vary one item, the background noise is so high that you cannot expect to detect any difference.

It is most unfortunate that the cereal industry has used the results of those studies and misinterpreted them to say that sweetened cereals are safe. That claim cannot be made for obvious reasons.

In the cereal study that was done at the Eastman Dental Center (Choung et al. J. Dent. Res. 52:504, 1973), the results also were inconclusive because of the high mortality of those rats, which were fed the cereals as their only source of nutrients. One could design a study with adequate supplements and demonstrate that any sucrose sweetened food fed rats will cause caries. The main point, that has to be recognized in this conference, is that it is futile for the food industry, and similarly for the consumer representatives, to target on only one food item. We have to look at the whole spectrum of foods.

There is a hierarchy of cariogenic foods. We can attempt to sweeten those foods high on the list (sticky and snack foods) with non-cariogenic sweetening agents. There will not be one food or one sweetener that will serve or will solve all those problems.

DR. MACKAY: Dr. Cantor pointed out yesterday that the roots of the revolution (rebellion might have been the better word) 200 years ago lay in the creation of the Caribbean sugar cane industry. This is another way of saying the food industry has had 200 years in which to find out things to do with sucrose.

In addition to the six things Mr. Hoskins mentioned, I could probably give you another 20 reasons for using sucrose. Sucrose is used not just because it's sweet. It's a monumental task to consider how you can--this would be a true revolution--do away with a major ingredient which has so many uses in the food industry. Some of these are trivial, some of them are highly specialized. When you have had this substance around for this length of time (and it's been so important as a preservative), it's extremely difficult to replace it. From the point of view of national health, I can't think of anything else you could use to replace this enormous tonnage and not then be concerned about what might happen with the substitutes. This is a very difficult problem indeed.

Mr. Choate does make some very valuable points, for which I'd like to commend him. I'd also like to suggest that there is a difference between a stick and a carrot. I don't think the food industry is particularly trying to increase the consumption of a particular category of foodstuff, though individually a manufacturer may be trying to increase consumption of his product within his category. TV has been with us for a few years out of the 200. Caries and the tremendous sugar consumption in the United Kingdom occurred long before commercial television ever came along. Since Mr. Choate does agree that private enterprise and capitalism should depend more on the carrot than the stick, if only there were ways in which manufacturers could be permitted to call attention to the fact that they are trying to do something about the problem.

As I said at the ADA meeting in Chicago about a month ago, I don't know any way in which a manufacturer of a product can call attention to the dental benefits of a product with less sugar without triggering a whole set of requirements. Dr. Gilkes pointed out yesterday that the word "non-cariogenic" would require substantiation by FDA. This is going to cost half a million dollars probably for a couple of clinical trials.

There's got to be some way in which the manufacturer can do what the dental profession wants, and perhaps what the medical profession wants. Somehow you have to ease the way, by cooperating with the industry rather than belaboring it with a stick and saying in effect go out of business.

MR. CHOATE: I'd like to respond with delight to this suggestion that industry help us to convey more information to the public about the products that are being so highly touted. Those of us who have studied mass communication of food information are quite aware that one individual advertisement for Life Savers or for M&M candies or for a sweet cereal containing 58% sucrose in and of itself is not going to ruin the nation's teeth. But you have the en masse impact of food advertising. When you have children seeing 8,500 to 13,000 food ads a year, the majority of which mention sweetness or sweet characteristics, you start to have a force at work on their imagination about what they should be eating that transcends any single product message. We do need the cooperation of the food industry as well as of the medical profession in convincing the conveyors of all of these messages, which are generally the networks, that they have a responsibility to carry many, many more messages that awaken the American public to the nutritional content of the foods that happen to get a large advertising budget, as well as to the dental implications of their heavy consumption.

DR. NAVIA: I want to emphasize here something that is very important. The food industry researchers and consumer advocates have to get together to do one thing that is very important, but very difficult, and that is to isolate and define the degree of cariogenicity of different foods. We don't have enough information on this one critical aspect. We really have to know whether food snacks such as Twinkies or Winkies or whatever are cariogenic or not, and to what extent they are, and whether we can modify them in some way to make them less caries-promoting.

The problem is not of one product, like Dr. Newbrun was saying. The problem is a large number of different products about which we need to obtain more information. Just a few weeks ago, at the meeting held under the auspices of ADA and NIDR there was a beginning of a concerted effort from industry, universities and others to try to look together into the problem, which I think is critical.

Dr. Bibby's remarks are directed to a problem that should be clearly recognized, and that is that there are modifiers of cariogenicity, and that just the presence of a sugar in general does not say anything absolute about the caries-promoting properties of a food. The other factors have to be recognized, evaluated and understood better.

DR. HERBERT: On the precedent that we have added a halogen, iodide, to a major food additive, salt, in the processing in order to reduce goiter, is it reasonable to add the halide fluoride in the processing of sugar to reduce cavities?

DR. NIZEL: We aren't interested in adding a halogen, fluoride, to sugar, but we have suggested the addition of a phosphate to foods that have a concentration of sugar which imparts an overt sweet taste like cookies, cakes, etc. The proof that phosphate additives can exert a significant anticaries effect has been clearly demonstrated in the experimental animal. The real problem is the funding of the type and numbers of clinical studies required by the F.D.A. before they give their approval that a dietary supplement is GRAS and that a health claim can be made for this additive. In addition, we have the problem of not being able to get the approval of university human investigation committees for a purposefully produced sugar control. So we have all of these kinds of problems.

DR. SHAW: I'd like to comment on Dr. Herbert's question. Fluoride has its major benefit during the development and the maturation of the tooth so that fluoride is in the crystal structure of the tooth at every level of the enamel and dentin. Fluoride works through modification of the host. We have no evidence to suggest that sucrose supplementation by a fluoride salt would result in the modification of the oral environment sufficiently to reduce caries. In addition, I would be concerned about the safety of fluoridation of sugar because of the variability in sugar use from one individual to another.

SESSION VI.

ASSESSMENTS AND RECOMMENDATIONS FOR THE
FUTURE

Moderator:

William E. Rogers

National Caries Program
National Institute of Dental Research
Westwood Building, Room 549
Bethesda, Maryland 20014

DESCRIPTION OF RESPONSIBILITIES OF THE FOUR TASK FORCES

Your responsibilities in the next 24 hours are going to be the hardest and yet the most challenging and interesting. During your discussions and the preparation of the task force reports, we are asking for complete freedom of thinking, lack of bias and prejudice in a cooperative group endeavor and venture, to help the National Caries Program and the National Institutes of Health to plan wisely for the future in this area of Sweeteners and Dental Caries.

I would like to amplify what you will be doing for us and to establish in your mind how important this task is. About five years ago the National Caries Program of the National Institute of Dental Research was established with the charge to develop the means, through preventive techniques, to eliminate dental caries as close to 100 percent as we possibly could. Of course, when you are told to carry out a research and development activity, you accept the responsibility to set objectives, to plan a course of action, and indeed to identify and implement specific projects. Priorities are set, currently in the context of expending approximately $10 million of your tax money per year. We would be derelict and careless in carrying out this responsibility if we did not ask for and try to attain the greatest degree of competence, the greatest inputs of information and wisdom in the decisions that we have to make, from the scientists and public interest groups in the United States and the world at large.

In addition to that philosophical statement, indeed we are required by regulation as implemented through the policy of the National Institutes of Health to review with peer review groups, such as this group, all projects which we are going to carry on and which we are going to solicit from the public. Earlier this policy did not extend into the grants area in that they were freely submitted to the National Institutes of Health and reviewed by peer groups and funded on the basis of scientific and technical matters. However, in the areas of collaborative research carried on through contracts and indeed extending into the area of grants at the present time, when we request research in particularly broad areas and sometimes allocate money to support those activities ahead of time, it is an NIH policy that the broad concepts of these projects be reviewed and approved by peer groups. The group at this conference is as beautiful a peer group for the evaluation of the sweetener area as one could possibly bring together.

In practice we have utilized the technique of workshops or symposia over the last two years to obtain these peer reviews and to obtain input from scientists and experts in the field and from people who know the public needs. In addition, through use of symposia and workshops we not only obtained these concepts for our immediate needs, but we have distributed the information that was developed from an integrated coordinated group to a wide public. In other words, this information has served not only our needs as we might utilize it with regard to issuing requests for proposals (RFP's) for contracts and other purposes within the National Caries Program. However, in addition, this information has been helpful in general purposes of advancement of science at the National Institutes of Health, and even in projects that might not be supported at the National Institutes of Health, but by such other agencies as the National Science Foundation.

You have an opportunity in the next 24 hours to generate ideas that will be useful to yourselves in terms of support of collaborative activities, to generate suggestions that are going to be specifically useful to the National Caries Program, and to help science move ahead in the sweetener area by integrating your best ideas and practical insights into solution of the problems that have been discussed these last two days. Even if you identify worthy projects that are a little removed from caries, but you think are highly important to conduct, here is an opportunity to present them in print to a wide readership in science where they might be put into practice by somebody else.

Our speakers have pinpointed all sorts of questions and projects that might be worked on profitably. I have been jotting down some of these projects and questions. I am going to mention them to refresh your minds, and trigger off thoughts again about things which have been said. I am sure that you will be bringing these and others up again in your discussion groups.

To study the binding of sweeteners and taste modifiers to the taste bud. To study the material in the taste bud that is bound by the taste modifier or the sweetener. What differences in response to sweeteners are there among such species as the rat, the dog, the monkey, and the human? What happens in utero in the ultimate development of the taste bud in terms of its functioning? Can these functions be conditioned either in utero or neonatally? What investigation needs to be done on aversion stimuli? Can these be used in an applied sense for aversion to particular types of sweeteners, perhaps? Can taste preference be modified by experience?

Is the consumption of sweets and other foods related to energy balance? Do people have a taste sensitivity early in life that predicts or predicates whether they are going to be obese later in life?

What is the maximum wholesale price for sweeteners if they are to be competitive with sucrose? What are the sources of supply of other sweeteners than the sugars? What economic studies need to be carried out about other sweeteners? Are there studies that could be done to change the current philosophy of the Food and Drug Administration that non-nutritive sweeteners cannot replace more than a small share of the total calories?

Would it be better for industry to continue looking for a new sweetener that has none of the ascribed defects of cyclamate, aspartame, or saccharin than put more dollars on the ones which have ascribed deficiencies or defects? Since sucrose is used in many ways because it is such a good bulking agent and is inexpensive, should we be looking for cheaper non-nutritive bulking agents?

Are studies needed on the effects of sucrose and other sugars on the development of teeth and enamel resistance? We know that there are effects; yet practically no one is looking at them. At what levels of carbohydrates do some of these developmental effects start? Using radiotelemetry could we do a vast number of studies on the effects of alternative sugars and sugar alcohols on pH changes in the mouth? Would these studies be useful in lieu of clinical trials?

We need studies on the absorption of polyols from the intestine and the adaptation process in the human. Is adaptation related to the microorganisms that live in the gut? Also, we need to study whether oral microorganisms have the capability to adapt to and metabolize polyols. Clinical studies need to be devised with suitable controls to assess the possible cariostatic effects of polyols. What other xylitol studies need to be carried out to corroborate or to test the hypotheses and results which have been achieved so far? What are some of the acceptable protocols which would be used in test situations for measuring reversal of incipient caries? What is the normal level of reversal in society?

What protocols can be developed for acceptable animal and clinical trials to evaluate the cariogenicity of sucrose and other specific sugars, of various foods and snack items as well as the possible cariostatic effects of various kinds of food additives? Perhaps you can make suggestions of how such evaluations can be done in foreign countries, where sugar substitutes are still utilized legally.

What toxicity, metabolism, and mutagenicity studies might the National Caries Program support that will be useful to industry? Is there research that can be carried out that demonstrates the value of non-nutritive sweet-

eners in the control of diabetes and obesity? Obviously many studies are
needed on the effects of different sugars and sugar alcohols on the metabolism and phenotype of various cells, both mammalian and microbial. We
also need to know more about the metabolism of polyols by these cells. With
regard to this, perhaps more studies are needed on some of these "vitamin
sparing" effects which we have seen of the polyols.

EDITORS' EXPLANATORY NOTE

The report of each task force is preceded by the charge and the list
of participants who took part in the discussion. Each report is recorded
on subsequent pages in the form produced as the result of the discussions
among the members of an individual task force. As such, each report represents the majority opinion of the members of that task force and not
necessarily the opinion of every member of the group. Each report was
read by the Moderator of that task force at the final plenary session of
all participants; numerous items of the four reports evoked extensive and
often heated discussion. In view of the size of the whole group, the nature of some recommendations, and the limited amount of time available, it
was not possible to review every point raised during the discussion or to
arrive at a list of recommendations that could be supported by all participants. As a result, some redundancy is present from one task force
to another and numerous participants of the conference will take exception to one or more recommendations.

REPORT OF TASK FORCE 1

<u>Charge</u>: Assess the current status of the numerous sweet compounds that are
available at various stages of development and recommend the research that
needs to be done to evaluate their potential usefulness and their relationship to human health including their cariogenicity.

> Moderator:L. M. Beidler
> Recorder: P. A. Swango
> Members: B. G. Bibby, J. G. Brand, G. Ev. Demetrakopoulos,
> L. K. Hiller, R. M. Horowitz, M. R. Kare, K. K. Mäkinen,
> P. P. Noznick, S. Tarka, and S. Weiss

1. The members of this task force encourage the development of new sweeteners and studies of their effect on human health.
2. Studies of the potential health benefits of sweeteners should include
   assessments of their effect on diabetes and obesity as well as on dental caries.
3. The effect of sweeteners on growth and development should be studied.
4. Evaluations of the safety of sugars and sweeteners should include assessments of their influence on learning behavior and related characteristics.
5. Studies of sweeteners and sugars should include assessments of risk-
   benefit ratios.
6. Studies on the physiological and psychological effects of the prolonged
   use of non-caloric sweeteners need to include evaluations of the influence of their use on the intake and distribution of other dietary components and on the consumption of calories.
7. Is there a role of sweetness in the functioning of the digestive system?
8. The possibility of immunogenic effects of high molecular weight sweeteners should be examined.
9. Origin of the preference for sweetness in both pre- and post-natal life
   needs to be studied with evaluation of the ways to modify it as well as
   changes in sweet preference that occur with age.
10. The possibility of using non-caloric sweeteners to enhance the accept-
    ability of nutritional supplements for undernourished populations needs
    to be assessed.
11. The effect of sweeteners on gastrointestinal tissues and endocrine

and exocrine functions needs to be determined in comparison to the effects of sucrose.

12. Investigations of the toxicology of saccharin should be continued including evaluation of the possibility that impurities are responsible for the adverse effects that have been reported.

13. The toxicology of glycyrrhizin from the standpoint of electrolyte balance, corticoid activity, and estrogenic properties should be clarified.

14. Reports of testicular atrophy in rats and dogs following the use of neohesperidine dihydrochalcone need to be evaluated further.

15. The toxicology of monellin, stevioside, dihydrochalcone analogs, and miraculin needs to be determined.

16. The status of sorbitol, mannitol, and xylitol, and possibly of lactitol and maltitol with regard to caries production needs to be evaluated further.

17. Microbial adaptation to polyols and their utilization by oral pathogens needs further study.

18. The effects of the polyols on oral mucosa with respect to penetration characteristics and ability to stimulate axon reflexes merit investigation.

19. The level of consumption of polyols resulting in osmotic diarrhea in humans, and the mechanism of adaptation need further study.

20. In replacement of sugars in foods, efforts should be concentrated on those foods that have the highest suspected cariogenic potential.

21. The possible cariogenicity of natural foods (fruits, etc.) that may be suggested as alternative snack foods needs to be determined.

22. In addition to replacing sucrose with other sweeteners, ways need to be sought for modifying sucrose and other sugars to reduce their cariogenicity.

23. Procedures are needed for reducing the sugar content of foods by incorporating taste-modifiers, particularly to reduce bitterness.

24. The receptor mechanism for the recognition of sweetness merits additional investigation.

REPORT OF TASK FORCE 2

Charge: What are the most fruitful avenues to pursue in the search for and evaluation of new, noncariogenic, commercially utilizable sweeteners?

Moderator: K. M. Beck
Recorder: H. van der Wel
Members: K. Arvidson, G.N. Bollenback, R. G. Bost, G. A. Crosby, J. R. Fordham, K. F. Gey, W. C. Griffin, M. Gumbmann, G. E. Inglett, D. A. M. Mackay, C. S. Nevin, E. Newbrun, P. A. Rossy, J. Soeldner, J. M. Talbot, and F. B. Zienty

The most fruitful avenues to pursue in the search for and the evaluation of new, non-cariogenic, commercially utilizable sweeteners are:

1. to consider that prevention of caries, control of obesity, and treatment of diabetes constitute separate needs for sweeteners.

2. to encourage the development of theories on how a sweetener interacts with the receptor and studies on the mechanism of sweet taste perception.

3. to provide a sound fundamental basis for designing new sweeteners, or modifying the structure of existing sweeteners, the biochemical basis of sweet and other taste sensations needs to be elucidated. This should include detailed studies of the interaction between sweet taste receptors and monellin, miraculin, and thaumatin as well as other sweet molecules as probes to aid in the isolation and characterization of taste receptor macromolecules. Such studies would provide a sound fundamental basis for rational structure–activity development of new sweeteners.

4. to find additional sweeteners

   (a) by preparation of new sweeteners based on a theoretical model for sweeteners, such as the dihydrochalcones, and preferably a sweetener that also has a bacteriostatic action for S. mutans.

   (b) by searching among botanicals for naturally occurring sweeteners that are metabolized with a capacity that does not cause any metabolic disturbance.

5.   to study sucrose-phosphate mixtures and develop commercial products of this type.

6.   to develop sweet tasting, non-fermentable products.

7.   to find a sweetener with a high intensity, fast impact as a flavoring agent for toothpaste.

8.   to provide funds for basic research on taste and sweeteners.

9.   to establish an independent test center to evaluate new products by various methods including preclinical testing in humans.

10. to develop acceptable tests for the evaluation of non-cariogenicity and anti-cariogenicity of products.

11. to establish a governmental committee to set standards for non-cariogenic and anti-cariogenic claims.

12. to develop a regulation to allow for non-cariogenic sweeteners to be used in cough syrups and other pharmaceuticals.

13. to monitor Congressional action on revision of the Food and Drug Act to insure that the development and approval of non-cariogenic products are readily possible.

14. to develop in consultation with appropriate members of the Food and Drug Administration and their consulting agencies acceptable procedures for the evaluation of the safety and effectiveness of products developed to be non-cariogenic or anti-cariogenic.

REPORT OF TASK FORCE 3

Charge: What currently-available sweeteners could be tested to determine their applicability for use in sucrose-free, processed foods and beverages? What would be the most appropriate investigations to carry out to
evaluate the usefulness in test situations of products sweetened by agents
other than cariogenic sugars?

> Moderator: W. A. Hoskins
> Recorder: R. H. Anderson
> Members: M. Brin, R. C. Cagan, D. Coursin, A. L. Coy-
> kendall, J. F. Emele, C. C. Gilkes, J. Rozanis, G. G.
> Roussos, and G. H. Schrotenboer

The members of the task force recommend the following:
1. that a suitable number of snack and/or dessert type foods
   believed to be highly cariogenic in their present forms
   need to be reformulated with non-cariogenic sweeteners
   and their cariogenicity determined in animal and clinical
   tests. Such food items would be selected on the basis of
   their total sugar content, retention properties in the
   oral cavity, frequency of consumption, etc. This list
   would include items such as: chewing gum; carbonated or
   non-carbonated beverages; hard candies; cookies; chocolate covered candy bars; caramels, toffees, and taffy
   type confections; pastries (pies, cakes, or other sweet
   baked goods); and frozen desserts.

   In this proposed study to investigate the feasibility and applicability of developing foods and beverages
   free of added sugars, it is recommended that all available non-cariogenic sweeteners and/or combinations of
   sweeteners whose functional properties are compatible
   with the food items being developed be utilized. Such
   sweeteners may include but should not be limited to:
   cyclamate, saccharin, dihydrochalcones, aspartame, xylitol, sorbitol, mannitol, monellin, thaumatin, stevioside, phyllodulcin, non-cariogenic "non-coupling" sugars, and miraculin.
2. Possibly the composition and texture of foods (fat, protein,

moisture, and carbohydrate) play a role in determining the cariogenicity of foods in addition to the influence of the sugars. Therefore it is recommended that a study be initiated to define the role of a food's total composition and texture such that the information resulting could provide a guideline to development of less cariogenic foods by the food industry.

3. To accelerate the development of less cariogenic foods, a simpler, less time-consuming, and less costly method or group of methods is needed to determine the cariogenicity of a food than the current clinical procedures. Six or more such procedures have been published but need to be validated for use on foods by a clinical study. Therefore, it is recommended that such laboratory evaluations be performed in conjunction with a clinical evaluation of a food or several foods' cariogenicity.

4. It is recommended that a study be initiated to define, develop, and evaluate the efficiency of non-conventional interventions such as gum, mouth rinses, chewable tablets, etc. Such items may include compounds such as phosphates, xylitol, bicarbonates, etc.

5. Consideration needs to be given to the overall physiological effect of changes in the food system as a result of reducing the cariogenicity of certain broad classes of foods. Consideration must be particularly given to such effects on special groups such as pregnant women and children.

6. The need for non-cariogenic sweeteners, particularly non-caloric types, is great. However, due to the very high risk/cost ratio, it is recommended that the National Institute of Dental Research and/or other branches of Health, Education and Welfare consider providing assistance in the development and evaluation of such agents.

7. It is recommended that current guidelines regarding the ethical limitations placed on selection of proper contols in double blind clinical studies be continually reviewed to develop procedures that are safe and yield appropriate evaluations of the products being tested.

REPORT OF TASK FORCE 4

Charge: Identify the social, legal, economic, educational, health, and regulatory considerations in the acceptance of products containing replacements
for sugar.

> Moderator: Juan M. Navia
> Recorder: Richard J. Jones
> Members: S. M. Cantor, J. P. Carlos, R. B. Choate, W. Cooley,
> V. Herbert, K. K. Krueger, H. Moskowitz, A. E. Nizel, E. Rat-
> ner, W. E. Rogers, C. D. Stone, B. J. Walter, and J. M. Weif-
> fenbach

The members of the task force considered that sugar included sucrose,
fructose, glucose, and other readily fermentable caries-promoting carbohydrates, and that substitutes included non-nutritive sweeteners, and such
nutritive, but non-cariogenic agents, as sorbitol, xylitol, and aspartame.
The members of the task force concluded that further research in the areas
of social, educational, and regulatory issues would benefit from research
that makes available accurate estimates of food consumption and dietary
patterns at the individual level, rather than having to depend upon commercial sources that reflect national averages based on disappearance from
the marketplace. They anticipated that the method of obtaining 24-hour
dietary recall as now practiced by the Division of Economic Research of the
U.S.D.A. will contribute to this end. It was also suggested that, in such
individual surveys, information would be desirable regarding the incidence
of caries among the people for whom consumption data were collected.

The members of the task force decided to pursue their charge by stating and then answering questions regarding the need for, the economic impact of, the educational initiatives for, and the safety considerations and
regulatory issues that relate to the use of sugar substitutes. These substitutes might be used in any food, but would perhaps have greater impact
on caries, and their impact would be easier to assess, when used in chewing
gum, soft drinks, and between-meal snacks. The increasing trend toward the
use of more snack foods in place of the traditional table foods at meal
time in the United States was noted. Questions regarding the anecdotal importance of sugar in the etiology of heart disease, obesity, and diabetes
mellitus were, for the most part, not debated.

In considering the need for sugar substitutes, the members of the task force observed that a non-nutritive sweetener, saccharin, has been in wide-spread use in this country since W. W. II, and that Congress was moved to extraordinary measures at the threat of its complete removal. The "need" in terms of reducing morbidity, and loss of work time due to caries was agreed, but there was no agreement that this could offset or compensate for the possible risk of subsequent bladder cancer as suggested by the Canadian saccharin study in rats. The perceived need for sweetness in chewing gum or soft drinks, for example, is probably based on acculturation: people could certainly survive as well, and probably with as well balanced psyches, without sugar, much less its substitutes. Nevertheless, our society has come to consider confections, desserts, and soft drinks to be just rewards for withstanding life's privations and stresses.

The economic impact of partial replacement of sugars in soft drinks, chewing gum, and snacks would be minor compared to many economic disloca-tions seen in many other business endeavors. The economy of sugar produc-ing areas might suffer temporarily, if sugars were to be totally replaced. Such total replacement seems unlikely. On the other hand, the added cost of certain substitutes for sugars must be considered as a burden upon the consumer and might cause economic hardship for some segments of the popul-ation. For example, saccharin has been much cheaper than the sugars it has replaced in many products in recent years, but the addition of xylitol to chewing gum at present cost would perhaps handicap it in the marketplace.

The educational initiatives required by the widespread introduction of substitutes for sugars should not be limited to commercial product promo-tion, but should include the efforts of the dentist, the dental hygienist, and the nutritionist, with perhaps advanced new teaching techniques employ-ing group dynamics and behavioral modification, as well as the traditional schoolroom instruction in dental hygiene. Any campaign undertaken on be-half of a substitute for sugars should be more educational and information-al than the usual promotion. However, if the substituted product lives up to its promise, it will promote itself. The different attributes of sugar and its substitutes should be stressed. Such a campaign should be modeled after the multi-media techniques used in Stanford's three-township study on the reduction of coronary risk factors and should benefit from the les-sons learned there. This study demonstrated that proper advertising mes-sages can be used to reach the citizenry and effect a measurable behavioral change. An even greater change is, of course, effected by group and indi-vidual instruction of people at high risk. In considering claims that such an advertiser might make regarding a chewing gum free of sugars, the members of the task force recommended that manufacturers should be allowed to promote gum substituting non-nutritive sweeteners for sugars as being non-cariogenic.

There was considerable discussion over the safety issue. It was pointed out that we are far from knowing all the consequences of completely replacing sugars in all foods. This consideration is an element in the present uncertainty about the safety of saccharin. Hence, it was suggested that limiting widespread sugar substitution to soft drinks, chewing gum, and snacks should have minimal impact on and could best be advocated as a part of improving the overall balance of daily nutrients. It was realized that we do not yet know the ultimate effect upon even caloric intake, much less distribution of nutrients, that is to be expected from full substitution of a non-nutritive sweetener for sugars in the normal individual. For the purposes of this discussion, sugar containing "food snacks" are defined as those foods (1) normally consumed apart from breakfast, lunch, or dinner, (2) designed for consumption without preparation, or (3) promoted for consumption at other occasions than at a planned meal or in lieu of a meal.

It was suggested that the use of multiple sugar substitutes would not only be desirable in providing a broader range of options for technical reasons, but might also reduce the mean exposure of the population to any one, thereby limiting any dose-related hazards. On the other hand, unanticipated adverse effects would perhaps be multiplied by the number of sweeteners available, even though each had passed the full amount of tests for mutagenicity, carcinogenicity, and acute toxicity.

The task force considered the problem posed by the absoluteness of the Delaney Amendment and its recent invocation in the case of saccharin. It did not wish to make specific recommendations, but it did suggest that regulatory authorities should be allowed some latitude for the application of scientific judgment in the determination of an acceptable risk by the Commissioner of FDA. In the case of a proposed substitute for sugar, allowance should be made of its benefits as well as the risks.

It was noted by the members of the task force that sugar/sucrose is on the GRAS list at present levels of consumption, yet the sugars are promoting factors in dental caries. Rather than advocating its removal from that list, the majority of the members of the task force recommended that there should be new labelling requirements especially applicable to snack foods containing added sugar. These foods containing added sugar should be required to state on the label the levels of total sugars, and also should be required to include a statement indicating that <u>frequent use of the product by itself between meals may be detrimental to dental health</u>. This is not considered to be an indictment of any one food, but of the misuse of the food by consuming it frequently between meals. These statements should also be reflected in advertising and promotion of the product. If any products containing sugar are shown not to be caries promoting, they should not be subject to this regulation and their manufacturers should have the option to state their non-cariogenicity on the labels.

This task force considered several areas of possible investigation that included the following:

1. Develop and test suitable human, animal and <u>in vitro</u> caries assays to identify the caries promoting properties of the most commonly used snacks available on the USA market as a means to resolve the dilemma about which ones are highly cariogenic.

2. Evaluate phosphates and other possible additives that might ameliorate the caries promoting property of foods containing sugars.

3. Study the effects upon the balance of other dietary nutrients imposed by extensively substituting a non-caloric sweetener for sugars in snacks and (carbonated) beverages.

4. Evaluate consumer acceptability of snacks using non-nutritive and nutritive sweeteners vs. unsweetened snacks.

5. Continue to search for, and develop applications for substitutes for sugar which can obviate the need for extensive safety testing, i.e., those which are either (a) not absorbed, or (b) broken down into metabolizable units on digestion, such as proteins and peptides.

6. Investigate the degree of success of nutrition counseling, as it relates to caries prevention, by individual and classroom instruction, group techniques, behavior modification or other behavioral science methods. The role of public service media advertising needs investigation.

7. Study the application of behavior modification to nutrition counseling of children.

8. Comprehensively evaluate the impact of introducing dental and nutritional advice into the television program/commercial mix; study the effect of selected and en masse

commercial food and beverage messages seen and heard on television by children under 12.

9. Study the acquisition of food habits and preferences through observation of human infants in transition from lactation to the adult dietary patterns.